About the Author

John T. Moore, Ed.D, grew up in the foothills of western North Carolina. He attended the University of North Carolina-Asheville where he received his bachelor's degree in chemistry. He earned his master's degree in chemistry from Furman University in Greenville, South Carolina. After a stint in the United States Army, he decided to try his hand at teaching. In 1971, he joined the chemistry faculty of Stephen F. Austin State University (SFASU) in Nacogdoches, Texas, where he still teaches chemistry. In 1985, he started back to school part-time, and in 1991 received his Doctorate in Education from Texas A&M University.

John's area of specialty is chemical education. He has developed several courses for students planning on teaching chemistry at the high school level. In the early 1990s, he shifted his emphasis to training elementary education majors and in-service elementary teachers in hands-on chemical activities. He has received four Eisenhower grants for professional development of elementary teachers and has served as co-editor (along with one of his former students) of the "Chemistry for Kids" feature of *The Journal of Chemical Education.* He is Director of SFASU's Teaching Excellence Center and is a Co-Director of SFA's Science, Technology, Engineering and Mathematics (STEM) Research Center. He is the author of several books on chemistry and is co-author on several more, including *Chemistry Essentials For Dummies, Biochemistry For Dummies,* and *Organic Chemistry II For Dummies.*

Although teaching has always been foremost in his heart, John found time to work part-time for almost five years in the medical laboratory of the local hospital and has been a consultant for a textbook publisher. He is active in a number of local, state, and national organizations.

John lives in the Piney Woods of East Texas with his wife Robin and their two dogs and two cats. He enjoys brewing his own beer and mead and making custom knife handles and pens from exotic woods. And he loves to cook. His two boys, Jason and Matt, along with his daughter-in-law Sarah and two grandchildren Zane and Sadie remain in the mountains of North Carolina.

Dedication

This book is dedicated to those children, past, present, and future, who will grow to love chemistry, just as I have done. You may never make a living as a chemist, but I hope that you will remember the thrill of your experiments and will pass that enjoyment on to your children. This book is also dedicated to my family: my wife Robin, who encouraged me and put up with my foul moods close to deadlines; my two sons, Jason and Matthew; Jason's wife Sarah; and to the two most wonderful grandchildren in the world, Sadie and Zane. It is also dedicated to Drs. Dexter Squibb and Lloyd Remington of Asheville-Biltmore College who turned me on to the wonders of chemistry and encourages me to continue my education.

Author's Acknowledgments

I would not have had the opportunity to write this book without the encouragement of my agent, Grace Freedson. She took the time to answer my constant e-mails. I owe many thanks to the staff at John Wiley & Sons, Inc., especially executive editor Lindsey Lefevere. Thanks to project editor Chad Sievers for his edits and wit. Thanks also to Rich Langley, my friend and writing partner. And let me offer many thanks to all my students over the past thirty years, especially the ones who became teachers. I've learned from you, and I hope that you've learned from me.

Publisher's Acknowledgments

We're proud of this book; please send us your comments at http://dummies.custhelp.com. For other comments, please contact our Customer Care Department within the U.S. at 877-762-2974, outside the U.S. at 317-572-3993, or fax 317-572-4002.

Some of the people who helped bring this book to market include the following:

Acquisitions, Editorial, and Vertical Websites

Project Editor: Chad R. Sievers

Executive Editor: Lindsay Lefevere

Copy Editor: Chad R. Sievers

Assistant Editor: David Lutton

Editorial Program Coordinator: Joe Niesen

Technical Editors: Michael R. Mueller, PhD, and Patti Smykal

Editorial Manager: Carmen Krikorian

Editorial Assistant: Rachelle Amick

Art Coordinator: Alicia B. South

Cover Photos: © iStockphoto.com/Gunnar Pippel

Cartoons: Rich Tennant (www.the5thwave.com)

Composition Services

Project Coordinator: Patrick Redmond

Layout and Graphics: Carrie A. Cesavice, Sennett Vaughn Johnson, Laura Westhuis

Proofreaders: John Greenough, Lisa Young Stiers

Indexer: Potomac Indexing, LLC

Publishing and Editorial for Consumer Dummies

Kathleen Nebenhaus, Vice President and Executive Publisher

Kristin Ferguson-Wagstaffe, Product Development Director

Ensley Eikenburg, Associate Publisher, Travel

Kelly Regan, Editorial Director, Travel

Publishing for Technology Dummies

Andy Cummings, Vice President and Publisher

Composition Services

Debbie Stailey, Director of Composition Services

Contents at a Glance

Introduction 1

Part I: A Basic Review of Chemistry I 7

Chapter 1: I Passed Chem I, But What About Chem II? 9

Chapter 2: Math for the (Chemistry) Masses 15

Chapter 3: Atomic Structure, the Periodic Table, and Bonding 27

Chapter 4: Digging up the Mole Concept and Stoichiometry 47

Chapter 5: Grasping Solutions and Intermolecular Forces 61

Chapter 6: Not Full of Hot Air: Gases and Gas Laws 81

Part II: Diving Into Kinetics and Equilibrium 97

Chapter 7: The Lowdown on Kinetics: Tortoise or the Hare? 99

Chapter 8: All Present in the Same State: Homogeneous Equilibrium 125

Chapter 9: Neutralizing Effects: Acid-Base Equilibrium 145

Chapter 10: Taking On Solubility and Complex Ion Equilibrium 179

Part III: A Plethora of Chemistry II Concepts 189

Chapter 11: Getting Hot with Thermodynamics 191

Chapter 12: Causing Electrons to Flow: Electrochemistry 207

Chapter 13: Going the Carbon Route: Organic Chemistry 231

Chapter 14: Pondering Polymers 247

Chapter 15: Bringing Biology into the Lab: Biochemistry 261

Part IV: Describing Descriptive Chemistry 277

Chapter 16: Examining the Ins and Outs of Petroleum 279

Chapter 17: Feeling the Power of Nuclear Chemistry 289

Chapter 18: Chemistry in the Home 309

Part V: The Part of Tens 327

Chapter 19: Ten Terrific Tips for Passing Chem II 329

Chapter 20: Ten Must-Know Formulas for Chem II 333

Chapter 21: Ten Great Chemistry Careers 339

Index 345

Table of Contents

Introduction 1

About This Book 1
Foolish Assumptions 1
What Not to Read 2
How This Book Is Organized 2
Part I: A Basic Review of Chemistry I 2
Part II: Diving Into Kinetics and Equilibrium 3
Part III: A Plethora of Chemistry II Topics 3
Part IV: Describing Descriptive Chemistry 4
Part V: The Part of Tens 4
Icons Used in This Book 4
Where to Go from Here 5

Part I: A Basic Review of Chemistry I 7

Chapter 1: I Passed Chem I, But What About Chem II? 9

Grasping the Nature of Chemistry II 10
Recapping general Chemistry I 10
Looking to where you are now: General Chemistry II 11
Examining the Branches of Chemistry 12
Comparing Macroscopic versus Microscopic Viewpoints 13

Chapter 2: Math for the (Chemistry) Masses 15

Digging Through the SI Measurement System 15
Figuring out what SI prefixes mean 16
Length 16
Mass 17
Volume 18
Temperature 18
Pressure 18
Energy 19
Dealing With Numbers, Both Really Big or Small 19
Using exponential and scientific notation 19
Using addition and subtraction 19
Multiplication and division 20
Raising a number to a power 20
Using a calculator 20

Solving the Quadratic Equation 21
Mastering the Unit Conversion Method 22
Working with Significant Figures 23
Understanding number basics: Exact and counted versus measured 23
Figuring out the number of significant figures in a measured number 24
Reporting the correct number of significant figures 24

Chapter 3: Atomic Structure, the Periodic Table, and Bonding. 27

Peering Inside the Atom: Subatomic Particles 27
Examining the Nucleus: At the Heart of It All 28
Modeling Electrons 29
Considering the not-so Bohr(ing) model 29
Looking at the quantum mechanical model 30
Trying to Find Those Electrons 30
Diagramming energy levels 31
Configuring electrons: Easy and space efficient 32
Peering at Patterns of Periodicity 33
Using the Periodic Table 36
Organizing by metals, nonmetals, and metalloids 36
Organizing by periods and families 38
Reviewing Bonding 38
Attracting opposites: Ionic bonding 39
Comprehending covalent bonding basics 43

Chapter 4: Digging up the Mole Concept and Stoichiometry. 47

Crossing the Mighty Mole Bridge 47
Getting a firm grip on the mole 48
Going from grams to moles and back again 48
Converting particles to moles and back again 50
Going from grams to particles and back again 51
Reacting to Reaction Stoichiometry 52
Calculating how many reactants and products 52
Percent yield: Something's missing 56
Limiting reactants: Running out of reactant(s) 57

Chapter 5: Grasping Solutions and Intermolecular Forces 61

Considering Solutes, Solvents, and Solutions 61
Discussing dissolving 62
A lowdown on saturated facts 63
Reconciling Solution Concentration Units 64
Percent composition 64
Molarity is numero uno 66
Defining molality 69
Especially for pollution: Parts per million and parts per billion 69

Contemplating Colligative Properties 70
Vapor-pressure lowering 70
Boiling-point elevation 71
Freezing-point depression 72
Osmotic pressure 73
Grasping the Types of Intermolecular Forces 74
London (dispersion) forces 75
Induced dipole 75
Dipole-dipole 76
Hydrogen bonding 76
Ion-dipole 76
Looking at the Properties of Liquids 77
Heat capacity 77
Capillary action 77
Viscosity 78
Surface tension 79

Chapter 6: Not Full of Hot Air: Gases and Gas Laws 81

Reviewing the Kinetic Molecular Theory 81
Obeying Laws: Gases, That Is 83
Boyle's law 84
Charles's law 86
Gay-Lussac's law 87
The combined gas law 89
Avogadro's law 90
The ideal gas equation 91
Dalton's and Graham's laws 92
Putting Gases Together with Reaction Stoichiometry 95

Part II: Diving Into Kinetics and Equilibrium 97

Chapter 7: The Lowdown on Kinetics: Tortoise or the Hare? 99

Comprehending Reaction Rates 100
Getting to Know Rate Laws 102
Slowing down the reaction 102
Identifying orders of reaction 103
Finding the rate constant 106
Grasping Integrated Rate Laws 107
First order 107
Second order 109
Zero order 109
Half-life ($t_{½}$) 109
Tackling the Collision Theory 111
Exothermic reactions 112
Endothermic reactions 113

Understanding Activation Energy 114
Recognizing Mechanisms 117
Getting elementary with reactions 117
Conforming the rate-determining step 120
Keeping Tabs on Catalysts 121
Heterogeneous catalysis 121
Homogeneous catalysis 122

Chapter 8: All Present in the Same State: Homogeneous Equilibrium 125

Considering Chemical Equilibrium 125
Looking at reaction (equilibrium) quotients: The Law of Mass Action 127
Transforming into an equilibrium constant 128
K_c' and K_p 129
Working Equilibrium Problems 131
Making Sense of LeChatelier's Principle 137
Altering the concentration 138
Altering the temperature 139
Altering the pressure 140
Looking at the effect of a catalyst 141
Putting It All Together: The Haber-Bosch Process 141
Concentration 143
Temperature 143
Pressure 143
Catalyst 143
Other considerations 144

Chapter 9: Neutralizing Effects: Acid-Base Equilibrium 145

Considering the Macroscopic Properties of Acids and Bases 145
Looking Closely at the Microscopic Properties of Acids and Bases 146
Dissolving in water: The Arrhenius theory 146
Accepting hydrogen: The Bronsted-Lowry theory 148
Taking and giving electrons: The Lewis acid-base theory 148
Examining Strong and Weak Acids and Bases 148
Disclosing the truth about acid strength 149
Uncovering the basic truth about base strength 150
Strong acids 151
Strong bases 152
Weak acids 153
Weak bases 154
Acidic/basic oxides 154
Bronsted-Lowry acid-base reactions 155
Determining pH 156
Calculating the pH of a solution 157
Using the pOH to calculate the pH 158
Determining the antilog relationship 158

Accepting Water's Autoionization 159
Tackling More Equilibrium Problems 160
K_a Problems 160
K_b Problems 163
Believing in Buffers 165
Working a basic buffer problem 165
Introducing the Henderson-Hasselbalch equation 167
Grasping buffer capacity 171
Tackling Titration Curves and Indicators 171

Chapter 10: Taking On Solubility and Complex Ion Equilibrium 179

Solving Solubility Equilibriums 179
Understanding the solubility product constant 180
Calculating the concentration of dissolved ions 180
Calculating solubility — molar and otherwise 181
Predicting precipitation 182
Eyeing the common-ion effect 184
Following the Formation of Complex Ions 185
Calculating the formation constant 186
Solving them from the dissociation 188

Part III: A Plethora of Chemistry II Concepts 189

Chapter 11: Getting Hot with Thermodynamics 191

Determining the Change in Energy 192
Discussing Enthalpy 193
Investigating Entropy 195
Defining entropy 195
Determining spontaneity 196
Tackling the Laws of Thermodynamics 197
Remaining constant: The first law 197
Checking for spontaneity: The second law 197
Zeroing in: The third law 199
Predicting Spontaneity for Enthalpy and Entropy Changes 199
Grasping Gibb's Free Energy 200
Checking Out Nonstandard Conditions 202
Revisiting the Haber-Bosch Process 203

Chapter 12: Causing Electrons to Flow: Electrochemistry 207

Following Those Pesky Electrons: Redox Reactions 208
Losing electrons: Oxidation 208
Finding electrons: Reduction 209
One's loss is the other's gain 210
Playing the numbers: Oxidation numbers 211
Balancing redox equations 212

Going Indirect: Clarifying Cells and Cell Potentials 216
Looking at the Daniell cell 216
Writing cell notation 218
Getting a grip on standard reduction potentials 219
Tackling the Nernst Equation 222
Solving a basic problem with this equation 222
Calculating equilibrium constants 223
Considering other uses for the equation 224
Checking Out Electrolysis 224
Power on the Go: Common Electrochemical Cells 226
Let the light shine: Flashlight cells 227
Starting your engine: Automobile batteries 228
Burning the Bacon! Combustion of Fuels and Foods 229

Chapter 13: Going the Carbon Route: Organic Chemistry 231

Hustling Hydrocarbons — the Simplest Organic Compounds 232
From gas grills to gasoline: Alkanes 232
Unsaturated hydrocarbons: Alkenes 238
A triple play: alkynes and their triple bond 240
Benzene and other aromatic (smelly) compounds 241
Following Functional Groups 241
Alcohols (make mine ethanol): R-OH 242
Bad odor carboxylic acids: R-COOH 243
Good odor esters: R-COOR' 244
Aldehydes (R-COH) and ketones (R-CO- R'): Related to alcohols 244
Sleepy ethers: R-O-R' 245
The organic bases: amines ($R-NH_2$) and amides ($R-CO-NH_2$) 245

Chapter 14: Pondering Polymers 247

Examining Natural Monomers and Polymers 247
Classifying Unnatural (Synthetic) Monomers and Polymers 249
Providing a little structure 249
Feel the heat 249
Used and abused 250
Chemical process 250
Reducing, Reusing, Recycling Plastics 258

Chapter 15: Bringing Biology into the Lab: Biochemistry 261

Understanding the Basic Elements Associated with Biochemistry 261
Contemplating Carbohydrates 263
Sticking with one sugar: Monosaccharides 263
Joining monosaccharides: Disaccharides and polysaccharides 265
Getting the Lowdown on Amino Acids 265

Piecing Together Proteins 267
Forming the sequence: Primary structure 267
Bonding hydrogen: Secondary structure 268
Interacting with side chains: Tertiary structure 268
Having more than one primary: Quaternary structure 269
Looking at Lipids 269
Influencing bodily functions: Steroids 269
Storing energy: Fats 271
Observing Nucleic Acids 273
Eyeing the parts of nucleic acids 273
Seeing what nucleic acids do 274
Defecting molecules leads to mutations 275

Part IV: Describing Descriptive Chemistry 277

Chapter 16: Examining the Ins and Outs of Petroleum 279
Understanding How Crude Oil Is Refined 279
Separating chemicals: Fractional distillation 280
Cracking up: Catalytic cracking 282
Moving molecular parts around: Catalytic reforming 283
Telling the Gasoline Story 284
Converting gasoline into a vapor: Volatility 284
Determining how good gas is: Octane ratings 285
Put the lead in, get the lead out: Additives 286
Eyeing Alternatives to the Internal Combustion Engine 288

Chapter 17: Feeling the Power of Nuclear Chemistry 289
Tracing Radioactivity and Transmutation 290
Radioactivity: An unstable nucleus decays 290
Transmutation: Under humanity's control 291
Examining How Nature Does It: Natural Radioactive Decay 292
Identifying stability 292
Decaying in a natural way 293
Decaying in less common ways 295
Determining Half-Lives and Radioactive Dating 296
Figuring out half-lives 298
Introducing radioactive dating 299
Clarifying Nuclear Fission 300
Dealing with critical mass and chain reactions 300
Powering civilization: Power plants 302
Breeder reactors: Making more nuclear stuff 304
Using the Sun's Power: Nuclear Fusion 305
Figuring out whether a fusion reactor is possible 306
Considering fusion's future 307

Chapter 18: Chemistry in the Home 309

Coping with Chemistry in the Laundry Room 309
Reducing the water tension: Surfactants 310
Make it soft: Water softeners 313
Make it whiter: Bleach 314
Cooking Up Chemistry in the Kitchen 314
Clean it all: Multipurpose cleaners 315
Wash those pots: Dishwashing products 315
Checking out Chemistry in the Bathroom 315
Detergent for the mouth: Toothpaste 315
Phew! Deodorants and antiperspirants 316
Skin care chemistry: Keeping it soft and pretty 317
Clean it, color it, curl it: Hair care chemistry 322
Chemistry in the Medicine Cabinet 324
Not tonight, I have a headache: The aspirin story 325
Introduction to serendipity: Minoxidil and Viagra 326

Part V: The Part of Tens 327

Chapter 19: Ten Terrific Tips for Passing Chem II 329

Developing and Sticking to a Regular Study Schedule 329
Striving For Understanding — Don't Just Memorize 330
Doing the Homework 330
Using Additional Resources 330
Skimming the Material before Going to Class 331
Taking Good Notes 331
Recopying Your Lecture Notes 331
Asking Questions 332
Getting a Good Night's Sleep before Exams 332
Paying Particular Attention to Charges and Significant Figures on Exams 332

Chapter 20: Ten Must-Know Formulas for Chem II 333

Molarity Equation 333
Rate Law Equation 333
Homogeneous Equilibrium Constant Expression 334
Acid and Base Equilibrium Constant Expressions 334
pH/pOH Equations 335
Henderson-Hasselbalch Equation 336
Solubility Product Equation 336
Gibbs Free Energy Equation 336
Nernst Equation 337
Half-Life Equations 337

Chapter 21: Ten Great Chemistry Careers 339

Patent Attorney.....339
Pharmaceutical/Chemical Sales.....340
Forensic Chemist.....340
Biochemistry/Biotechnology.....341
Agricultural Chemist.....341
Material Science.....341
Food and Flavor Chemist.....342
Water Quality Chemist.....342
Cosmetic Chemists.....343
Chemistry Teaching.....343

Index.....345

Introduction

Congratulations. You jumped the first hurdle in understanding the basics of chemistry by passing Chemistry I. Perhaps you even used my book, *Chemistry For Dummies, Second Edition* (John Wiley & Sons, Inc.). If you did, thank you. If you didn't, I'm glad you've entrusted me with your Chemistry II endeavors. The very fact that you're at least looking at this book indicates that you feel you may need a little help in your Chemistry II class. Chem I (believe it or not) isn't as mathematical as Chem II. In Chem I you had a lot of descriptive material; Chem II is all about solving problems, so get ready.

About This Book

My goal with this book is not to make you into a chemistry major. My goal is simply to give you a basic understanding of some chemical topics that commonly appear in the second half of a university introductory chemistry course or the second year in a high school chemistry course. If you're taking a course, use this book as a reference in conjunction with your notes and textbook.

Simply watching people play the piano, no matter how intently you watch them, doesn't make you a musical expert. You need to practice. And the same is true with chemistry. It's *not* a spectator sport. You probably figured that out in Chem I; you need to practice and work on problems. Chemistry II adds a lot more math problems, which may be challenging for some people. Sharpen up your calculator skills — you'll need them. I show you how to work certain types of problems in this book — homogeneous equilibrium, for example — but use your textbook for practice problems. It's work, yes, but it really can be fun. This book is for those of you who want some additional help with Chem II topics.

Foolish Assumptions

When I wrote this book, I made a few assumptions about you. Those assumptions include the following:

- You're taking (or retaking) a second-semester college chemistry course or preparing to take a second-semester college chemistry course.
- You're taking (or retaking) a second-year high school chemistry course or preparing to take a second-year high school chemistry course.

- You at least passed the first-year high school chemistry course and are wondering whether you want to take the next class.
- You at least passed the first-semester college chemistry course and are wondering whether you want to take the next class.
- You feel relatively comfortable with arithmetic and know enough algebra to solve for a single unknown in an equation.
- You have a scientific calculator capable of doing exponents and logarithms.

If you're buying this book just for the thrill of finding out about something different — with no plan of ever taking a chemistry course — I applaud you and hope that you enjoy this adventure. Feel free to skip those topics that don't hold your interest; for you, there will be no tests, only the thrill of increasing your knowledge about something new.

What Not to Read

I know you're a busy person and want to get just what you need from this book. Although I want you to read every single word I've written, I understand you may be on a time crunch. If so, feel free to skip the sidebars, the gray-shaded boxes that appear here and there. These interesting bits of info aren't essential to understanding the stuff you need to know.

I mark some paragraphs with Technical Stuff icons. What I tell you in these paragraphs is more than you need to know, strictly speaking, but it may give you helpful or interesting detail about the topic at hand. If you want just the facts, you can skip these paragraphs.

How This Book Is Organized

I've organized the topics in a logical progression — basically the same way I organize my courses for science and nonscience majors. Following is an overview of each part of the book.

Part I: A Basic Review of Chemistry I

In this part, I give you a basic review of those topics commonly found in a Chem I course that I feel are critical to your progression through the Chem II concepts. I review the simple concepts of chemistry in Chapter 1, and then in Chapter 2, I give you a quick review of chemical calculations. I show you how to use the factor label method of calculations, along with an introduction to the SI (metric) system.

In Chapter 3, I give a review of atomic structure, the periodic table, and the different types of bonding. I don't cover topics in a lot of depth, but just enough to jog your memory about energy level configurations, periodicity, and bonding. In Chapter 4, I give you a good review of reaction stoichiometry because you will really need these mole-related concepts in Chem II.

In Chapter 5, I review solutions and solution concentration units. I also review the different types of intermolecular forces and the properties of liquids. In Chapter 6, I review the properties of gases including the gas laws (Boyle's law, Charles's law, Gay-Lussac's law, the combined gas law, the ideal gas law, Avogadro's law, and more). That's it! Six chapters of review of a course it took half a year (or a full year) to complete.

Part II: Diving Into Kinetics and Equilibrium

In this part, you get into the real meat of Chemistry II. In Chapter 7, I discuss the factors associated with the speed of a reaction. I show you how to determine the rate law for a reaction. A rate law relates the changes in concentrations of reactants to the overall speed of reaction. I also discuss the current model on how reactions occur and end up with a discussion of catalysts.

The rest of this part focuses on equilibrium — three chapters worth. The study of equilibrium is probably the most important topic in a second-semester college chemistry (or a second-year high school) course. First, I introduce you to the basic concepts of equilibriums, and then I apply these basic concepts to homogeneous equilibrium systems. I also cover acid-base equilibrium systems, heterogeneous equilibrium systems (solubility), and complex ion equilibrium. Lots of different types of equilibriums, yet all remarkably similar.

Part III: A Plethora of Chemistry II Concepts

In this part I start off by examining thermodynamics, building on that little taste of thermochemistry you studied in Chem I. I talk about enthalpy and entropy and Gibbs Free Energy, as well as the three laws of thermodynamics. Then it's off to electrochemistry for a discussion of redox reactions. I show you how to balance redox reactions, which can be a bane of second-semester chemistry students, and then show how redox reactions are related to electrochemical cells — batteries and electroplating.

The last three chapters of this part give you a rest from calculations. I start off by giving you a glimpse into the world of organic chemistry, the chemistry of carbon. I discuss hydrocarbons in a little detail and give you a brief introduction to other functional groups, such as the alcohols. Then I show you an application of organic chemistry — polymers. I discuss some of the different types of *polymers* (plastics) in terms of their structure and usage. Finally, I introduce you to the world of biochemistry, the chemistry of living things.

Part IV: Describing Descriptive Chemistry

In this part, I start by discussing the chemistry of petroleum. I introduce terms like *cracking* and *reforming around,* as well as discussing what that octane rating really means. Then I leave an old fuel for a new fuel — nuclear power. I show you the different types of nuclear decays, discuss fission and fusion, and show how to deal with half-life problems.

I finish up this part, with a discussion of chemistry in the home. That's not covered much in general chemistry, but I believe that you deserve an opportunity to explore the practical side of chemistry a little. I discuss the chemical nature of soaps and detergents, deodorants and antiperspirants, aspirin and Viagra.

Part V: The Part of Tens

In this part, I introduce you to ten terrific tips for passing Chem II. They really do work. Then I give you the top ten mathematical formulas that you will be using in Chem II, and wind up with ten chemical careers, for those of you who are dreaming about graduating and getting a job in chemistry.

Icons Used in This Book

If you've read other *For Dummies* books, you recognize the icons used in this book, but here's the quickie lowdown for those of you who aren't familiar with them:

This icon gives you a tip on the quickest, easiest way to perform a task or conquer a concept. This icon highlights stuff that's good to know and stuff that'll save you time and/or frustration.

The Remember icon is a memory jog for those really important things you shouldn't forget.

I use this icon when I describe safety in doing a particular activity, especially mixing chemicals.

This icon points out different example problems you may encounter with the respective topic. I walk you through them step by step to help you gain confidence.

I don't use this icon very often because I keep the content pretty basic. But in those cases where I expand on a topic beyond the basics, I use this icon. You can safely skip this material, but you may want to look at it if you're interested in a more in-depth description.

Where to Go from Here

I present this book's content in a logical (at least to me) progression of topics. But this doesn't mean you have to start at the beginning and read to the end of the book. Each chapter is self-contained, so feel free to skip around. Sometimes, though, you'll get a better understanding if you do a quick scan of a background section as you're reading. To help you find appropriate background sections, I've placed "see Chapter XX for more information" cross-references here and there throughout the book.

Because I'm a firm believer in concrete examples, I also include lots of illustrations and figures with the text. They really help in the understanding of chemistry topics. And to help you with the math, I break up problems into steps so that you can easily follow exactly what I'm doing.

If you're trying to clarify something specific, go right to that chapter and section. If you're a real novice, start with Chapter 1 and go from there. If you did okay in Chemistry I, I suggest quickly reviewing Part I and then going on to Part II. Chapter 7 on kinetics is essential, and so are Chapters 8–10 on equilibrium.

You really can't go wrong. Whether taking a course, reviewing for a professional exam, or just wanting to know a little more about chemistry, I think you will get your money's worth. I hope that you enjoy your chemistry trip.

Part I
A Basic Review of Chemistry I

In this part . . .

In this part, I give you a basic review of those topics commonly found in a Chem I course that I feel are critical to your progression through the Chem II concepts. I review the really basic concepts of science and chemistry in Chapter 1. In Chapter 2, I give you a quick review of chemical calculations. I show you how to use the factor label method of calculations, along with an introduction to the SI (metric) system.

In Chapter 3, I give a review of atomic structure, the periodic table, and the different types of bonding. I don't cover topics in great depth here, but just enough to jog your memory about energy level configurations, periodicity, and bonding. In Chapter 4, I provide you a good review of reaction stoichiometry because you'll really need these mole-related concepts in Chem II.

In Chapter 5, I review solutions and solution concentration units. I also touch on the different types of intermolecular forces and the properties of liquids. In the last review chapter, I review the properties of gases including the gas laws (Boyle's law, Charles's law, Gay-Lussac's law, the combined gas law, the ideal gas law, Avogadro's law, and more). That's it — six chapters of review of a course it took you full year if you're in high school or a full semester if you're in college to complete.

Chapter 1

I Passed Chem I, But What About Chem II?

In This Chapter

- Comprehending chemistry
- Discovering science and technology
- Examining the general areas of chemistry

You already know what chemistry is. You passed your first year of high school or your first semester of college chemistry. Now you're ready to take on your second year or second semester, and you want a resource to help you explain concepts in plain English. This chapter sets the stage for the rest of the book by showing you what the differences are between Chem I and Chem II so that you can relate better to this new material. It also relates some of the major areas of chemistry to the topics you'll be studying in Chemistry II. If you're already in the midst of a Chem II college or high school course, you may want to skim over this chapter for a quick review of some basic concepts and then go right to the subject area in the book that is troubling you.

If you bought this book just to have fun discovering something new and aren't taking a chemistry course, you may need a little refresher on the really fundamental chemical topics. I suggest buying a copy of the first book in this series, *Chemistry For Dummies*. That book, now in its second edition, can give you the basics and make this book more meaningful.

Teaching chemistry is very enjoyable. For me, it's more than just a collection of facts and a body of knowledge. Although I wasn't a chemistry major when I entered college, I quickly became hooked when I took my first chemistry course. The subject seemed so interesting and so logical. Watching chemical changes take place, figuring out unknowns, using instruments, extending my senses, and making predictions to figure out why they were right and wrong all seems so fascinating. Your journey into Chem II starts here.

Grasping the Nature of Chemistry II

Chem I, in most schools, is a mixture of a lot of different topics. You naturally find some carryover between topics; you finish the chapter on gases and only briefly cover those topics again, until you hit the final exam. Your Chemistry II class is more consistent in these topics. Chem II is also much more mathematical than Chem I, which was great for me because I always enjoyed the quantitative aspects of chemistry more than the descriptive part. That's why I am an analytical chemist instead of an organic chemist. I enjoy working with numbers.

The following sections give you a quick reminder at the content in a typical Chem I course and then show you what to expect in a typical Chem II class that you are or might be taking.

Recapping general Chemistry I

In your first couple of weeks in your Chemistry II class, you probably will review the basics of what you covered in your Chemistry I class. I dedicate the chapters in Part I of *Chemistry II For Dummies* to these topics to help you review these important topics. Here are the topics you can find:

- **Problem solving:** The metric or SI system is essential to studying chemistry at any level. You need to be able to use the factor-label method of problem solving, also called *unit analysis.* This method allows you to manipulate units to generate the set-up for a particular problem. About this same time you become proficient in determining the number of significant figures you should report in your final answer. Refer to Chapter 2 for more information.
- **Atomic structure:** Having a firm understanding of subatomic particles (protons, electrons, and neutrons), the nucleus, and the electron clouds is important when taking a chemistry course. Chapter 3 gives you an overview of these topics. You can also find information on *electron configurations* (the way to represent the various electrons in an atom), average atomic masses, and the mole concept. For an overview of these topics, see Chapters 3 and 4.
- **The Periodic Table and periodic properties:** Chemistry I gave you the basics on electron configurations, ionization energies, sizes of atoms, and a host of other topics related to the periodic table. You definitely need this knowledge when studying Chem II. Chapter 3 gives you a brief overview.
- **Bonding:** Chemical bonding, both ionic and covalent, form an important part of Chem I. Having a firm foundation on these topics is also important in Chem II. See Chapter 3 for a review.
- **Molecules, compounds, and chemical equations:** Here is where chemical nomenclature was first introduced in your Chem I class. You may remember discussing chemical formulas, chemical names, and vice versa. Calculating molar masses and determining the empirical formula

from percentage data is also important. You also figured out how to balance chemical equations. Chemical nomenclature is an absolute necessity of Chemistry II as well as the balancing of chemical equations and the determination of molar masses. For a review, refer to Chapter 4.

- **Reaction stoichiometry:** You probably remember that this topic was a main crux of your Chem I course. You learned how to calculate how much — how much reactant, how much produce, how many moles, how many grams, and how many particles. Balanced chemical equations go hand in hand to allow you to do these calculations. You also focused on the basic reaction types and sometimes even a little solution stoichiometry. The reaction stoichiometry and the mole concept are of primary importance in Chem II. Flip to Chapter 4 to ensure you have a good understanding of these topics.
- **Solutions:** More than likely you studied solution concentration units, especially molarity and molality, in your Chem I class. Solution concentrations are extremely important in Chem II. Refer to Chapter 5 for a review.
- **Gas properties:** Many textbooks and Chem I instructors cover the properties of gases, including numerous gas laws and the kinetic molecular theory. Understanding the kinetic molecular theory also makes it easier to see how the various factors affect the kinetics of a reaction in Chem II. Check out Chapter 6 for more info.
- **Nuclear chemistry:** Some instructors cover nuclear chemistry as part of the Chem I curriculum; some cover it in Chem II. Chapter 17 touches on what you need to know.

If you want more in-depth explanation of these topics, you can check out my book, *Chemistry For Dummies,* 2nd Edition (John Wiley & Sons, Inc.).

Looking to where you are now: General Chemistry II

In Chem II you can expect to encounter the following topics, but not necessarily in this exact order:

- **Chemical kinetics:** A Chemistry II class usually covers this topic early after you finish reviewing the topics of Chem I. *Kinetics* is the study of the speed of reactions. Along with kinetics reaction mechanisms, the series of steps a reaction proceeds through in going from reactants to products is included. Chapter 7 covers kinetics.
- **Chemical equilibrium:** This is the largest topic in most Chem II classes. An *equilibrium* is established when a chemical reaction goes from reactants to products and at the same time is also proceeding from products to reacts. These two reactions occur at the same reaction rate (speed). You can uncover all the different types of equilibriums: homogeneous,

heterogeneous, acid-base, solubility, and complex-ion. You can also find out about ways to manipulate the equilibrium system so as to form as much product as possible. I discuss equilibrium in Chapters 8, 9, and 10.

- **Thermodynamics:** Thermodynamics is another important topic of Chemistry II. *Thermodynamics* is basically the study of energy transfer. It builds on the thermochemistry concepts of Chem I, but it has the goal of being able to predict under what conditions a reaction is spontaneous. Chapter 11 covers thermodynamics.
- **Electrochemistry:** The study of batteries and cells also appears in Chem II. You figure out how to balance redox reaction and then move on to electrochemical cells. You discover all about cells and batteries, including automobile batteries and flashlight cells. Chapter 12 explains electrochemistry in more depth.
- **Radioactivity:** Chemistry II classes sometimes cover this topic. Sometimes Chemistry I classes cover it. *Radioactivity* essentially is the spontaneous decay of an unstable nucleus to a more stable one. This is the stuff of atomic bombs and nuclear power plants. Chapter 17 discusses radioactive decay, half-lives, fission, and fusion.
- **Other topics:** Some instructors also cover organic chemistry and biochemistry. I cover these topics in Chapters 13 through 15.

Examining the Branches of Chemistry

As you go through your Chemisty II course, you may actually start to wonder what chemists do all day. Well, some make things (synthesis), others examine the properties of things (analysis), and others explain things (teach). But all chemists have a specialty area in which they have received more training. The following describes the general areas of chemistry.

- **Physical chemistry:** This branch figures out how and why a chemical system behaves as it does. Physical chemists study the physical properties and behavior of matter and try to develop models and theories that describe this behavior. Especially keep this branch in mind when you're studying thermodynamics in Chapter 11.
- **Analytical chemistry:** This branch is highly involved in the determination of the properties of a substance (analysis). Chemists from this field of chemistry may be trying to find out what substances are in a mixture *(qualitative analysis)* or how much of a particular substance is present *(quantitative analysis)* in something. Analytical chemists typically work in industry in product development or quality control. If a chemical manufacturing process goes wrong and is costing that industry hundreds of thousands of dollars an hour, that quality control chemist is under a lot of pressure to fix it and fix it fast. A lot of instrumentation is

used in analytical chemistry. Chapter 12, electrochemistry, is a typical topic studied by analytical chemists.

- **Inorganic chemistry:** This branch is involved in the study of inorganic compounds such as salts. It includes the study of the structure and properties of these compounds. It also commonly involves the study of the individual elements of the compounds. Inorganic chemists probably say that this field is the study of everything except carbon, which they leave to the organic chemists. Inorganic chemists are interested in the descriptive chemistry of the elements.
- **Organic chemistry:** This field is the study of carbon and its compounds. It's probably the most organized of the areas of chemistry — with good reason. There are millions of organic compounds, with thousands more discovered or created each year. Industries such as the polymer industry, the petrochemical industry, and the pharmaceutical industry depend on organic chemists. Chapters 13 and 14 describe aspects of organic chemistry. Much more about organic chemistry can be found in *Organic Chemistry II For Dummies* (John Wiley & Sons, Inc.).
- **Biochemistry:** This branch specializes in living organisms and systems. Biochemists study the chemical reactions that occur at the *molecular level* of an organism — the level where items are so small that people can't directly see them. Biochemists study processes such as digestion, metabolism, reproduction, respiration, and so on. Sometimes, distinguishing between a biochemist and a molecular biologist is difficult because they both study living systems at a microscopic level. However, a biochemist really concentrates more on the reactions that are occurring. Check out Chapter 15 for a taste of biochemistry, but for a full meal see my book *Biochemistry For Dummies* (John Wiley & Sons, Inc.).
- **Biotechnology:** This is a relatively new area of science that is commonly placed with chemistry. It's the application of biochemistry and biology when creating or modifying genetic material or organisms for specific purposes. It's used in such areas as cloning and the creation of disease-resistant crops, and it has the potential for eliminating genetic diseases in the future. I also suggest you check out my book *Biochemistry For Dummies* (John Wiley & Sons, Inc.) for more information.

Comparing Macroscopic versus Microscopic Viewpoints

As you go through your chemistry course, pay attention to the way your instructor shifts from talking about matter in terms of atoms and molecules and then shifts very naturally into the concrete world of grams and kilograms. These two viewpoints are called the *microscopic* viewpoint and the *macroscopic* viewpoint. Nearly all chemists, no matter what field they study, study the world around them in two ways:

- **Macroscopic view:** This view is what you see, feel, and touch. This is the world of dirty lab coats — of mixing solutions and weighing out elements. This viewpoint is the world of experiments, or what some nonscientists call the real world.
- **Microscopic view:** This view focuses on work with models and theories. Chemists may describe a chemical reaction, such as the Haber reaction to produce ammonia, in terms of individual atoms and molecules. This is the microscopic world.

Scientists are often so used to going back and forth between the two views that they don't even realize that they're doing so. An occurrence or observation in the macroscopic world generates an idea related to the microscopic world, and vice versa. You may find this flow of ideas disconcerting at first. You may have noticed this back and forth some in your Chemistry I studies; you'll notice it more in your Chemistry II studies. You may need some adjusting to it before moving back and forth becomes second nature to you.

Contrasting pure and applied chemistry

In *pure chemistry,* chemists are free to carry out whatever research interests them — or whatever research they can get funded. There is no real expectation of practical application at this point. The researcher simply wants to know for the sake of knowledge. This type of research (often called *basic research*) is most commonly conducted at colleges and universities. The chemist uses undergraduate and graduate students to help conduct the research. The work becomes part of the students' professional training. The researcher publishes his or her results in professional journals for other chemists to examine and attempt to refute. Funding is almost always a problem, because the experimentation, chemicals, and equipment are quite expensive.

In *applied chemistry,* chemists normally work for private corporations. Their research is directed toward a very specific short-term goal set by the company — product improvement or the development of a new plastic or medicine, for example. Normally, more money is available for equipment and instrumentation with applied chemistry, but the chemists also have the pressure of meeting the company's goals.

These two types of chemistry, pure and applied, share the same basic differences as science and technology. In *science,* the goal is simply the basic acquisition of knowledge. There doesn't need to be any apparent practical application. Science is simply knowledge for knowledge's sake. *Technology* is the application of science toward a very specific goal.

Society has a place for science *and* technology — likewise for the two types of chemistry. The pure chemist generates data and information that is then used by the applied chemist. Both types of chemists have their own sets of strengths, problems, and pressures. In fact, because of the dwindling federal research dollars, many universities are becoming much more involved in gaining patents, and they're being paid for technology transfers into the private sector.

Chapter 2

Math for the (Chemistry) Masses

In This Chapter

- Feeling out the SI measurement system
- Working with numbers, both really big or small
- Using the quadratic equation to solve problems
- Figuring out unit conversions
- Getting comfortable with significant figures

Congratulations. You made it through Chemistry I. You probably remember that chemistry has a lot of calculations. Guess what? Chemistry II has a lot more calculations than Chemistry I. The good news: I am confident you can handle these calculations, which include more arithmetic and simple algebra. You also need to be able to use the quadratic equation and be able to do a few more calculator functions.

Because you rely heavily on the SI system, exponential notation, and the unit conversion method, I focus on reviewing these concepts in this chapter. I also touch on a quick review of significant figures and rounding off. Most of these concepts should be review. Reading this chapter or parts of this chapter that you need to review can bring you up to speed for Chemistry II.

Digging Through the SI Measurement System

The *SI system* is a decimal system. It has base units for mass, length, volume, and so on. You use the SI system in nearly all calculations you make in chemistry. You probably used it a lot in your Chemistry I class, and you'll use it probably even more in your Chemistry II class. This section lists the most common SI prefixes, base units for physical quantities in the SI system, and some useful SI-English conversions.

The English used the English system of weights and measures. The US colonies and then the United States adopted it. The US system uses pounds and ounces, gallons and quarts, and miles and yards with all kinds of weird conversions between them. The metric system (SI system) is much easier to use, so much in fact that the English have abandoned their own system and use the metric system now. The so-called English system is now called the *US customary system,* because the United States is just about the only country that still uses it. The decimal-based SI system is much easier for scientists to use and is understood worldwide.

Figuring out what SI prefixes mean

The prefixes modify the base units and tell you how much of an item is in question. For example, *kilo-* means 1,000; a kilogram is 1,000 grams, and a kilometer is 1,000 meters.

Use Table 2-1 as a handy reference for the abbreviations and meanings of some of the most commonly used SI prefixes.

Table 2-1 Selected SI Prefixes

Prefix	*Abbreviation*	*Meaning*
kilo-	k	1,000 or 10^3
hecto-	h	100 or 10^2
deka-	da	10 or 10^1
deci-	d	0.1 or 10^{-1}
centi-	c	0.01 or 10^{-2}
milli-	m	0.001 or 10^{-3}
micro-	µ	0.000001 or 10^{-6}
nano-	N	0.000000001 or 10^{-9}

Length

The base unit for length in the SI system is the *meter.* The exact definition of meter has changed over the years, but it's now defined as the distance that light travels in a vacuum in 1/299,792,458 of a second. The most common SI units of length that you will encounter are:

millimeter (mm)

centimeter (cm)

meter (m)

kilometer (km)

Some common length conversions from the English system to the SI system are

1 mile (mi) = 1.61 kilometers (km)

1 yard (yd) = 0.914 meters (m)

1 inch (in) = 2.54 centimeters (cm)

I find the inch/cm conversion to be the most useful because many of the problems I deal with fall into that length range. I suggest you find the one that works best for you.

Mass

The base unit for mass in the SI system is the *kilogram*. It's the weight of the standard platinum-iridium bar found at the International Bureau of Weights and Measures. Here are the most common SI units of mass you will encounter:

milligram (mg)

gram (g)

kilogram (kg)

Some common English to SI system mass conversions are

1 pound (lb) = 453.6 grams (g)

1 ounce (oz) = 28.4 grams (g)

I find the lb/g conversion to be the most useful because many of the problems I work with fall into this range.

Volume

The base unit for volume in the SI system is the *cubic meter;* however, chemists normally use the liter. They do so because it's customary for graduated glassware used in chemistry to be in milliliters or liters, unlike medical instruments such as syringes that are in cc's (cm^3). A liter is 0.001 m^3. Here are the most common SI units of volume:

1 milliliter (mL) = 1 cubic centimeter (cm^3 or cc)

1 liter (L) = 1,000 milliliters (mL)

Some common English to SI system volume conversions are

1 quart (qt) = 0.946 liters (L)

1 fluid ounce (fl oz) = 29.6 milliliters (mL)

1 gallon (gal) = 3.79 liters (L)

I find the qt/L conversion to be the most useful because again it fits better with most metric to English conversion that I do.

Temperature

The base unit for temperature in the SI system is *Kelvin*. Here are the three major temperature conversion formulas:

Celsius to Fahrenheit: $°F = (9/5)°C + 32$

Fahrenheit to Celsius: $°C = (5/9)(°F–32)$

Celsius to Kelvin: $K = °C + 273$

Pressure

The SI unit for pressure is the *pascal*, where 1 pascal equals 1 newton per square meter. But pressure can also be expressed in a number of different ways, so here are the most common pressure conversions:

1 millimeter of mercury (mm Hg) = 1 torr

1 atmosphere (atm) = 760 millimeters of mercury (mm Hg) = 760 torr

1 atmosphere (atm) = 101 kilopascals (kPa)

1 barr = 10^5 Pa

Energy

The SI unit for energy (such as heat) is the *joule*, but many chemists and chemistry professors still use the metric unit of heat, the *calorie*, because it still is used extensively in the popular and chemical literature. Here are some common energy conversions:

1 calorie (cal) = 4.184 joules (J)

1 Nutritional (food) Calorie (Cal) = 1 kilocalorie (kcal) = 4,184 joules (J)

Dealing With Numbers, Both Really Big or Small

Chemists work with very large and very small numbers on a daily basis. For example, when chemists talk about the number of ions in a gram of table salt, they're talking about a very, very large number. But when they talk about the diameter of a single sodium cation, they're talking about a very, very small number. As you found in Chem I, chemists can use exponential or scientific notation to represent these large or small numbers. These sections review the ways you can handle very large and small numbers so you're ready for them in Chem II.

Using exponential and scientific notation

In *exponential notation*, a number is represented as a value raised to a power of ten. The decimal point can be located anywhere within the number as long as the power of ten is correct. In *scientific notation*, the decimal point is always located between the first and second digit — and the first digit must be a number other than zero.

For example, the number 328,000 could be represented a 3.28×10^5 while 0.0054 would equal 5.4×10^{-3}.

Using addition and subtraction

To add or subtract numbers in exponential or scientific notation, both numbers must have the same power of ten. If they don't, you must convert them to the same power. Here's a subtraction example:

$$(2.5 \times 10^5 \text{ cm}) - (2.2 \times 10^4 \text{ cm}) = (25 \times 10^4 \text{ cm}) - (2.2 \times 10^4 \text{ cm}) =$$

$$22.7 \times 10^4 \text{ cm (exponential notation)} = 2.27 \times 10^3 \text{ cm (scientific notation)}$$

You perform the addition exactly the same way.

Multiplication and division

To multiply numbers expressed in exponential notation, multiply the *coefficients* (the numbers) and add the *exponents* (powers of ten):

$$(2.25 \times 10^{-2} \text{ cm}) \times (3.37 \times 10^{-5} \text{ cm}) = (2.25 \times 3.37) \times 10^{(-2 + -5)} \text{ cm}^2$$

$$= 7.58 \times 10^{-7} \text{ cm}^2$$

To divide numbers expressed in exponential notation, divide the coefficients and subtract the exponent of the denominator from the exponent of the numerator:

$$(6.27 \times 10^5 \text{ g}) \div (1.25 \times 10^3 \text{ mL}) = (6.27 \div 1.25) \times 10^{5-3} \text{ g/mL} = 5.02 \times 10^2 \text{ g/mL}$$

Raising a number to a power

To raise a number in exponential notation to a certain power, raise the coefficient to the power and then multiply the exponent by the power:

$$(2.33 \times 10^{-5}\text{cm})^3 = (2.33)^3 \times 10^{-5 \times 3} \text{ cm}^3 = 12.6 \times 10^{-15} \text{ cm}^3 = 1.26 \times 10^{-14} \text{ cm}^3$$

Using a calculator

Scientific calculators make doing calculations much easier. You don't have to focus as much time on the actual calculations and can spend more time on the problem itself.

You can use a calculator to add and subtract numbers in exponential notation without first converting them to the same power of ten. Just be careful that you correctly enter the exponential numbers.

For example, assume that your calculator has a key labeled *EXP*. The EXP stands for $\times 10$. After you press the EXP key, you enter the power. For example, to enter the number 6.25×10^3, you type 6.25, press the EXP key, and then type 3.

What about a negative exponent? If you want to enter the number 6.05×10^{-12}, you type 6.05, press the EXP key, type 12, and then press the +/– key.

When using a scientific calculator, *don't* enter the × *10* part of your exponential number. Press the EXP key to enter this part of the number.

Solving the Quadratic Equation

When you get into the sections on equilibriums, you may need to solve the quadratic equation. The *quadratic equation* is a way for solving second degree equations of the form $ax^2 + bx + c = 0$. If you're not familiar with it from your algebra class, you may want to do some reviewing using your textbook or the Internet.

Here I give you some tips so that you can minimize this experience, but you won't be able to avoid it entirely. If it has been a while since you worked with this equation, use the following example as a refresher to brush up on this mathematical solution process.

The quadratic equation is useful in solving problems like this:

$2x^2 + 5x = 52$ can be rearranged to $2x^2 + 5x - 52 = 0$

The quadratic equation is of the form:

$$x = \frac{-b \pm \sqrt{b^2 - 4ac}}{2a}$$

where in this case $a = 2$, $b = 5$, and $c = -52$. Substituting the values into the equation gives:

$$\begin{aligned} x &= \frac{-5 \pm \sqrt{5^2 - 4(2)(-52)}}{2(2)} \\ &= \frac{-5 \pm \sqrt{25 - (-416)}}{4} \\ &= \frac{-5 \pm \sqrt{441}}{4} \\ &= \frac{-5 \pm 21}{4} \\ &= 4 \text{ or } -6.5 \end{aligned}$$

Only one value is going to have any significance in the real world. Many times x is a concentration and you can't have a negative concentration. Concentrations might be small, but not less than zero.

Mastering the Unit Conversion Method

You have probably found that actually setting up chemistry problems to solve them is sometimes hazy or unclear. A scientific calculator can help you by handling the math, but a calculator can't tell you what mathematical operations you need to perform.

That's where the *unit conversion method*, sometimes called the *factor label method*, comes into play. This method can help you set up chemistry problems and calculate them correctly. Hopefully you got familiar with it in Chem I, but a brief review shouldn't hurt.

Two basic rules are associated with the unit conversion method:

- **Rule 1:** Always write the unit and the number associated with the unit. Rarely in chemistry will you have a number without a unit. Pi is the major exception that comes to mind.
- **Rule 2:** Carry out mathematical operations *with* the units, canceling them until you end up with the unit you want in the final answer. In every step, you must have a correct mathematical statement.

This example may stimulate your recall of the unit conversion method. A *firkin* is a little known unit of volume in the U.S. Customary System. A firkin is equal to 9.0 gallons. How many liters are in 1 firkin? You need to solve for liters/firkin, so follow these steps:

1. **Write down what you start with:**

   $$\frac{9.0 \text{ gallons}}{\text{firkin}}$$

   Note that per Rule #1, the equation shows the unit and the number associated with it.

2. **Convert gallons to quarts, canceling the unit of gallons per Rule #2:**

   $$\frac{9.0 \text{ gallons}}{\text{firkin}} \times \frac{4 \text{ quarts}}{1 \text{ gallon}}$$

3. **Convert quarts to liters:**

   $$\frac{9.0 \text{ gallons}}{\text{firkin}} \times \frac{4 \text{ quarts}}{1 \text{ gallon}} \times \frac{0.946 \text{ L}}{1 \text{ quart}}$$

4. **Now that you have the units of liters per firkin, do the math to get the answer:**

   34 L/firkin

5. **Stop and ask yourself whether the answer is reasonable.**

   Nine gallons would contain 36 quarts and a liter is approximately a quart, so that the answer should be around 36 L. The answer is reasonable and has the correct units.

   Note that the answer has been rounded off to the correct number of significant figures. If you're a little rusty with significant figures, the next section gives you details on how to do so.

Note that although the setup of the preceding example is correct, it's certainly not the only correct setup. Depending on what conversion factors you know and use, there may be many correct ways to set up a problem and get the correct answer. With a little practice, you can really appreciate and like the unit conversion method. It got me through my introductory physics course.

Working with Significant Figures

Significant figures (no, I'm not talking about Donald Trump's net worth) are the number of digits that you report in the final answer of the mathematical problem you are calculating. The number of the significant figures is limited by the accuracy of the measurement. The following sections explain how to determine the number of significant figures in a number, how to determine how many significant figures you need to report in your final answer, and how to round your answer off to the correct number of significant figures.

Understanding number basics: Exact and counted versus measured

If I ask you to count the number of television sets that you and your family own, you can do it without any guesswork involved. Your answer might be 0, 1, 2, or 10, but you would know exactly how many televisions you have. Those numbers are what are called *counted numbers.* If I ask you how many feet are in a mile, your answer will be 5,280, which is an *exact number.* Another exact number is the number of centimeters per inch — 2.54. This number is exact by definition. In both exact and counted numbers, there is no doubt what the answer is. When you work with these types of numbers, you don't have to worry about significant figures.

Now suppose that I ask you and four friends to individually measure the length of an object as accurately as you possibly can with a ruler. You then report the results of your measurements: 4.87 centimeters, 4.85 centimeters,

4.88 centimeters, 4.81 centimeters, and 4.83 centimeters. Which of you is right? You are all within experimental error. These measurements are measured numbers, and measured values always have some error associated with them. You determine the number of significant figures in your answer by your least reliable *measured* number.

Figuring out the number of significant figures in a measured number

When determining the number of significant figures, or *sig. figs.*, in a measured number, you need to keep the following rules in mind to help:

- **Rule 1:** All nonzero digits are significant. All numbers, one through nine, are significant, so 523 contains three sig. figs., 2.7×10^5 contains two, and 0.4327 contains four. The zeroes are the only numbers that you have to worry about.
- **Rule 2:** All of the zeroes between nonzero digits are significant. For example, 705 contains 3 sig. figs., 386,004,605 contains nine, and 9.023×10^{-6} contains four.
- **Rule 3:** All zeros to the left of the first nonzero digit are *not* significant. For example, 0.0032 contains two sig. figs. and 0.0000070057 contains five (expressed in scientific notation it would be 7.0057×10^{-6}).
- **Rule 4:** Zeroes to the right of the last nonzero digit are significant if a decimal point is present. For example, 5,040.0 contains five sig. figs., 0.000680620 contains six, and 5.40600×10^7 also contains six sig. figs.
- **Rule 5:** Zeroes to the right of the last nonzero digit are *not* significant if no decimal point is present. (Actually, a more correct statement is that I really don't know about those zeroes if no decimal point is present. I would have to know something about how the value was measured. But most scientists use the convention that if no decimal point is present, the zeroes to the right of the last nonzero digit aren't significant.) For example, 68,000 would contain two sig. figs and 70,200 would contain three.

Reporting the correct number of significant figures

For the most part, the *least* precise measured value determines the number of significant figures that you report in your calculation. What values qualify as the least precise measurement varies depending on the mathematical operations involved.

Addition and subtraction

In addition and subtraction, you should report your answer to the number of decimal places used in the number that has the fewest decimal places. For example, suppose you're adding the following amounts:

3.543 mL + 4.27 mL + 7.278 mL + 5.7654 mL

Your calculator will show 20.8564, but you're going to round off to the hundredths place based on the 4.27, because it has the fewest number of decimal places. You then round the figure off to 20.86.

Multiplication and division

In multiplication and division, you can report the answer to the same number of significant figures as the number that has the *least* significant figures. Remember that counted and exact numbers don't count in the consideration of significant numbers. For example, suppose that you're calculating the density in grams per liter of an object that weighs 42.3864 (6 sig. figs.) grams and has a volume of 20.30 milliliters (4 sig. figs.). The setup looks like this:

$$\frac{42.3864\text{g}}{20.30\text{ mL}} \times \frac{1{,}000\text{ mL}}{1\text{ L}}$$

Your calculator will read 2087.014778. You have six significant figures in the first number and four in the second number (the 1000 mL/L does not count because it is an exact conversion). You should have four significant figures in your final answer, so round the answer off to 2087 g/L. Only round off your final answer. Don't round off any intermediate values.

Rounding off numbers

When rounding off numbers, use the following rules:

- **Rule 1:** Look at the first number to be dropped; if it is 5 or greater, drop it and all the numbers that follow it, and increase the last retained number by 1.

  For example, suppose that you want to round off 675.784 to four significant figures. You drop the 8 and the 4. The 8, the first dropped number, is greater than 5, so you increase the retained 7 to 8. Your final answer is 675.8.

- **Rule 2:** If the first number to be dropped is less than 5, drop it and all the numbers that follow it, and leave the last retained number unchanged.

  If you're rounding 6.58457 to three significant figures, you drop the 4, the 5, and the 7. The first number to be dropped is 4, which is less than 5. The 5, the last retained number, stays the same. So you report your answer as 6.58.

Chapter 3

Atomic Structure, the Periodic Table, and Bonding

In This Chapter

- Looking closer at the atom
- Exploring the nucleus
- Understanding electron modeling
- Locating electrons
- Gazing at periodicity
- Utilizing the periodic table
- Identifying bonding, both ionic and covalent

This chapter reviews a plethora of topics, including atomic structure, the periodic table, and ionic and covalent bonding. You're probably thinking, "Wow, what a chapter! It took about a half of a semester in Chem I to learn all of that." This chapter isn't meant to re-teach this material; all I do is to touch upon these important topics briefly and hope that they come back to you. If you're struggling, I suggest you grab your textbook and a copy of my book, *Chemistry For Dummies,* 2nd edition (John Wiley & Sons, Inc.), for more detailed review.

Peering Inside the Atom: Subatomic Particles

As you have been told ever since elementary school, the *atom* is the smallest part of matter that represents a particular element. After the atom was discovered, scientists then found that atoms are composed of certain subatomic particles. It didn't matter what the element, the same *subatomic particles* make up the atom. The number of each of the various subatomic particles is the only thing that varies.

Scientists now recognize that many subatomic particles exist. But in order to be successful in chemistry, you really only need to be concerned with the three major subatomic particles:

- **Protons:** The subatomic particle found in the atom's dense central core that has a positive charge.
- **Neutrons:** The subatomic particle found in the atom's dense central core that has no charge.
- **Electrons:** The subatomic particle found outside the atom's dense central core that has a negative charge.

Table 3-1 summarizes the characteristics of these three subatomic particles.

Table 3-1 The Three Major Subatomic Particles

Name	*Symbol*	*Charge*	*Mass (g)*	*Mass (amu)*	*Location*
Proton	p^+	+1	1.673×10^{-24}	1	Inside the nucleus
Neutron	n^o	0	1.675×10^{-24}	1	Inside the nucleus
Electron	e^-	–1	9.109×10^{-28}	0.0005	Outside the nucleus

Examining the Nucleus: At the Heart of It All

The *nucleus* is very small and very dense when compared to the rest of the atom and serves as the atom's center. The nucleus contains both the protons and neutrons. To give you an idea how small an atom's diameter is, it's measured at around 10^{-10} meters. *Nuclei* (the plural form of nucleus) are around 10^{-15} meters in diameter. For example, if a hydrogen atom were the size of a typical professional football stadium, the nucleus would be about the size of a small marble and would weigh more than 100 million tons.

The protons of an atom are all crammed together inside the nucleus. Forces in the nucleus counteract the repulsion that the protons have for each other and hold the nucleus together.

The nucleus is very small, and it also contains most of the atom's mass. In fact, for all practical purposes, the mass of the atom is the sum of the masses of the protons and neutrons. (Ignoring the minute mass of the electrons is okay unless you're doing very, very precise calculations, which you probably won't be doing in your Chemistry II class.)

The number of protons found in a particular atom is called the *atomic number* and the sum of the number of protons plus the number of neutrons in an atom is called the *mass number.*

An element may be composed of atoms containing different numbers of neutrons. Atoms of the same element (therefore the same number of protons and electrons) that have differing numbers of neutrons are called *isotopes*. Chemists have adopted a certain symbolization as a way of identifying a particular isotope of an element. This symbolization is shown in Figure 3-1.

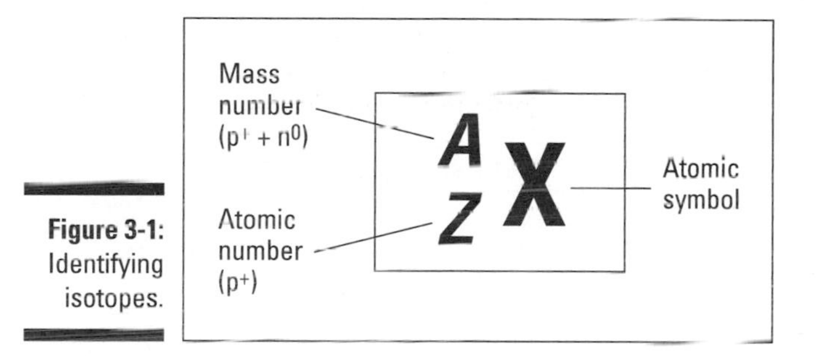

Figure 3-1: Identifying isotopes.

Look at Figure 3-1. *X* represents the chemical symbol as found on the periodic table. *Z* represents the atomic number — the number of protons in the nucleus. *A* represents the mass number, the sum of the number of protons plus neutrons. The mass number (also called the *atomic weight*) is listed in amu.

Modeling Electrons

When scientists first developed early models of the atom, they had electrons spinning around the nucleus in a random fashion or similar to planets orbiting the sun. Scientists today have a much better idea of the atomic structure. They use two models: the Bohr model and the quantum mechanical model. Both are helpful in understanding the atom. The following sections give you a brief overview of these two models.

A model is useful because it helps you understand what's observed in nature. Having more than one model represent and help people understand a particular topic isn't unusual.

Considering the not-so Bohr(ing) model

The Bohr model, sometimes called the *planetary model,* is simple and relatively easy to understand. This model shows that the electrons in atoms are in orbits of differing energy around the nucleus. Niels Bohr, a Danish scientist, used the term *energy levels* (or *shells*) to describe these orbits of differing energy. He determined that the energy of an electron is *quantized,* meaning electrons can be in one energy level or another but not in between.

An electron's energy level is normally called its *ground state.* But the electron can move to a higher-energy, less-stable level, or shell, by absorbing energy. This higher-energy, less-stable state is the electron's *excited state.* After the electron is done being excited, it can return to its original ground state by releasing the energy it has absorbed.

He discovered that the closer an electron is to the nucleus, the less energy it possesses. The farther away it is, the more energy it possesses. Hence, Bohr assigned numbers to the electron's energy levels. The higher the energy-level number is, the farther away the electron is from the nucleus — and the higher the energy.

Bohr also uncovered something else interesting. The various energy levels can hold differing numbers of electrons: energy level 1 may hold up to two electrons; energy level 2 may hold up to eight electrons, and so on.

This model works well for very simple atoms such as hydrogen (which has one electron). However it doesn't work the best for more complex atoms, such as carbon or even helium with its two electrons. Although some entry-level textbooks still use this model today, a more sophisticated (and complex) model — the quantum mechanical model — is used much more frequently (check out the next section).

Looking at the quantum mechanical model

The quantum mechanical model is based on *quantum theory,* which says matter also has properties associated with waves. This model can explain observations made on complex atoms. According to quantum theory, knowing the exact position and *momentum* (speed and direction) of an electron at the same time is impossible, which is called the *uncertainty principle.* So scientists had to replace Bohr's simple circular orbits with complex shapes *orbitals* (sometimes called *electron clouds*), volumes of space in which an electron is *likely* to be. In other words, probability replaced certainty.

Trying to Find Those Electrons

Both chemists and chemistry students find it useful to represent which energy level, subshell, and orbital are occupied by electrons in any particular atom. These representations are also useful in showing why certain elements behave in similar ways. The two most useful representations are

- Energy-level diagrams
- Electron configurations

In this section, I show you how to use an energy-level diagram and write electron configurations.

Diagramming energy levels

One way chemists and chemistry students can do things like figure out bonding or molecular geometry is by diagramming energy levels. Figure 3-2 is a blank energy-level diagram you can use to depict electrons for any particular atom. This figure doesn't show all the known orbitals and subshells, but with this diagram, you can do just about anything you need to. If you don't remember what orbitals, subshells, or all those numbers and letters in the figure have to do with the price of beans, check out *Chemistry For Dummies,* 2nd Edition (John Wiley & Sons, Inc.) for a refresher.

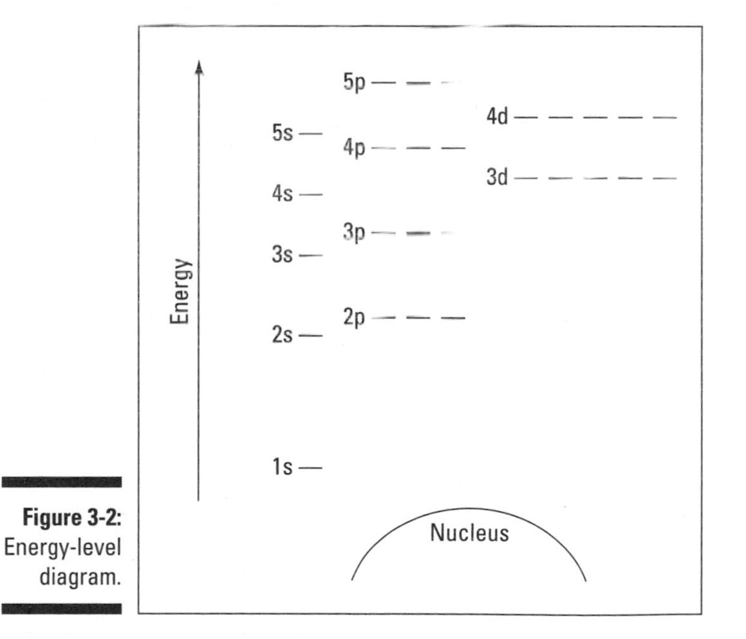

Figure 3-2: Energy-level diagram.

Orbitals are represented with dashes in which you can place a maximum of two electrons. The 1s orbital is closest to the nucleus and has the lowest energy. It's also the only orbital in energy level 1. At energy level 2, both s and p are orbitals, with the 2s having lower energy than the 2p. Three dashes of the same energy represent the three 2p subshells. This figure also shows energy levels 3, 4, and 5. Notice that the 4s has lower energy than the 3d. This exception isn't what you may have thought, but it's what's observed in nature.

Go figure. Speaking of which, Figure 3-3 shows the *aufbau principle,* a method for remembering the order in which orbitals fill the vacant energy levels.

Figure 3-3: Aufbau filling chart.

In using the energy-level diagram (Figure 3-2), remember two things:

- Electrons fill the lowest vacant energy levels first.
- When more than one subshell lies at a particular energy level, such as at the 3p or 4d levels, only one electron fills each subshell until each subshell has one electron. Then electrons start pairing up in each subshell. This rule is named *Hund's rule.*

Figure 3-4 is the energy-level diagram for sodium.

Configuring electrons: Easy and space efficient

When you have to figure out chemical reactions and bonding, using energy-level diagrams is helpful. The downside: They're bulky. The *electron configuration* gives just about the same information but in a much more concise, shorthand-notation form.

The electron configuration for sodium is $1s^2 2s^2 2p^6 3s^1$. Compare that notation with the energy-level diagram for sodium in Figure 3-4. You can derive the electron configuration from the energy-level diagram. The first two electrons in sodium fill the 1s orbital, so you show it as $1s^2$ in the electron configuration. The 1 is the energy level, the s represents the type of orbital, and the superscript 2 represents the number of electrons in that orbital. The next two electrons are in the 2s orbital, so you write $2s^2$. Then you show the 6 electrons in the 2p orbital as $2p^6$. Finally, you have a single electron in the 3s orbital, shown as $3s^1$. Put it all together, and you get $1s^2 2s^2 2p^6 3s^1$.

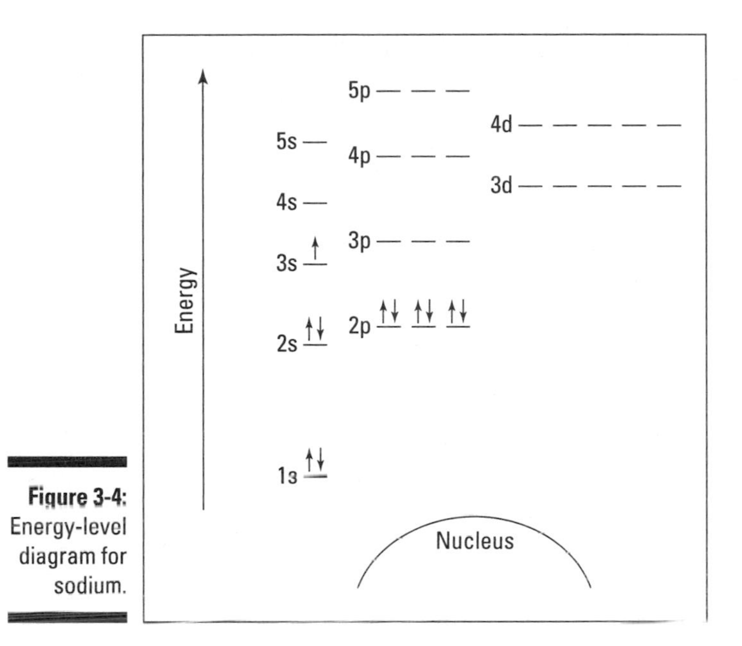

Figure 3-4: Energy-level diagram for sodium.

The sum of the superscript numbers equals the atomic number, or the number of electrons in the atom. Also recall that those electrons in the outermost energy level are called *valence electrons.* These electrons are gained, lost, or shared during a chemical reaction.

Peering at Patterns of Periodicity

In the mid-1800s, Dmitri Mendeleev, a Russian chemist, identified a repeating pattern of chemical properties in the elements that were known at the time. He arranged the elements in order of increasing atomic mass to form something that fairly closely resembles the modern periodic table. He even predicted the properties of some of the then-unknown elements. Later, the elements were rearranged in order of increasing *atomic number,* the number of protons in the nucleus of the atom.

Today chemistry students and chemists can't fathom doing much in chemistry without the periodic table. Of course mastering the properties of all 118+ elements isn't the best use of your time because more are created almost every year; however, you, and other chemists, can grasp the properties of families of elements. Finding the relationships among elements and figuring out the formulas of many different compounds is easy by simply referring to the periodic table. The table readily provides atomic numbers, mass numbers, and information about the number of valence electrons, outermost s and p electrons. Figure 3-5 shows the periodic table.

PERIODIC TABLE OF THE ELEMENTS

Period	1 IA	2 IIA	3 IIIB	4 IVB	5 VB	6 VIB	7 VIIB	8 VIIIB	9 VIIIB
1	1 H Hydrogen 1.00797								
2	3 Li Lithium 6.939	4 Be Beryllium 9.0122							
3	11 Na Sodium 22.9898	12 Mg Magnesium 24.312							
4	19 K Potassium 39.102	20 Ca Calcium 40.08	21 Sc Scandium 44.956	22 Ti Titanium 47.90	23 V Vanadium 50.942	24 Cr Chromium 51.996	25 Mn Manganese 54.9380	26 Fe Iron 55.847	27 Co Cobalt 58.9332
5	37 Rb Rubidium 85.47	38 Sr Strontium 87.62	39 Y Yttrium 88.905	40 Zr Zirconium 91.22	41 Nb Niobium 92.906	42 Mo Molybdenum 95.94	43 Tc Technetium (99)	44 Ru Ruthenium 101.07	45 Rh Rhodium 102.905
6	55 Cs Cesium 132.905	56 Ba Barium 137.34	57 La Lanthanum 138.91	72 Hf Hafnium 179.49	73 Ta Tantalum 180.948	74 W Tungsten 183.85	75 Re Rhenium 186.2	76 Os Osmium 190.2	77 Ir Iridium 192.2
7	87 Fr Francium (223)	88 Ra Radium (226)	89 Ac Actinium (227)	104 Rf Rutherfordium (261)	105 Db Dubnium (262)	106 Sg Seaborgium (266)	107 Bh Bohrium (264)	108 Hs Hassium (269)	109 Mt Meitnerium (268)

Lanthanide Series	58 Ce Cerium 140.12	59 Pr Praseodymium 140.907	60 Nd Neodymium 144.24	61 Pm Promethium (145)	62 Sm Samarium 150.35	63 Eu Europium 151.96
Actinide Series	90 Th Thorium 232.038	91 Pa Protactinium (231)	92 U Uranium 238.03	93 Np Neptunium (237)	94 Pu Plutonium (242)	95 Am Americium (243)

Figure 3-5: The periodic table.

10 VIIIB	11 IB	12 IIB	13 IIIA	14 IVA	15 VA	16 VIA	17 VIIA	18 0
								2 He Helium 4.0026
			5 B Boron 10.811	6 C Carbon 12.01115	7 N Nitrogen 14.0067	8 O Oxygen 15.9994	9 F Fluorine 18.9984	10 Ne Neon 20.183
			13 Al Aluminum 26.9815	14 Si Silicon 28.086	15 P Phosphorus 30.9738	16 S Sulfur 32.064	17 Cl Chlorine 35.453	18 Ar Argon 39.948
28 Ni Nickel 58.71	29 Cu Copper 63.546	30 Zn Zinc 65.37	31 Ga Gallium 69.72	32 Ge Germanium 72.59	33 As Arsenic 74.9216	34 Se Selenium 78.96	35 Br Bromine 79.904	36 Kr Krypton 83.80
46 Pd Palladium 106.4	47 Ag Silver 107.868	48 Cd Cadmium 112.40	49 In Indium 114.82	50 Sn Tin 118.69	51 Sb Antimony 121.75	52 Te Tellurium 127.60	53 I Iodine 126.9044	54 Xe Xenon 131.30
78 Pt Platinum 195.09	79 Au Gold 196.967	80 Hg Mercury 200.59	81 Tl Thallium 204.37	82 Pb Lead 207.19	83 Bi Bismuth 208.980	84 Po Polonium (210)	85 At Astatine (210)	86 Rn Radon (222)
110 Ds Darmstadtium (281)	111 Rg Roentgenium (286)	112 Cn Copernicium (285)	113 Uut Ununtrium (284)	114 Uuq Ununquadium (289)	115 Uup Unupentium (288)	116 Uuh Ununhexium (298)	117 Uus Ununseptium (294)	118 Uuo Ununoctium (294)

64 Gd Gadolinium 157.25	65 Tb Terbium 158.924	66 Dy Dysprosium 162.50	67 Ho Holmium 164.930	68 Er Erbium 167.26	69 Tm Thulium 168.934	70 Yb Ytterbium 173.04	71 Lu Lutetium 174.97
96 Cm Curium (247)	97 Bk Berkelium (247)	98 Cf Californium (251)	99 Es Einsteinium (254)	100 Fm Fermium (257)	101 Md Mendelevium (258)	102 No Nobelium (259)	103 Lr Lawrencium (260)

Using the Periodic Table

In your Chemistry I course, you probably used the periodic table with different experiments and exercises. In Chemistry II, you'll use the periodic table even more, so get used to it. Spend time with it and treat it like a good friend.

The elements are arranged in order of increasing atomic number on the periodic table. Here is a review of what you'll find on the periodic table:

- **Atomic number:** The atomic number (number of protons) is located right above the element symbol.
- **Atomic mass:** Under the element symbol is the atomic mass, or *atomic weight* (sum of the protons and neutrons). Atomic mass is a *weighted average* of all naturally occurring isotopes. (And if that's Greek to you, just flip to Chapter 4 for tons of fun with atomic mass and isotopes.)

  Notice also that two rows of elements — Ce–Lu (commonly called the *lanthanides*) and Th–Lr (the *actinides*) — have been pulled out of the main body of the periodic table. If they were included in the main body of the periodic table, the table would be much wider.

You can classify the elements in many ways, which the following sections describe. Two quite useful ways are

- **Metals, nonmetals, and metalloids:** These categories refer to large groupings of elements based on certain physical properties such as conductivity.
- **Families and periods:** These categories refer to the columns and rows, respectively, of the periodic table.

Organizing by metals, nonmetals, and metalloids

One way you can use the periodic table is by categorizing the elements in three of the following ways:

Metals

Most of the elements in the periodic table are considered *metals.* All the elements on the left-hand side and in the middle of the periodic table are metals, with one notable exception — hydrogen, which is the first element and is unique in its properties. Scientists stick it above lithium, but it doesn't react like a metal.

The metals have properties that you normally associate with the metals you encounter in everyday life. They are solid (with the exception of mercury, Hg, a liquid), shiny, good conductors of electricity and heat, *ductile* (they can be drawn into thin wires), and *malleable* (they can be easily hammered into very thin sheets). And all these metals easily tend to lose electrons.

Nonmetals

Except for a few elements that border the metals on the right, the elements to the right of the periodic table are classified as *nonmetals* (along with hydrogen).

Nonmetals have properties opposite those of the metals. The nonmetals are brittle, not malleable or ductile, poor conductors of both heat and electricity, and tend to gain electrons in chemical reactions. Some nonmetals are liquids; some are gases.

Metalloids

The elements that border the metals and the nonmetals are *metalloids.* The metalloids, or *semimetals,* have properties that are somewhat of a cross between metals and nonmetals. They tend to be economically important because of their unique conductivity properties (they only partially conduct electricity), which make them valuable in the semiconductor and computer chip industry.

Hydrogen: The most common element

In many ways, hydrogen is a unique element. Hydrogen is the key element to the acids and is also important to hydrogen bonding. About 90 percent of all the atoms in the universe are hydrogen atoms (nearly 10 percent are helium atoms).

Hydrogen exists as a colorless, odorless, diatomic gas at room temperature and is extremely flammable. Flammable gases are only flammable within certain concentration limits. The limits for hydrogen are wider than for any other flammable material, making it more hazardous than many other flammable gases. Even very low amounts of hydrogen in air will burn.

Having an understanding of the element hydrogen can help you grasp its reactions (in acid compounds and so on) and in its intermolecular interactions, as in hydrogen bonding, which plays such a big role in aqueous chemistry and in biochemistry.

The placement of hydrogen in the IA family on the periodic table is a consequence of its electron configuration, $1s^1$. Its placement in the IA family is somewhat misleading, because in many ways hydrogen doesn't behave like other elements in the same family. Different periodic tables treat hydrogen slightly different as a result. Wherever hydrogen is placed, make sure to remember that hydrogen is an exception to nearly every rule or trend that you have learned. For example, the electronegativity of hydrogen places it between boron and carbon.

Some periodic tables attempt to circumvent this problem by placing hydrogen at the top of both Group 1 and Group 17. Like the elements in Group 1, hydrogen can form a +1 ion. Like the elements in Group 17, hydrogen can form a –1 ion. Therefore, the dual placement helps to illustrate the ability of hydrogen to form both cations and anions.

Organizing by periods and families

Another way you can categorize elements with the periodic table is through the use of the following two methods:

- **Periods:** These are the horizontal rows on the periodic table. The periods are numbered 1 through 7 on the left-hand side of the table. The atomic numbers increase from left to right in each period.

  Even though they're in the same period, these elements have chemical properties that aren't all that similar. Consider the first two members of period 4: potassium (K) and calcium (Ca). In reactions, they both tend to lose electrons (after all, they are metals), but potassium loses one electron, whereas calcium loses two. Bromine (Br), down near the end of the period, tends to gain an electron (it's a nonmetal).

- **Families:** These are the vertical columns. The families, also called *groups,* may be labeled at the top of the columns in one of two ways:
  - **Roman numerals and letters:** Many chemists (especially old ones like me) prefer this older method and still use this method.
  - **Numbers 1 through 18:** The newer method simply uses 1–18.

    I use the older method in describing the features of the table because relating the position of an element on the periodic table with its number of valence electrons is much easier than the newer method.

  The members of a family do have similar properties. Consider the IIA family, starting with Beryllium (Be) and going through Radium (Ra). All these elements tend to lose two electrons in reactions. And all the members of the VIIA family tend to gain one electron.

Reviewing Bonding

If chemistry was simply about elements, I would have chosen a more exciting career, like accounting. But chemistry is far more than the properties of elements — it's about chemical reactions and chemical bonding. During chemical reactions, bonds are broken and new bonds are formed. Electrons are lost and gained or shared. New substances are formed; old substances disappear. Heat is given off or absorbed. Gases may be formed and, yes, even fires and explosions may occur. Chemistry is the breaking and making of bonds.

Two basic types of bonding — ionic and covalent — happen in chemistry. In the next few sections I review these types of bonding.

Attracting opposites: Ionic bonding

Ionic bonding is the type of bonding that holds salts together. Here I briefly discuss simple ions and polyatomic ions: how they form and how they combine. I also show you how to predict the formulas of ionic compounds and how chemists detect ionic bonds.

Sodium + Chlorine = Table Salt

The process of creating table salt is pretty amazing. You take two very hazardous substances, and from them you make a substance that's necessary for life.

Sodium is an alkali metal, a member of the IA family on the periodic table. The Roman numerals at the top of the A families show the number of valence electrons (s and p electrons in the outermost energy level) in the particular element. So sodium has 1 valence electron and 11 total electrons because its atomic number is 11. Its electron configuration would be:

Sodium (Na)→$1s^2 2s^2 2p^6 3s^1$

Sodium has one valence electron; by the octet rule, it becomes stable when it has eight valence electrons. Two possibilities exist for sodium to become stable: It can gain seven more electrons to fill energy level 3, or it can lose the one 3s electron so that energy level 2 (which is already filled with eight electrons) becomes the valence energy level. In general, the loss or gain of one, two, or sometimes even three electrons can occur, but an element doesn't ordinarily lose or gain more than three electrons. So to gain stability, sodium loses its 3s electron. At this point, it has 11 protons (11 positive charges) and 10 electrons (10 negative charges). The once-neutral sodium atom now has a single positive charge [11(+) plus 10(–) equals 1+]. It's now an *ion,* an atom that has a charge due to the loss or gain of electrons. And ions that have a positive charge (such as sodium) due to the loss of electrons are called *cations.* You can write an electron configuration for the sodium cation:

Na^+ $1s^2 2s^2 2p^6$

The sodium ion (cation) has the same electron configuration as neon, so it's *isoelectronic* with neon.

On the other hand, chlorine is a member of the halogen family — the VIIA family on the periodic table. It has 7 valence electrons and a total of 17 electrons. Its electron configuration would be:

Chlorine (Cl) $1s^2 2s^2 2p^6 3s^2 3p^5$

Chlorine has 7 valence electrons. To obtain its full octet, it must lose the 7 electrons in energy level 3 or gain one at that level. Because elements don't generally gain or lose more than three electrons, chlorine must gain a single electron to fill energy level 3. At this point, chlorine has 17 protons (17 positive charges) and 18 electrons (18 negative charges). So chlorine becomes an ion with a single negative charge (Cl^-). The neutral chlorine atom becomes the chloride ion. Ions with a negative charge due to the gain of electrons are called *anions.* The electronic configuration for the chloride anion is

Cl^- $1s^2 2s^2 2p^6 3s^2 3p^6$

The chloride anion is isoelectronic with argon.

Sodium achieves its full octet and stability by losing an electron. Chlorine fills its octet by gaining an electron. If the two are in the same container, then the electron sodium loses can be the same electron chlorine gains. The transfer of an electron creates ions — cations (positive charge) and anions (negative charge) — and opposite charges attract each other. The Na^+ cation attracts the Cl^- anion and forms the compound NaCl, or table salt. This is an example of an *ionic bond,* which is a *chemical bond* (a strong attractive force that keeps two chemical elements together) that comes from the *electrostatic attraction* (attraction of opposite charges) between cations and anions.

The compounds that have ionic bonds are commonly called *salts.* Cations and anions can have more than one unit of positive or negative charge if they lose or gain more than one electron. In this fashion, many different kinds of salts are possible.

Ions: Cations, anions, and polyatomics

Other salts are formed in the same fashion as sodium chloride. Some metal loses one or more electrons, and some nonmetal gains those electrons. Cations and anions are formed, and the attraction between the positive charge of the cation and negative charge of the anion brings the particles together (this is called an *electrostatic attraction*) and creates the ionic compound.

A metal reacts with a nonmetal to form an ionic bond.

You can often use the element's position on the periodic table to determine the ion's charge. For example, all the alkaline earth metals (the IIA elements) lose two electrons to form a cation with a 2+ charge. In the same way, the alkali earth metals (IA elements) lose a single electron to form a 1+ cation. Aluminum, a member of the IIIA family, loses three electrons to form a 3+ cation.

By the same reasoning, the members of the oxygen family (VIA elements) all have six valence electrons. All these elements gain two electrons to fill their valence energy level. And all of them form an anion with a 2– charge. The VIIA elements (the halogens) gain one electron to form anions with a 1– charge, and the VA elements gain three electrons to form anions with a 3– charge.

The electrical charge that an atom achieves is sometimes called its *oxidation state or oxidation number.* Many of the ions derived from the transition metals have more than one oxidation states.

Determining the number of electrons that members of the transition metals (the B families) lose is more difficult. In fact, many of these elements lose a varying number of electrons so that they form two or more cations with different charges.

Ions aren't always *monoatomic,* composed of just one atom. Ions can also be *polyatomic,* composed of a group of atoms. The nitrate ion, for example, is NO_3^-.

The symbol for the nitrate ion, NO_3^-, indicates that one nitrogen atom and three oxygen atoms are bonded together and that the whole polyatomic ion has one extra electron (as evidenced by the single negative charge).

For tables of the common monoatomic cations and anions, as well as polyatomic ions, see my book, *Chemistry For Dummies,* 2nd Edition, (John Wiley & Sons, Inc.).

Ionic compounds

Forming an ionic compound consists of the cation and anion attracting to each other, which results in a salt's formation. Don't forget: The compound must be *neutral* — it must have equal numbers of positive and negative charges. In this section I explain how to predict the formula of an ionic compound simply by considering the electronic configurations and/or the charges on the cations and anions.

Suppose you want to know the *formula,* or composition, of the compound that results from reacting calcium with iodine.

The electron configurations for calcium and iodine are

Calcium (Ca) $1s^2 2s^2 2p^6 3s^2 3p^6 4s^2$

Iodine (I) $1s^2 2s^2 2p^6 3s^2 3p^6 4s^2 3d^{10} 4p^6 4d^{10} 5s^2 5p^5$

Calcium, a member of the alkaline earth metal family, loses two valence electrons to form a cation with a 2+ charge. The electron configuration for the calcium cation is

Ca^{2+} $1s^2 2s^2 2p^6 3s^2 3p^6$

Iodine is a halogen that has seven valence electrons, so it gains one to complete its octet (eight valence electrons) and form the iodide anion with a 1– charge. The electron configuration for the iodide anion is

I^{1-} $1s^2 2s^2 2p^6 3s^2 3p^6 4s^2 3d^{10} 4p^6 4d^{10} 5s^2 5p^6$

Note that if an ion simply has 1 unit of charge, positive or negative, you don't have to write the 1; you just use the plus or minus symbol, with the 1 being understood. But I like to use the 1 in addition to the charge for clarity (so that people understand that I just didn't forget to write the number).

The compound must be neutral; it must have a zero charge. It achieves this by having equal numbers of positive and negative charges. The calcium ion has a 2+, so it requires 2 iodide anions, each with a single negative charge, to balance the 2 positive charges of calcium. So the formula of the compound that results from reacting calcium with iodine is CaI_2.

The *crisscross rule* is a quick way to determine the formula of an ionic compound. The crisscross rule uses the ionic charges of the ions to predict the formula of the ionic compound. It doesn't work all the time, but it is a good way of checking your result using the previous method.

Look at Figure 3-6 for an example of using this rule. Take the numerical value of the metal ion's superscript (forget about the charge symbol) and move it to the bottom right-hand side of the nonmetal's symbol — as a subscript. Then take the numerical value of the nonmetal's superscript and make it the subscript of the metal. (Note that if the numerical value is 1, it's just understood and not shown.) So in this example, you make aluminum's 3 a subscript of sulfur and make sulfur's 2 a subscript of aluminum, and you get the formula Al_2S_3.

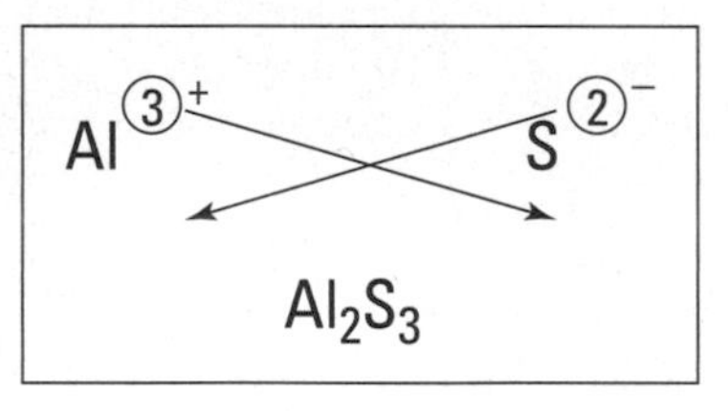

Figure 3-6: Figuring the formula of aluminum sulfide.

Compounds involving polyatomic ions work exactly the same way as with the monatomic ions. For example, here's the compound made from the ammonium cation (NH_4^+) and the carbonate anion (CO_3^{2-}):

$(NH_4)_2CO_3$

Notice that because two ammonium ions (two positive charges) are needed to neutralize the two negative charges of the carbonate ion, the ammonium ion is enclosed in parentheses and a subscript 2 is added.

The crisscross rule works very well, but you have to be careful if both ions have the same numeral in the superscript. Suppose that you want to write the compound formed when strontium reacts with sulfur. Strontium, an alkaline earth metal, forms a 2+ cation, and sulfur forms a 2– anion. So you might predict that the formula is:

Sr_2S_2

But this formula is incorrect. After you use the crisscross rule, you need to reduce all the subscripts by a common factor, if possible. In this case, you divide each subscript by 2 and get the correct formula:

SrS

Comprehending covalent bonding basics

Covalent bonding occurs between two nonmetals. The properties of these two types of compounds are different; covalently bonded compounds can be solids, liquids, or gases. Furthermore covalent compounds have a much lower melting point than ionic compounds (salts), and covalent compounds tend to be nonelectrolytes.

Understanding multiple bonds

Covalent bonding is the sharing of one *or more* electron pairs. In hydrogen and most other diatomic molecules, only one electron pair is shared. But in many covalent bonding situations, more than one electron pair is shared.

It takes energy to break a covalent bond, which is called its *bond strength.* More energy generally is needed to break a double bond than a single bond if the same elements are involved. For example, a carbon-to-oxygen double bond (two shared pairs of electrons) has a higher bond strength (requires more energy to break the bond) than a carbon-to-oxygen single bond. The double bond isn't twice as strong as a single bond, but is considerably greater. And a triple bond is stronger yet. Chemists also observe that multiple bonds are shorter in bond length (the distance between the nuclei of the bonded atoms) than single bonds — double bonds shorter than single bonds and triple bonds shorter than double bonds.

Nitrogen (N_2) is a diatomic molecule in the *VA family* on the periodic table, meaning that it has five valence electrons. So nitrogen needs three more valence electrons to complete its octet. A nitrogen atom can fill its octet by sharing three electrons with another nitrogen atom, forming three covalent bonds, a so-called *triple bond.* A triple bond isn't quite three times as strong as a single bond, but it's a very strong bond. In fact, the triple bond in nitrogen is one of the strongest bonds known. This strong bond is what makes nitrogen gas very stable and resistant to reaction with other chemicals. It's also why many explosive compounds (such as TNT and ammonium nitrate) contain nitrogen.

Seeing structural formula

If you want to write a formula to stand for the exact compound you have in mind, you often must write the structural formula. The *structural formula* shows the elements in the compound, the exact number of each atom in the compound, and the bonding pattern for the compound. Two examples of structural formulas are the electron-dot formula and Lewis formula.

The following steps explain how to write the electron-dot formula for a simple molecule. The electron-dot formula allows you to keep track of where the valence electrons are located. Doing so helps in following chemical reactions and determining molecular geometry. The Lewis structural formula is similar, except that it replaces the dots representing a bonding pair of electrons with a dash.

1. **Write a skeletal structure showing a reasonable bonding pattern using just the element symbols.**
2. **Take all the valence electrons from all the atoms and throw them into an imaginary container that I call an *electron pot.***
3. **Use the $N - A = S$ equation to figure the number of bonds in this molecule. In this equation,**

   N equals the sum of the number of valence electrons needed by each atom. N has only two possible values — 2 or 8.

   A is the number of valence electrons in your electron pot — the sum of the number of valence electrons available for each atom.

   S equals the number of electrons *s*hared in the molecule. And if you divide S by 2, you have the number of covalent bonds in the molecule.
4. **Distribute the electrons from your electron pot to account for the bonds.**
5. **Distribute the rest of the electrons (normally in pairs) so that each atom achieves its full octet of electrons.**

For a more in depth discussion of writing electron-dot (Lewis structures), see my book, *Chemistry For Dummies,* 2nd Edition (John Wiley & Sons, Inc.).

Notice that this structural formula actually shows two types of electrons: *bonding electrons,* the electrons that are shared between two atoms, and *nonbonding electrons,* the electrons that are not being shared.

Attracting bonding electrons: Electronegativities

What happens when the two atoms involved in a bond aren't the same? The two positively charged nuclei have different attractive forces; they "pull" on the electron pair to different degrees. The end result is that the electron pair is shifted toward one atom. But the question is, "Which atom does the electron pair shift toward?" Electronegativities provide the answer.

Electronegativity is the strength an atom has to attract a bonding pair of electrons to itself. The larger the value of the electronegativity, the greater the atom's strength to attract a bonding pair of electrons. With a few exceptions, the electronegativities increase from left to right, in a period, and decrease down a family.

Electronegativities are useful because they give information about what will happen to the bonding pair of electrons when two atoms bond. Basically three types of bonds can be formed.

- **Ionic bonding:** In this case the bonding electrons are totally removed from one of the atoms and ions are formed. Look at a case in which the two atoms forming a compound have extremely different electronegativities — potassium bromide (KBr). Potassium bromide is ionically bonded (metal + nonmetal). An electron has transferred from potassium to bromine. Potassium has an electronegativity of 0.8, and bromine has an electronegativity of 2.8. That's an electronegativity difference of 2.0 (2.8 – 0.8), making the bond between the two atoms very, very polar.

- **Nonpolar covalent bond:** This bond has an electron pair that is equally shared. You have a nonpolar covalent bond anytime the two atoms involved in the bond are the same or anytime the difference in the electronegativities of the atoms involved in the bond is very small. For example, look at the O_2 molecule. Oxygen has an electronegativity value of 3.5. Each oxygen atom attracts the bonding electrons with a force of 3.5. Because there's an equal attraction, the bonding electron pair is shared equally between the two oxygen atoms and is located halfway between the two atoms.

- **Polar covalent bond:** The electron pair in this bond is shifted toward one atom. The atom that more strongly attracts the bonding electron pair is slightly more negative, while the other atom is slightly more positive. The larger the difference in the electronegativities, the more negative and positive the atoms become.

  Consider hydrogen bromide (HBr). Hydrogen has an electronegativity of 2.1, and bromine has an electronegativity of 2.8. The electron pair that is bonding HBr together shifts toward the bromine atom because it has a larger electronegativity value.

Considering polar covalent bonding

If the two atoms involved in the covalent bond aren't the same, the bonding pair of electrons is pulled toward one atom, with that atom taking on a slight (partial) negative charge and the other atom taking on a partial positive charge. In many cases, the molecule then has a positive end and a negative end. This molecule can be called a *dipole.*

In hydrogen fluoride (HCl), the bonding electron pair is pulled much closer to the fluorine atom than to the hydrogen atom, so the fluorine end becomes partially negatively charged and the hydrogen end becomes partially positively charged. The same thing takes place in ammonia (NH_3); the nitrogen has a greater electronegativity than hydrogen, so the bonding pairs of electrons are more attracted to it than to the hydrogen atoms. The nitrogen atom takes on a partial negative charge, and the hydrogen atoms take on a partial positive charge.

Having a polar covalent bond present explains why some substances act the way they do in a chemical reaction. Because this type of molecule has a positive end and a negative end, it can attract the part of another molecule with the opposite charge.

Hydrogen fluoride (HF) has many unusual properties because of its polar covalent bond. Fluorine has a larger electronegativity than hydrogen, so the electron pair are pulled in closer to the fluorine atom, giving it a partial negative charge. Subsequently, the hydrogen atom takes on a partial positive charge.

Hydrogen fluoride is a dipole and acts like a magnet, with the fluorine end having a negative charge and the hydrogen end having a positive charge. These charged ends can attract other hydrogen fluoride molecules. The partially negatively charged fluorine atom of one hydrogen fluoride molecule can attract the partially positively charged hydrogen atom of another hydrogen fluoride molecule. This attraction between the molecules occurs frequently and is a type of *intermolecular force* (force between different molecules).

Intermolecular forces can be of three different types, listed from the weakest to the strongest:

- **London force:** Also called the *dispersion force,* this force is a very weak type of attraction that generally occurs between nonpolar covalent molecules, such as oxygen (O_2), nitrogen (N_2), or hydrogen (H_2). It results from the ebb and flow of the electron orbitals, giving a very weak and very brief charge separation around the bond.
- **Dipole-dipole interaction:** This intermolecular force occurs when the positive end of one dipole molecule is attracted to the negative end of another dipole molecule. It's much stronger than a London force, but it's still pretty weak.
- **Hydrogen bond:** The third type of interaction is really just an extremely strong dipole-dipole interaction that occurs when a hydrogen atom on one molecule is bonded to one of three extremely electronegative elements — O, N, F, and sometimes Cl — on another molecule. These four elements have a very strong attraction for the bonding pair of electrons, so the atoms involved in the bond take on a large amount of partial charge. This bond turns out to be highly polar — and the higher the polarity, the more effective the bond. It's only about 5 percent of the strength of an ordinary covalent bond, but still very strong for an intermolecular force. The hydrogen bond is the type of interaction that's present in water.

Chapter 4

Digging Up the Mole Concept and Stoichiometry

In This Chapter

- Getting acquainted with the mole
- Understanding reaction stoichiometry

When chemists make new substances, a process called *synthesis*, a number of questions come to mind. These questions include "How much?", "How much of this reactant do I need to make this much product?", and "How much product can I make with this much reactant?" To answer these questions, chemists take a balanced chemical equation, expressed in terms of atoms and molecules, and convert it to grams or pounds or tons — some type of unit they can actually weigh in the lab. The mole concept enables chemists to move from the microscopic world of atoms and molecules to the real world of grams and kilograms and is one of the most important central concepts in chemistry. In this chapter, I remind you of the mole concept.

Crossing the Mighty Mole Bridge

When humans deal with objects, they often think in terms of a convenient amount. For example, when you buy socks, you buy a pair. When you go to the hardware store, you buy a dozen bolts. And when I go the office supply store, I buy copy paper by the ream.

People use words to represent numbers all the time — a pair (2) of earrings, a dozen (12) eggs, and a ream (500) of paper. All these words represent units of measure, and they represent a convenient number for the objects they're used to measure. Rarely would you buy a ream of eggs or a pair of bolts.

Likewise, when chemists deal with atoms and molecules, they need a unit that takes into consideration the very small size of atoms and molecules, a unit that is convenient for these small objects. The *mole* is that unit. The mole is central to understanding the quantitative aspects of chemistry. Without it, chemistry students and chemists could never answer those "how much" questions. The following sections show you how to count particles with a mole.

Getting a firm grip on the mole

The world *mole* stands for a number — 6.022×10^{23}. A mole is commonly called *Avogadro's number,* named for Amedeo Avogadro, the scientist who laid the groundwork for the mole principle.

If you had a mole of unpopped popcorn and made them into a ball, it would be a sphere of about 600,000 meters. A mole of moles, those furry little creatures that make such a mess of my yard, would have a mass approximately equal to the moon; laid end-to-end (not a pretty sight) they would stretch about 65 trillion miles. Now that is a mole bridge to the stars!

Avogadro's number is a number that represents a very large number of very small particles, especially atoms and molecules. But how does it relate to the macroscopic world of grams and pounds and tons in which you and I live?

The answer to that question is that in the macroscopic world a *mole* (abbreviated as *mol*) is the number of particles in exactly 12 grams of carbon-12. So if you have exactly 12 grams of ^{12}C, you have Avogadro's number (6.022×10^{23}) of carbon atoms. This is also a mole of ^{12}C atoms. For any other element, a mole is the atomic mass (weight) of that element expressed in grams. For a compound, a mole is the formula (or molecular) mass of that compound expressed in grams.

The weight of a carbon dioxide (CO_2) molecule is 44.01 amu. Because a mole is the formula (or molecular) mass of a compound expressed in grams, you can now say that the mass of a mole of carbon dioxide is 44.01 grams. You can also say that 44.01 grams of carbon dioxide contains 6.022×10^{23} CO_2 molecules, or a mole of carbon dioxide. And the mole of carbon dioxide is composed of one mole of carbon and two moles of oxygen.

REMEMBER

The mole is the bridge between the microscopic world of atoms and molecules and the macroscopic world in which you live.

$$6.022 \times 10^{23} \text{ particles} = 1 \text{ mole} = \text{atomic/formula/molecular mass in grams}$$

If you have any one of those three things — particles, moles, or grams — then you can calculate the other two.

Going from grams to moles and back again

In chemistry the relationship between the mole and grams is very important. It allows you to answer those "how much" questions. It allows you to consider a balanced chemical equation in terms of grams (or any other mass units) and not just individual particles. This section provides some examples to illustrate this relationship.

Suppose you want to know how many moles of ammonia are in 50.0 grams of ammonia. You probably immediately thought, "I need the formula and molar mass of ammonia." And you would be correct. The formula of ammonia is NH_3. It has a molar mass of 14.01g/mole + 3(1.010g/mole) = 17.04 g/mole. That means that if you had 17.04 grams of ammonia, you would have 1 mole of ammonia (and 6.022×10^{23} ammonia molecules). Now you can use the unit conversion method to go from grams of ammonia to moles of ammonia:

$$\frac{50.0\text{ g NH}_3}{1} \times \frac{1\text{ mole}}{17.04\text{ g}} = 2.93\text{ moles of NH}_3$$

Finally, make sure that you have rounded to the correct number of significant figures and that the answer is reasonable. Refer to Chapter 2 for help in rounding.

Or suppose that you want to know how many grams are in 2.75 moles of sodium sulfate. You again have to know the formula and formula mass of sodium sulfate. (It's a salt, so it is more correct to use the term formula mass than molecular mass, because no molecules are present.) The formula is Na_2SO_4. Its formula mass is 142.04 g/mole. Now you apply the unit conversion process:

$$\frac{2.75\text{ moles of sodium sulfate}}{1} \times \frac{142.04\text{ g}}{1\text{ mole}} = 391\text{ grams of sodium sulfate}$$

Be sure to check that you have rounded off your answer to the correct number of significant figures and that your answer is reasonable.

Before I move on to the particles to mole conversion, I also want to provide an example of a gram to mole to gram conversion.

Calculate the number of grams of nitrogen in 100.0 grams of ammonium phosphate [$(NH_4)_3PO_4$]. The formula mass of ammonium phosphate is 149.12 g/mole.

To solve, follow these steps:

1. **Convert grams of ammonium phosphate to moles.**

$$\frac{100.0\text{ g}(\text{NH}_4)_3\text{PO}_4}{1} \times \frac{1\text{ mole}(\text{NH}_4)_3\text{PO}_4}{149.12\text{ g}}$$

2. **Convert from moles of ammonium phosphate to moles of nitrogen.**

$$\frac{100.0\text{ g}(\text{NH}_4)_3\text{PO}_4}{1} \times \frac{1\text{ mole}(\text{NH}_4)_3\text{PO}_4}{149.12\text{ g}} \times \frac{3\text{ mole N}}{1\text{ mole}(\text{NH}_4)_3\text{PO}_4}$$

3. **Convert from moles of nitrogen to grams of nitrogen.**

$$\frac{100.0\ g(NH_4)_3PO_4}{1} \times \frac{1\ mole(NH_4)_3PO_4}{149.12\ g} \times \frac{3\ mole\ N}{1\ mole(NH_4)_3PO_4} \times \frac{14.01\ g\ N}{1\ mole\ N}$$

$$= 28.19\ g\ N$$

4. **Check that you have rounded off your answer to the correct number of significant figures and that your answer is reasonable.**

Converting particles to moles and back again

You can use a similar procedure to convert from particles to moles that you used to convert grams to moles.

You want to figure out how many moles of oxygen are in 1.00×10^{25} molecules of carbon dioxide, CO_2.

Follow these steps to calculate:

1. **Convert from molecules of carbon dioxide to moles of carbon dioxide.**

$$\frac{1.00 \times 10^{25}}{1} \times \frac{1\ mole\ CO_2}{6.022 \times 10^{23}}$$

2. **Convert from moles of carbon dioxide to moles of oxygen.**

$$\frac{1.00 \times 10^{25}}{1} \times \frac{1\ mole\ CO_2}{6.022 \times 10^{23}} \times \frac{2\ moles\ O}{1\ mole\ CO_2} = 33.2\ moles\ O$$

3. **Check that you have rounded off your answer to the correct number of significant figures and that your answer is reasonable.**

If you can go from particles to moles, you should be able to do the reverse. The following walks you through an example.

How many hydrogen atoms are contained in 3.55 moles of water, H_2O?

These three easy steps can help you solve this example.

1. **Convert from moles of water to moles of hydrogen.**

$$\frac{3.55\ moles\ H_2O}{1} \times \frac{2\ moles\ H}{1\ mole\ H_2O}$$

2. **Convert from moles of hydrogen to atoms of hydrogen.**

$$\frac{3.55\text{ moles }H_2O}{1}\times\frac{2\text{ moles H}}{1\text{ mole }H_2O}\times\frac{6.022\times10^{23}\text{ atoms}}{1\text{ mole H}}=4.28\times10^{24}\text{ H atoms}$$

3. **Check that you have rounded off your answer to the correct number of significant figures and that your answer is reasonable.**

Going from grams to particles and back again

You can put this gram/mole/particles thing all together and do some gram-to-particle calculations.

How many ethane (C_2H_6) molecules are in 150.0 grams of ethane?

Follow these steps to calculate:

1. **Convert the grams to moles.**

   On any of these types of problems, if you can't think of anything else to do, convert to moles.

$$\frac{150.0\text{ g }C_2H_6}{1}\times\frac{1\text{ mole }C_2H_6}{30.08\text{ g}}$$

2. **Convert from moles to particles.**

$$\frac{150.0\text{ g }C_2H_6}{1}\times\frac{1\text{ mole }C_2H_6}{30.08\text{ g}}\times\frac{6.022\times10^{23}\text{ molecules}}{1\text{ mole}}=3.003\times10^{24}\text{ molecules}$$

3. **Check your significant figures and make sure that the answer is reasonable.**

In Chemistry II, you probably will encounter problems that are a bit more advanced. Here is a more involved problem going from particles to grams.

A certain sample of ethane (C_2H_6) contained 7.5×10^{24} atoms of hydrogen. How many grams of ethane did the sample contain?

In looking at the formula of ethane, notice that it has six hydrogen atoms for ethane molecule; you need this information to solve this problem. Stick to these steps to solve this problem.

1. **Convert from atoms of hydrogen to moles of hydrogen.**

$$\frac{7.5\times10^{24}\text{ H atoms}}{1}\times\frac{1\text{ mole H}}{6.022\times10^{23}\text{ atoms}}$$

2. **Convert from moles of hydrogen atoms to moles of ethane molecules.**

$$\frac{7.5\times10^{24}\text{ H atoms}}{1}\times\frac{1\text{ mole H}}{6.022\times10^{23}\text{ atoms}}\times\frac{1\text{ mole ethane molecules}}{6\text{ moles of H atoms}}$$

3. **Convert from moles of ethane to grams of ethane.**

$$\frac{7.5\times10^{24}\text{ H atoms}}{1}\times\frac{1\text{ mole H}}{6.022\times10^{23}\text{ atoms}}\times\frac{1\text{ mole ethane molecules}}{6\text{ moles of H atoms}}$$

$$\times\frac{30.08\text{ g}}{1\text{ mole }C_2H_6}=62\text{ g ethane}$$

4. **Make sure you check the significant figures and whether the answer is reasonable.**

Reacting to Reaction Stoichiometry

With weight relationships established, you're ready for some stoichiometry problems. *Stoichiometry* refers to the mass relationship in balanced chemical equations. In stoichiometry, the mole concept is essential. The coefficients in the balanced equation are not only the number of individual atoms or molecules but also the number of moles. The ratio of the moles of one substance to another substance in the balanced chemical equation is called a *stoichiometric ratio.* With this ratio, you can convert from the moles of one substance in a balanced chemical equation to the moles of another substance. And if you can get moles, you can easily generate grams.

In order to do any type of stoichiometry problem, you must have the balanced chemical equation.

Calculating how many reactants and products

Industrial chemists are interested in how much (grams, kilograms, tons, and so on) of a particular product can be formed with a certain amount (grams, kilograms, tons, and so on) of reactants in a certain chemical reaction. The mole concept allows these industrial chemists (and chemistry students) to do this. To help you understand how to make this calculation, I use the Haber process for the synthesis of ammonia gas from nitrogen and hydrogen gases:

$$N_2(g) + 3\,H_2(g) \rightarrow 2\,NH_3(g)$$

Suppose that you want to know how many grams of diatomic nitrogen gas will be needed to fully react with 150.00 grams of hydrogen gas.

$N_2(g) + 3\,H_2(g) \rightarrow 2\,NH_3(g)$

1 mole + 3 moles → 2 moles

To solve, work through these steps:

1. **Convert from grams of hydrogen gas to moles of hydrogen gas.**

   $$\frac{150.00\text{ g }H_2}{1} \times \frac{1\text{ mole }H_2}{2.016\text{ g}}$$

2. **Incorporate that stoichiometric ratio into our problem set-up.**

   Looking at the balanced chemical equation, you see that for every three moles of hydrogen used, it takes one mole of nitrogen gas.

   $$\frac{150.00\text{ g }H_2}{1} \times \frac{1\text{ mole }H_2}{2.016\text{ g}} \times \frac{1\text{ moles }N_2}{3\text{ moles }H_2}$$

3. **Convert from moles of nitrogen gas to grams of nitrogen gas.**

   $$\frac{150.00\text{ g }H_2}{1} \times \frac{1\text{ mole }H_2}{2.016\text{ g}} \times \frac{1\text{ moles }N_2}{3\text{ moles }H_2} \times \frac{28.02\text{ g}}{1\text{ mole }N_2} = 694.9\text{ g }N_2$$

Use the same information to work another problem.

Calculate the number of grams of ammonia gas produced when 150.00 grams of hydrogen gas reacts with excess ammonia.

Stick with the Haber process and follow these steps to solve:

1. **Convert from grams of hydrogen gas to moles of hydrogen gas.**

   $$\frac{150.00\text{ g }H_2}{1} \times \frac{1\text{ mole }H_2}{2.016\text{ g}}$$

2. **Convert from moles of hydrogen gas to moles of ammonia.**

   Looking at the balanced chemical equation, you see that for every three moles of hydrogen gas that react, two moles of ammonia gas are produced:

   $$\frac{150.00\text{ g }H_2}{1} \times \frac{1\text{ mole }H_2}{2.016\text{ g}} \times \frac{2\text{ moles }NH_3}{3\text{ moles }H_2}$$

3. **Convert from moles of ammonia to grams of ammonia.**

   $$\frac{150.00\text{ g }H_2}{1} \times \frac{1\text{ mole }H_2}{2.016\text{ g}} \times \frac{2\text{ moles }NH_3}{3\text{ moles }H_2} \times \frac{17.03\text{ g}}{1\text{ mole }NH_3} = 844.7\text{ g }NH_3$$

4. **Make sure you check the significant figures and whether the answer is reasonable.**

   Well, you started with 150.00 grams of hydrogen and you calculated that it would take 694.9 grams of nitrogen to react with it. If the *Law of Conservation of Matter* (matter is neither created nor destroyed in ordinary chemical reactions) holds, then the number of grams of ammonia produced should simply be the sum of the hydrogen plus nitrogen — 150.00g hydrogen + 694.9g nitrogen = 844.9g ammonia. You calculate 844.7g. The difference in these two values is a round-off error. They are certainly close enough for me!

Take a look at another reaction — the thermite reaction. In this reaction rust is reacted with aluminum metal producing aluminum oxide and iron metal. This thermite reaction was once used to join pieces of railroad tracks. The balanced chemical reaction looks like this:

$$Fe_2O_3(s) + 2\,Al(s) \rightarrow Al_2O_3(s) + 2\,Fe(s)$$

When you get ready to work stoichiometry problems, you *must* start with a balanced chemical equation. If you don't have it to start with, you have to go ahead and balance the equation.

In this example, the formula weights you need are

- **Fe_2O_3:** 159.70 g/mol
- **Al:** 26.98 g/mol
- **Fe:** 55.85 g/mol
- **Al_2O_3:** 101.96 g/mol

Suppose that you want to know how many grams of rust would react with 1.000 kilogram of aluminum.

You need to convert from kilogram of aluminum to grams and convert the grams to moles of aluminum. Then you can use a stoichiometric ratio to convert from moles of aluminum to moles of rust and finally to grams. To do so, stick to these steps:

1. **Convert to grams.**

   $$\frac{1.000 \text{ kg Al}}{1} \times \frac{1000 \text{ g}}{1 \text{ kg}}$$

2. **Convert moles of carbon.**

   $$\frac{1.000 \text{ kg Al}}{1} \times \frac{1000 \text{ g}}{1 \text{ kg}} \times \frac{1 \text{ mole Al}}{26.98 \text{ g}}$$

3. **Apply the stoichiometric ratio to convert from moles of aluminum to moles of rust.**

$$\frac{1.000 \text{ kg Al}}{1} \times \frac{1000 \text{ g}}{1 \text{ kg}} \times \frac{1 \text{ mole Al}}{26.98 \text{ g}} \times \frac{1 \text{ moles } Fe_2O_3}{2 \text{ moles Al}}$$

4. **Convert from moles of Fe_2O_3 to grams:**

$$\frac{1.000 \text{ kg Al}}{1} \times \frac{1000 \text{ g}}{1 \text{ kg}} \times \frac{1 \text{ mole Al}}{26.98 \text{ g}} \times \frac{1 \text{ moles } Fe_2O_3}{2 \text{ moles Al}} \times \frac{159.70 \text{ g}}{1 \text{ mole } Fe_2O_3}$$

$= 2960.$ g rust

That answer looks pretty reasonable and has been rounded off to the correct number of significant figures.

Making these calculations is so much fun. Here is another one for you.

Aluminum metal reacts with oxygen to produce aluminum oxide. How many grams of aluminum oxide will be produced if 500.0 grams of aluminum metal is reacted with excess oxygen gas?

Something is missing, which is the balanced chemical equation. Here is the unbalanced equation:

$$Al(s) + O_2(g) \rightarrow Al_2O_3(s)$$

The balanced equation looks like this:

$$4\ Al(s) + 3\ O_2(g) \rightarrow 2\ Al_2O_3(s)$$

Now you're ready to solve. Work through these steps:

1. **Convert from grams of aluminum to moles.**

$$\frac{500.0 \text{ g Al}}{1} \times \frac{1 \text{ mole Al}}{26.98 \text{ g}}$$

2. **Apply the reaction stoichiometry as given by the balanced chemical equation to convert to moles of aluminum oxide.**

$$\frac{500.0 \text{ g Al}}{1} \times \frac{1 \text{ mole Al}}{26.98 \text{ g}} \times \frac{2 \text{ moles } Al_2O_3}{4 \text{ moles Al}}$$

3. **Calculate the grams of aluminum oxide.**

$$\frac{500.0 \text{ g Al}}{1} \times \frac{1 \text{ mole Al}}{26.98 \text{ g}} \times \frac{2 \text{ moles } Al_2O_3}{4 \text{ moles Al}} \times \frac{101.96 \text{ g}}{1 \text{ mole } Al_2O_3} = 944.8 \text{ g } Al_2O_3$$

Again, the significant figures are correct and the answer looks reasonable.

In your Chemistry II class, you need to stay aware of little twists that may show up in problems, such as in the following:

> How many oxygen molecules will it take to react with 500.0 grams of aluminum metal?

You have been calculating mass in the previous problems, but now you are asked to calculate the number of molecules. Just be sure you don't confuse moles and molecules.

To figure out this problem, follow these steps:

1. **Convert the grams of Al to moles.**

   $$\frac{500.0 \text{ g Al}}{1} \times \frac{1 \text{ mole Al}}{26.98 \text{ g}}$$

2. **Convert from mole of Al to moles of oxygen using the coefficients in the balanced chemical equation.**

   $$\frac{500.0 \text{ g Al}}{1} \times \frac{1 \text{ mole Al}}{26.98 \text{ g}} \times \frac{3 \text{ moles } O_2}{4 \text{ moles Al}}$$

3. **Convert from moles of oxygen gas to molecules of oxygen gas.**

   $$\frac{500.0 \text{ g Al}}{1} \times \frac{1 \text{ mole Al}}{26.98 \text{ g}} \times \frac{3 \text{ moles } O_2}{4 \text{ moles Al}} \times \frac{6.022 \times 10^{23} \text{ molecules}}{1 \text{ mole } O_2}$$

   $$= 8.370 \times 10^{24} \text{ molecules}$$

Check the significant figures and the reasonableness of the answer. You should have a feeling that there will be a large number of molecules formed, so that your answer should be a large number — and it is.

In one of the previous examples, you predicted that you could form 944.8 g Al_2O_3 if you reacted 500.0 grams of aluminum with excess oxygen. What if, however, you carry out this reaction and only get 823.0 grams of aluminum oxide formed? Several reasons, such as sloppy technique or impure reactants, may cause you to produce less than you expect. The reaction may also quite likely be an equilibrium reaction, and you'll never get 100 percent conversion from reactant to products. (Equilibrium reactions don't use up all the reactants — see Chapters 8 to 10 for more than all you ever wanted to know about equilibrium reactions). Wouldn't it be nice if you had a way to label the efficiency of a particular reaction? You do. It's called the percent yield, which I discuss in the following section.

Percent yield: Something's missing

Rarely is anything 100 percent efficient. Your automobile isn't; your study habits aren't or you would be getting straight 100's; and neither are chemical reactions. In any reaction, there are going to be less products formed than expected. Several reasons may cause it, but it happens. Chemists can get an idea of the efficiency of a reaction by calculating the *percent yield* for the reaction using the following equation:

$$\frac{\text{Actual Yield}}{\text{Theoretical Yield}} \times 100\%$$

The *actual yield* is how much of the product you actually got when you carried out the reaction. The *theoretical yield* is how much of the product you *calculated* you would get. The ratio of these two yields gives you an idea about how efficient the reaction is. For the reaction of aluminum to aluminum oxide (see the preceding section), your theoretical yield was 944.8 grams of Al_2O_3; your actual yield was 823.0 grams. Therefore, the percent yield is

$$\frac{823.0\text{ g}}{944.8\text{ g}} \times 100\% = 87.11\%$$

A percent yield of about 87 percent is pretty good, but chemist and chemical engineers would rather see a lot higher, preferably more than 90%. The higher the percent yield, the more money the company makes.

Limiting reactants: Running out of reactant(s)

In a chemical reaction rarely are all the reactants present in stoichiometric ratios so that you use up all the reactants at the same time. Normally, you run out of one of the reactants and have some others left over. (In some of the problems sprinkled throughout this chapter, I told you which reactant is the limiting one by saying you have *an excess* of the other reactant(s).)

In this section, I show you how to calculate which reactant is the limiting reactant and how to base your stoichiometric calculations on it.

Consider the reaction between the hydrocarbon hexane (C_6H_{14}) and oxygen gas to produce carbon dioxide. The balanced equation is:

$$2\ C_6H_{14}(l) + 19\ O_2(g) \rightarrow 12\ CO_2(g) + 14\ H_2O(g)$$

Suppose you placed 500.0 grams each of hexane and oxygen gas in a reaction vessel and initiated the combustion reaction. You want to know how much (grams) of each product is formed and which and how much of the reactant is left over. You must determine the limiting reactant and base your stoichiometric calculations on it.

A sure way to know that you have a limiting reactant problem is when you are given amounts of more than one reactant.

In order to figure out which reactant is the limiting reactant, you calculate the mole-to-coefficient ratio. You calculate the number of moles of both hexane and oxygen, and then you divide each by their coefficient in the balanced chemical equation. The one with the smallest mole-to-coefficient ratio is the limiting reactant. For the reaction of hexane and oxygen, you can calculate the mole-to-coefficient ratio for the hexane and oxygen like this:

$$\frac{500.0\text{ g hexane}}{1}\times\frac{1\text{ mole hexane}}{86.17\text{ g}}\times\frac{1}{2\text{ coefficient}}=2.901\frac{\text{mole}}{\text{coefficient}}\text{hexane}$$

$$\frac{500.0\text{ g oxygen}}{1}\times\frac{1\text{ mole }O_2}{32.00\text{ g}}\times\frac{1}{19\text{ coefficient}}=0.8224\frac{\text{mole}}{\text{coefficient}}\text{oxygen}$$

Hexane has a mole-to-coefficient ratio of 2.901, and oxygen has a ratio of 0.8224. Because oxygen has the lowest mole-to-coefficient ratio, oxygen is the limiting reactant and determines how much carbon dioxide and water will be produced. You need to base your calculations on the limiting reactant, the oxygen. Now it's just another stoichiometry problem.

To calculate the number of grams of carbon dioxide that will be produced, follow these steps:

1. **Convert from grams of oxygen to moles.**

   $$\frac{500.0\text{g oxygen}}{1}\times\frac{1\text{ mole }O_2}{32.00\text{ g}}$$

2. **Apply the stoichiometric ratio between oxygen and carbon dioxide.**

   $$\frac{500.0\text{ oxygen}}{1}\times\frac{1\text{ mole }O_2}{32.00\text{ g}}\times\frac{12\text{ moles }CO_2}{19\text{ moles }O_2}$$

3. **Convert from moles of carbon dioxide to grams of carbon dioxide.**

   $$\frac{500.0\text{ oxygen}}{1}\times\frac{1\text{ mole }O_2}{32.00\text{ g}}\times\frac{12\text{ moles }CO_2}{19\text{ moles }O_2}\times\frac{44.01\text{ g}}{1\text{ mole }CO_2}=434.3\text{ g }CO_2$$

Significant figures and reasonableness looks good. In a similar fashion, you can calculate the amount of water produced. The only thing that will change will be the stoichiometric ratio and the molar mass:

$$\frac{500.0\text{ oxygen}}{1}\times\frac{1\text{ mole }O_2}{32.00\text{ g}}\times\frac{14\text{ moles }H_2O}{19\text{ moles }O_2}\times\frac{18.02\text{ g}}{1\text{ mole }H_2O}=207.5\text{ g }H_2O$$

Looks like a good answer; significant figures are fine and the answer is reasonable.

Finally, you can check how many grams of hexane were used and how much is left over. You can do so in a couple ways:

One way is the way we have been doing, using the limiting reactant and the stoichiometric ratio to calculate the amount of hexane used:

$$\frac{500.0\text{g oxygen}}{1} \times \frac{1 \text{ mole } O_2}{32.00 \text{ g}} \times \frac{2 \text{ moles hexane}}{19 \text{ moles } O_2} \times \frac{86.17 \text{ g}}{1 \text{ mole hexane}}$$

$$= 141.7 \text{ g hexane}$$

The amount of hexane left over would be the initial amount minus the amount used: 500.0 g – 141.7 g = 358.3 g excess hexane.

The other way is more of a reasoning approach. You can calculate the mass of total products: 434.3 g CO_2 + 207.5 g H_2O = 641.8 g. Of that total mass, 500.0 grams is oxygen (the limiting reactant is completely consumed). The rest, 141.8 g, must come from the hexane. Notice that this amount agrees within rounding error that you calculated using the stoichiometric ratio.

The advantage of calculating the amount of nonlimiting reactant consumed using the stoichiometric ratio is that if you have made a mistake in calculating either of the other products, it doesn't affect your calculation of the nonlimiting reactant.

Chapter 5

Grasping Solutions and Intermolecular Forces

In This Chapter

- Understanding solutes, solvents, and solutions
- Comprehending solution concentration units
- Tapping into colligative properties
- Identifying intermolecular forces
- Eyeing the properties of liquids

Solutions are all around you. The air you breathe, the sports drink (and other types of drinks) you drink, even your tap water is a solution. In this chapter, I remind you of some of the properties of solutions. I show you the different ways chemists represent a solution's concentration, and I tell you about *colligative properties,* the interaction between particles that have such an impact on solution behavior.

Considering Solutes, Solvents, and Solutions

A *solution* is a *homogeneous* mixture. A homogeneous mixture has the same composition and set of properties throughout the entire sample. For example, suppose you dissolve a tablespoon of sugar in a glass of water. Mix it up well. Now look at the solution. Do you see any difference in the solution close to the surface compared to the bottom? No, the sugar isn't settling out. If you were to sample a bit of the solution at the top of the glass and at the bottom of the glass, it would taste the same. It is a homogeneous mixture, the same throughout the entire sample. The same is true of solid solutions. Look at your gold ring. It's not pure gold; it's an alloy of gold and some other metal. It's also a solution, a homogeneous mixture.

A solution is made up of two components:

- **A solvent:** The *solvent* component is the part of the solution that doesn't change state and is in the largest amount (grams, moles, and so on). If you dissolve sugar in water, the water is the solvent. It's still in the liquid state, and you have a lot more water there than sugar.
- **One or more solutes:** The *solute* component of a solution is the substance that changes state and is present in the lesser amount. In the previous example, sugar would be the solute; it's gone from the solid state to the aqueous solution state and there is less sugar than water.

You can determine which component is which based upon the amounts most of the time, but in a few cases of extremely soluble salts, such as lithium chloride, more than 5 grams of salt can be dissolved in 5 milliliters of water. However, water is still considered the solvent, because it's the species that has not changed state.

You can have more than one solute in a solution. If you dissolve sugar in water to make a sugar-water solution and then dissolve some salt into the same solution, you have two solutes, sugar and salt. You still, however, have only one solvent, though — water.

Most people naturally think of liquids when the term *solution* is used. But there can also be solutions of gases. Our atmosphere, for example, is a solution. Because air is almost 79 percent nitrogen, it is considered the solvent, and the oxygen, carbon dioxide, and other gases are considered the solutes. And I've already mentioned that solids composed of one metal "dissolved" in another metal are considered solutions. Brass is a solution of zinc in copper.

Discussing dissolving

Why do some things dissolve in one solvent and not another? For example, oil and water don't mix to form a solution, but oil dissolves in gasoline. This question is complex and involves the interaction between the solute particles, the solvent particles, and the solute — solvent particles. Polarity of both the solute and solvent play a big role in the dissolving process. A general rule of solubility says that *like-dissolves-like* with respect to polarity of both the solvent and solutes.

A typical oil, for example, is a nonpolar material; it's composed of nonpolar covalent bonds. Oil will dissolve nonpolar solutes, such as gasoline or grease. Water, on the other hand is a polar solvent. It is composed of polar covalent bonds. It's a fine solvent for polar materials, such as salts and alcohols, but not for nonpolar materials such as grease and oil. Now you know why oil and water don't mix.

Only so much solute can be dissolved in a given amount of solvent. I'm bad about misjudging how much sugar will dissolve in my glass of iced tea. No matter how much I stir, there's some undissolved sugar at the bottom of the glass. This is because the sugar has reached its maximum solubility in water at that temperature. The term *solubility* is the maximum amount of solute that will dissolve in a given amount of a solvent at a specified temperature. Chemists normally express solubility in units of grams solute per 100 milliliters of solvent (g/100 mL).

The solubility of a solute is related to the temperature of the solvent. For solid solute dissolving in liquid solvents, the solubility of the solute normally increases if the temperature is increased. If you heat that iced tea, the sugar at the bottom readily dissolves. For gases dissolving in liquids, the solubility of the gaseous solute is less as the temperature increases. This is true for oxygen gas dissolving in lake water. As the temperature of the lake increases, less oxygen will dissolve. This is *thermal pollution;* the addition of heat to water that decreases the solubility of the oxygen and affects the aquatic life.

A lowdown on saturated facts

You can express the relative amount of solute and solvent in a solution in a couple of qualitative ways. One set of qualitative terms is saturated, unsaturated, or supersaturated. A *saturated* solution contains the maximum amount of dissolved solute possible at a certain temperature. If the solution has less than this maximum amount, it's called an *unsaturated* solution. However, under certain unusual circumstances, the solvent may actually dissolve more than its maximum amount and become *supersaturated.* This supersaturated solution is relatively unstable and sooner or later some solute will precipitate (form a solid) until the concentration is reduced to the saturation point.

If a solution is unsaturated, the amount of solute that is dissolved can vary over a wide range. A couple of other qualitative terms, concentrated and dilute, can be used to describe the relative amount of solute and solvent:

- A solution may be said to be *concentrated* if it contains a large amount of solute per given amount of solvent. If you dissolve 200 grams of sugar in a liter of water, for example, the resulting solution is concentrated.
- The solution is said to be *dilute* if, relatively speaking, it contains very little solute per given amount of solvent. If you dissolve 0.01 grams of sugar in a liter of water, for example, the solution is really dilute.

However, what if you dissolve 15 grams or 45 grams of sugar in a liter of water? Is the solution concentrated or is it dilute? These terms aren't very useful in many cases, such as in IV solutions, where each bag *must* have a very precise amount of solute in them, or the patient is in danger. In cases like this, you must have a precise quantitative way of describing the relative amount of solute and solvent in a solution; you must use solution concentration units.

Reconciling Solution Concentration Units

Chemists and chemistry students use a variety of solution concentration units to quantitatively describe the relative amounts of the solute(s) and the solvent. Many times in nonlaboratory situations, percentage is commonly used. However, in chemistry, *molarity* (the moles of solute per liter of solution) is the solution concentration unit of choice. In certain circumstances another unit, *molality* (the moles of solute per kilogram of solvent), is used. And I use parts-per-million or parts-per-billion when describing air and water pollutants. These sections take a quick look at some of the more important concentration units.

Percent composition

Percentage is the amount per one hundred. Depending on the way you choose to express the percentage, the units of amount per one hundred vary. Three different percentages are commonly used:

- Mass/mass (m/m) percentage
- Mass/volume (m/v) percentage
- Volume/volume (v/v) percentage

Unfortunately, although the percentage of solute is often listed, the method (m/m, m/v, v/v) is not. I normally assume that the method is weight/weight, but I'm sure you know about assumptions.

Here are some examples of solutions using percentage as a concentration unit. Most of the solutions are *aqueous*, solutions in which water is the solvent.

Mass/mass percentage

In *mass/mass percentage,* or *mass percentage,* the mass of the solute is divided by the mass of the solution and then multiplied by 100 to get the percentage. Mass percentage is sometimes referred to as *weight percentage*. Normally a gram is the weight unit. Mathematically, mass percentage looks like this:

$$\frac{\text{mass}}{\text{mass}}\% = \frac{\text{mass of solute}}{\text{mass of solution}} \times 100\%$$

If, for example, you dissolve 15.0 grams of sodium chloride in 150 grams of water, the mass percentage is:

$$\frac{\text{mass}}{\text{mass}}\% = \frac{15\text{ g solute}}{165\text{ g solution}} \times 100\% = 9.1\%$$

Therefore, the solution is a 9.1 percent (m/m) solution.

Suppose that you want to make 500.0 grams of a 10.0 percent (m/m) sucrose, or table sugar, solution. How many grams of sucrose will be needed and how shall you prepare the solution?

You know that 10.0 percent of the weight of the solution is sugar, so you can multiply the 500.0 grams by 0.100 to get the weight of the sugar:

$$500.0 \text{ grams} \times 0.100 = 50.0 \text{ grams of sugar}$$

The rest of the solution (500.0 grams – 50.0 grams = 450.0 grams) is water. You can simply weigh out 50.0 grams of sugar and add it to 450.0 grams of water to get your 10 percent (m/m or w/w) solution.

Mass percentage is the easiest percentage solution to make, just weigh out the solute and solvent. But sometimes it's inconvenient to weigh out large amounts of solvent. In this case, you can use the mass/volume percentage, which I discuss in the next section.

Mass/volume percentage

Mass/volume percentage is very similar to mass/mass percentage, but instead of using grams of solution in the denominator, it uses milliliters of solution:

$$\frac{\text{mass}}{\text{volume}}\% = \frac{\text{mass of solute}}{\text{volume of solution}} \times 100\%$$

Many times in chemistry the volume unit is milliliters.

Suppose that you want to make 250.0 milliliters of a 20.0 percent (m/v) potassium chloride solution. Calculate the grams of potassium chloride needed and describe how to make the solution.

In order to solve this problem, follow these steps:

1. **Calculate the grams of KCl needed.**

   A 20.0 percent (m/v) solution has 20.0 grams of potassium chloride per 100.0 mL of solution. But you're making 250.0 mL of solution, so that:

   $$\frac{20.0 \text{ g KCl}}{100.0 \text{ mL of solution}} \times \frac{250.0 \text{ mL solution}}{1} = 50.0 \text{ g KCl}$$

2. **Dissolve the 50.0 grams of KCl in a small amount of water and dilute it to exactly 250.0 milliliters in a volumetric flask and mix well.**

   In other words, you dissolve and dilute 50.0 grams of KCl to 250.0 milliliters. (I tend to abbreviate *dissolve* and *dilute* by writing *d & d*.) You won't know exactly how much water you put in, but it's not important as long as the final volume is 250.0 milliliters.

You can also use the percentage and volume to calculate the grams of solute in a certain amount of a solution like the following example shows.

Suppose you wanted to know how many grams of sodium hypochlorite (NaOCl) are in 300.0 milliliters of a 5.0 percent (m/v) solution of household bleach. Because it's a 5.0 percent (m/v), there are 5.0 grams of sodium hypochlorite per 100 milliliters of solution. You can set up the problem like this:

$$\frac{5.0\text{ g NaOCl}}{100.0\text{ mL of solution}} \times \frac{300.0\text{ mL solution}}{1} = 15.0\text{ g NaOCl}$$

You now know that you have 15.0 grams of sodium hypochlorite in the 300 milliliters of solution.

Volume/volume percentage

However, sometimes both the solute and solvent are liquids. In this case, using a volume/volume percentage is more convenient. With *volume/volume percentages,* both the solute and solution are expressed in milliliters:

$$\frac{\text{volume}}{\text{volume}}\% = \frac{\text{volume of solute}}{\text{volume of solution}} \times 100\%$$

You work volume/volume percentage problems in a similar fashion as the other percentage types. Just make sure that you don't assume that volumes of liquids are additive. For example, you can't simply add 25 milliliters of alcohol to 25 milliliters of water to get 50 milliliters of solution — you actually get a little less. The polar water molecules attract the polar alcohol molecules, which tends to fill in the open framework of water molecules and prevents the volumes from simply being added together.

Molarity is numero uno

Most chemists use *molarity* as the concentration unit of choice because it utilizes moles. The mole concept and chemistry go hand in hand like butter and toast. Molarity lets chemists easily work solutions into reaction stoichiometry. *Molarity (M)* is defined as the moles of solute per liter of solution.

$$\text{Molarity}(\text{M}) = \frac{\text{moles solute}}{\text{liter solution}}$$

When making molar (molarity) solutions (whether it be on paper or in the lab), calculating the amount of solute needed and then dissolving and diluting to the required volume is extremely important.

If 25.0 grams of $BaCl_2$ are dissolved and diluted to 450.0 milliliters, what is the molarity of the solution and how would you prepare it?

To solve this problem, stick to these steps:

1. **Take the grams of barium chloride and convert them to moles using the formula weight of $BaCl_2$ (208.24 g/mol).**
2. **Divide the moles by 0.4500 liters (450.0 milliliters).**

   You can set up the equation like this:

   $$\frac{25.0 \text{ g barium chloride}}{1} \times \frac{1 \text{ mole barium chloride}}{208.24 \text{ g}} \times \frac{1}{0.4500 \text{ L}} = 0.267 \text{ M}$$

3. **Take the 25.0 grams of barium chloride and dissolve and dilute to 450.0 mL.**

Suppose that you want to prepare 1.50 liter of a 0.250 M KCl solution. Describe the amounts used and how the solution is to be prepared.

You can use the molarity, volume, and molar mass of KCl to calculate the number of grams of KCl needed as this equation shows:

$$\frac{0.250 \text{ moles KCl}}{1 \text{ L}} \times \frac{1.50 \text{ L}}{1} \times \frac{74.56 \text{ g}}{1 \text{ mole KCl}} = 28.0 \text{ g KCl}$$

You weigh out 28.0 grams of KCl and dissolve and dilute it to 1.50 liters.

Deciding about dilution

Sometimes you're faced with making a solution from another solution. For example, suppose you need a 2 M bleach solution, but you only have a 5 M bleach solution. You can make a more dilute solution from a more concentrated one by a process called *dilution*. In dilution, you add more solvent, spreading the solute particles farther apart and lowering the concentration. You can use the following useful equation to figure how much of the concentrated solution you need or how much solvent is needed:

$$V_{old} \times C_{old} = V_{new} \times C_{new}$$

In this equation, V_{old} is the old volume, or the volume of the original solution, C_{old} is the concentration of the original solution, V_{new} is the volume of the new solution, and C_{new} is the concentration of the new solution. You can use any concentration unit but most of the time molarity is used, so we would use M_{old} and M_{new}.

Suppose you want to prepare 500.0 milliliters of 2.00 M HCl and all you have is some 12.0 M HCl. You can dilute some of the 12.0 M to 2.00 M, but how much of the 12.0 M HCl is needed (V_{old})?

To solve this problem, stick to these steps:

1. **Write down the mathematical relationship.**

   $V_{old} \times M_{old} = V_{new} \times M_{new}$

2. **Substitute in the known quantities.**

   $V_{old} \times 12.0\ M = 500.0\ mL \times 2.00\ M$

   $V_{old} = (500.0\ mL \times 2.00\ M)/12.0\ M = 83.3$ milliliters

3. **Put about 400 milliliters of water into a 500.0 mL volumetric flask, slowly add the 83.3 milliliters of the concentrated HCl as you stir, and then dilute to the final 500.0 milliliters with water and mix thoroughly.**

If you're actually doing a dilution of concentrated acids, be sure to *add the acid to the water* instead of the other way around! Adding the water to the concentrated acid can generate significant heat that may likely splatter the solution all over you.

Solutions and reaction stoichiometry

The molarity concentration unit is really useful when dealing with reaction stoichiometry. Recall that reaction stoichiometry is the quantitative relationships between chemicals. Using it you calculate how much reactant is consumed and how much product is formed during a reaction. Check out Chapter 4 for a quick review of stoichiometry.

Suppose that you want to know how many milliliters of 2.50 M sulfuric acid it takes to neutralize a solution containing 50.0 grams of sodium hydroxide.

To solve this problem, follow these steps:

1. **Write the balanced chemical equation.**

   $H_2SO_4(aq) + 2\ NaOH(aq) \rightarrow 2\ H_2O(l) + Na_2SO_4(aq)$

2. **Convert the mass of sodium hydroxide to moles.**

   $$\frac{50.0\ \text{g NaOH}}{1} \times \frac{1\ \text{mole NaOH}}{40.00\ \text{g}}$$

3. **Put in the stoichiometric ratio relating the number of moles of sodium hydroxide to mole of sulfuric acid.**

   $$\frac{50.0\ \text{g NaOH}}{1} \times \frac{1\ \text{mole NaOH}}{40.00\ \text{g}} \times \frac{1\ \text{mole}\ H_2SO_4}{2\ \text{moles NaOH}}$$

4. **Calculate the volume using the molarity of sulfuric acid and convert to liters.**

$$\frac{50.0 \text{ g NaOH}}{1} \times \frac{1 \text{ mole NaOH}}{40.00 \text{ g}} \times \frac{1 \text{ mole } H_2SO_4}{2 \text{ moles NaOH}} \times \frac{1 \text{ L}}{2.50 \text{ mole}} = 0.250 \text{ L}$$

It takes 250.0 milliliters (0.250 L) of the 2.50 M H_2SO_4 solution to completely react with the solution that contains 50.0 grams of NaOH.

Defining molality

Molality is another concentration term that involves moles of solute. It isn't used very much except in dealing with *colligative properties* (properties of solutions that simply depend on the number of particles, not their type). *Molality (m)* is defined as the moles of solute per kilogram of solvent. It's one of the few concentration units that doesn't use the solution's weight or volume. Mathematically, it looks like this:

$$\text{molality}(m) = \frac{\text{moles of solute}}{\text{kg of solvent}}$$

Suppose you want to dissolve 15.0 grams of KCl in 50.0 grams of water. What is the molality of the solution?

To calculate the molality, you can start with grams of KCl over the mass of water. Then convert the grams of KCl to moles and the grams of water to kilograms and finish the arithmetic.

$$\frac{15.0 \text{ g KCl}}{50.0 \text{ g water}} \times \frac{1 \text{ mole KCl}}{74.56 \text{ g}} \times \frac{1000 \text{ g}}{1 \text{ kg}} = 4.02 \text{ m}$$

Especially for pollution: Parts per million and parts per billion

Chemists tend to use percentage and molarity, and even molality, because they're convenient units for the solutions that chemists routinely make in the lab or the solutions commonly found in nature. However, these units aren't very convenient when dealing with the concentrations of certain pollutants in the environment. These pollutant concentrations are extremely small. In order to express the concentrations of very dilute solutions, scientists have developed another concentration unit — parts per million.

You might not have thought of it this way but percentage is parts per hundred, or grams solute per 100 grams of solution. *Parts per million (ppm)* is grams solute per one million grams of solution. However, it is commonly expressed as milligrams solute per kilogram solution, which is the same ratio. The reason it's expressed this way is that chemists can easily weigh out milligrams or even tenths of milligrams, and, if you're talking about aqueous solutions, a kilogram of solution is the same as a liter of solution. (The density of water is 1 gram per milliliter, or 1 kilogram per liter. The weight of the solute in these solutions is so very small that it's negligible when converting from the mass of the solution to the volume.)

By law, the maximum contamination level of lead in drinking water is 0.05 ppm. This number corresponds to 0.05 milligrams of lead per liter of water. That's pretty dilute. But mercury is regulated at the 0.002 ppm level. Sometimes, even this unit isn't sensitive enough, so environmentalists have resorted to the parts per billion (ppb) or parts per trillion (ppt) concentration units. Some neurotoxins are deadly at the parts per billion level.

Contemplating Colligative Properties

Some properties of solutions depend on the specific nature of the solute. For example, sugar solutions taste sweet, whereas salt solutions taste salty. Salt solutions conduct electricity (they're electrolytes), while sugar solutions don't (they're nonelectrolytes). Solutions containing the copper cation are commonly blue, while those containing the nickel cation are green.

However, some properties of solution don't depend on the specific type of solute — just the *number* of solute particles. Properties that simply depend on the relative number of solute particles are called *colligative properties*. The effect the solute has on the properties of the solution simply depends on the number of solute particles present. The following sections discuss in greater detail these colligative properties — these effects — including

- Vapor-pressure lowering
- Boiling-point elevation
- Freezing-point depression
- Osmotic pressure

Vapor-pressure lowering

A liquid that is contained in a closed container will eventually evaporate, and the gaseous molecules contribute to the pressure of the gas above the liquid.

The pressure due to the gaseous molecules of the evaporated liquid is called the liquid's *vapor pressure.*

But if you make that same liquid the solvent in a solution, the vapor pressure due to the solvent evaporation is lower, because the solute particles in the liquid take up space at the surface and the solvent can't evaporate as easily. Also many times the solute and solvent may have an attractive force that also makes it more difficult for the solvent to evaporate. That lowered vapor pressure is independent of what kind of solute you use. Instead, it just depends on the number of solute particles that are present in the solution.

For example, if you add one mole of glucose to a liter of water and add one mole of sucrose to another liter of water, the amount that the vapor pressure is lowered is the same because you're adding the same *number* of solute particles. If, however, you add a mole of potassium nitrate to a liter of water, the vapor pressure lowers by about twice the amount of the glucose or sucrose solutions. Potassium nitrate breaks apart into two ions, the potassium cation and the nitrate anion, so adding a mole of potassium nitrate yields two moles of particles (ions), and the greater number of solute particles leads to lower pressure.

Boiling-point elevation

Each liquid has a specific temperature at which it boils (at a given atmospheric pressure). This temperature is called the liquid's *boiling point.* If you use a particular liquid as a solvent in a solution, you find that the boiling point of the solution is always higher than the pure liquid. This is called the *boiling-point elevation.*

Boiling-point elevation explains why you use antifreeze in your automobile's radiator in the summer. You want the engine coolant to boil at a higher temperature so that it absorbs as much engine heat as possible *without* boiling. You also use a pressure cap on your radiator, because the higher the pressure, the higher the boiling point. This concept also explains why a pinch of salt in the cooking water causes foods to cook a little faster. The salt raises the boiling point so that more energy can be transferred to cooking the food during a given amount of time.

You can calculate the amount of boiling-point elevation by using this formula:

$$\Delta T_b = K_b m$$

ΔT_b is the *increase* in the boiling point, K_b is the boiling-point elevation constant (0.512°C/m for water), and m is the molality of particles. (For molecular substances, the molality of particles is the same as the molality of the substance; for ionic compounds, you have to take into consideration the formation of ions and calculate the molality of the ion particles.) Solvents other than water have a different boiling point elevation constant (K_b).

Determine the boiling point of 3.0 m aqueous KNO_3 solution.

KNO_3 is a salt, a strong electrolyte. In a 3.0 m solution it is 6.0 m in particles, because KNO_3 dissociates completely into K^+ and NO_3^-. Therefore:

$$\Delta T_b = K_b m$$

$$\Delta T_b = (0.512°C/m) \times (6.0\ m)$$

$$\Delta T_b = 3.1°C$$

The change in boiling point is 3.1 degrees Celsius. You know that the boiling point of a solution is always *higher* than the pure solvent, so the solution's boiling point is:

$$100.0°C + 3.1°C = 103.1°C$$

Freezing-point depression

Just as each liquid has a specific boiling point; it also has a specific temperature at which it freezes. If you use a particular liquid as a solvent in a solution, though, you find that the freezing point of the solution is always lower than the pure liquid. This is called the *freezing-point depression*, and it's a colligative property of a solution, meaning that it depends on the number of solute particles.

Why do you put rock salt in the ice/water mix when making homemade ice cream? The rock salt forms a solution with a freezing point lower than water (or the ice cream mix that's to be frozen). That is why you use antifreeze in your automobile's cooling system during the winter. The more you use (up to a concentration of 50/50), the lower the freezing point.

The freezing-point depression effect also explains why a salt (normally calcium chloride, $CaCl_2$) is spread on ice to melt it. The dissolving of calcium chloride is highly *exothermic* (it gives off a lot of heat). As the calcium chloride dissolves, it melts the ice and forms a solution in the resulting water. The salt solution that's formed when the ice melts has a lowered freezing point that keeps the solution from refreezing. Who knew that those dump trucks spreading calcium chloride on the icy roads were making use of a colligative property?

You can calculate the amount the freezing point will be depressed:

$$\Delta T_f = K_f m$$

ΔT_f is the amount the freezing point will be lowered, K_f is the freezing point depression constant (1.86°C /m for water), and m is the molality of the particles.

Suppose you want to calculate the freezing point of that 3.0 m aqueous KNO_3 solution used in the previous section.

You know that the solution is 6.0 m in particles, so

$$\Delta T_f = K_f m$$

$$\Delta T_f = (1.86°C/m) \times 6.0\ m$$

$$\Delta T_f = 11.2°C$$

You know that the freezing point of the solution is lower than the freezing point of water (the solvent), so the freezing point of the solution is

$$0.0°C - 11.2°C = -11.2°C$$

Osmotic pressure

What does an IV solution, a pickle, a bottle of sports drink, and the bottled water that I drink have in common? They all have a tie to the osmotic pressure of solutions. *Osmotic pressure* is prevalent in your everyday lives, but few people, outside of chemists, realize it. Here I take a look at this topic; and by the way it is another colligative property.

Take a container and divide it into two compartments with a thin membrane containing microscopic pores large enough to allow water molecules but not solute particles to pass through. This membrane is called a *semipermeable membrane*; it lets some small particles pass through but not other, larger particles.

Then add a concentrated salt solution to one compartment and a more dilute salt solution to the other. Initially, the two solution levels start out the same. But after a while, you observe that the level on the more concentrated side has risen, and the level on the more dilute side has dropped. This change in levels is due to the passage of water molecules from the more dilute side to the more concentrated side through the semipermeable membrane. This process is called *osmosis,* the passage of a solvent through a semipermeable membrane into a solution of higher solute concentration. The pressure that you have to exert on the more concentrated side in order to stop this process is called the osmotic pressure of the solution.

The solvent always flows through the semipermeable membrane from the more dilute side to the more concentrated side. In fact, you can have pure water on one side and any salt solution on the other, and water always goes from the pure-water side to the salt-solution side. The more concentrated the salt solution, the more pressure it takes to stop the osmosis (the higher the osmotic pressure).

But if you apply more pressure than is necessary to stop the osmotic process, exceeding the osmotic pressure, water is forced through the semipermeable membrane from the more concentrated side to the more dilute side. This process is called *reverse osmosis.* Reverse osmosis is a good, relatively inexpensive way of purifying water.

One of the most biologically important consequences of osmotic pressure involves the cells within your own body. You can look at red blood cells as an example. Inside the blood cell is an aqueous solution, and outside the cell is another aqueous solution (intercellular fluid). When the solution outside the cell has the same osmotic pressure as the solution inside the cell, it's said to be *isotonic.* Water can be exchanged in both directions, helping to keep the cell healthy. However, if the intercellular fluid becomes more concentrated and has a higher osmotic pressure *(hypertonic),* water flows primarily out of the blood cell, causing it to shrink and become irregular in shape. This is a process called *crenation.* The process may occur if the person becomes seriously dehydrated, and the crenated cells are not as efficient in carrying oxygen. If, on the other hand, the intercellular fluid is more dilute than the solution inside the cells and has a lower osmotic pressure *(hypotonic),* the water flows mostly into the cell. This process, called *hemolysis,* causes the cell to swell and eventually rupture

The processes of crenation and hemolysis explain why the concentration of IV solutions is so very critical. If they're too dilute, then hemolysis can take place, and if they're too concentrated, crenation is a possibility.

You can calculate the osmotic pressure (π) by using the following equation:

$$\pi = (nRT/V)i = iMRT$$

In this equation π is the osmotic pressure in atmospheres, n is the number of moles of solute, R is the ideal gas law constant (0.0821 L atm/K mol), T is the Kelvin temperature, V is the volume of the solution and i is the *van't Hoff factor* (the number of moles of particles that will be formed from 1 mole of solute). n/V may be replaced by M, the molarity of the solution.

Grasping the Types of Intermolecular Forces

The particles around us, in the air, in our glass of beverage, and in our bodies are all interacting with each other. The interactions are called intermolecular. *Intermolecular forces* are those attractive or repulsive forces (interactions) that take place between atoms, molecules, and ions. They are all related to charge, whether a full charge in the case of ions or a partial charge in the case of atoms or molecules.

Like charges repel; unlike charges attract.

Being able to recognize whether or not a molecule is polar is important in order to determine which intermolecular forces are important in a particular situation. *Polar* molecules have a partial positive and partial negative end and thus are a dipole. You may want to check out *Chemistry For Dummies,* 2nd edition (John Wiley & Sons, Inc.) for more on covalent bonding and molecular geometry. For a quick review, see the section on polar bonding in Chapter 3 of this book.

The following intermolecular forces are listed in order of increasing strength of interaction. Many of the properties of chemical substances, such as solubility, are related to intermolecular forces. Being able to recognize the intermolecular forces present many times allows you to predict some physical properties and even explain why some substances react in the way they do.

London (dispersion) forces

This intermolecular attraction occurs in all substances, but is usually only significant for nonpolar substances. It is created from the momentary distortion of the electron cloud in which the electron density flows to one side of the atom/molecule. That electron cloud is not fixed; I like to compare it to a ball of cotton candy, easily pushed around (but not as tasty).

This distortion causes a very weak temporary dipole. This weak dipole induces a dipole in another molecule. These weak dipoles lead to an attraction. The more electrons present, the larger the cloud and the greater the London force. This interaction is extremely weak and an individual one does not last very long. However, it is strong enough to allow chemists to liquefy nonpolar gases such as hydrogen, H_2, which would be impossible if intermolecular forces didn't attract these molecules.

Induced dipole

Induced dipole intermolecular forces occur when the charge on an ion or a dipole distorts the electron cloud of a nonpolar molecule. A cation attracts the electron cloud, whereas an anion repels it. Either force induces a temporary dipole in the nonpolar molecule. These interactions are fairly weak and tend to occur in solution.

Dipole-dipole

Dipole-to-dipole attraction is important when dipoles are present. The positive end of one dipole is attracted to the negative end of another. For example, in the extremely reactive chlorine monofluoride gas, the chorine has a partial positive charge, and the more electronegative fluorine has a partial negative charge, allowing a dipole-dipole intermolecular attraction between two of the molecules:

Cl-F — Cl-F

$\delta+\ \delta- — \delta+\ \delta-$

Dipole-dipole forces tend to be especially important in polar liquids and are considered a strong intermolecular force, but not as strong as ion-dipole.

Hydrogen bonding

Hydrogen bonding is a type of dipole-dipole interaction, but one in which a hydrogen is bonded to an extremely electronegative element (N, O, F). The covalent bond between the hydrogen and these three elements is extremely polar. The hydrogen on one molecule can interact with the O, N, or F on another molecule. This intermolecular force is much stronger than other dipole-dipole forces and so therefore is given its own special name. The strength of this interaction explains a number of unusual properties of substances, such as water. The reason that water has such relatively high boiling and melting points, among other properties, is related to its hydrogen bonding.

In order for an intermolecular force to be hydrogen bonding, a hydrogen atom must be bonded to a N, O, or F, and this hydrogen must interact with an O, N, or F on another molecule.

Ion-dipole

Sometimes an ion is attracted to a molecule that is a dipole. If the ion is a *cation* (positive charge), it's attracted to the negative end of the dipole, and if the ion is an *anion* (negative charge), it's attracted to the positive end of the dipole. This interaction occurs quite commonly in aqueous solution in which an ion attracts water molecules.

Suppose, for example, you dissolve some white, crystalline $AlCl_3$ in water. It dissolves and the Al^{3+} cation attracts water molecules because it is a small ion with a large amount of positive charge. (Chemists say it has a large charge density.) Because of its high charge density, it attracts the partial negative end of the water molecule, the oxygen. In fact, it attracts a total of six water molecules, creating the hydrated aluminum ion, $Al(H_2O)_6^{3+}$.

The chloride ions attract the partial positive ends of water molecules, the hydrogens. However, because chloride ions are large ions with only a single negative charge, their charge density is low and so the attractive force is much weaker.

If you carefully evaporate the solution to dryness, you can recover $Al(H_2O)_6Cl_3$, the hexahydrated aluminum chloride. This substance is a hydrate; it has six waters of hydration incorporated into its crystalline structure. Careful heating of most hydrates results in the waters being driven out of the crystalline structure, leaving the anhydrous form.

Looking at the Properties of Liquids

A liquid is a phase in which the randomly orientated particles are in contact. The particles may clump together to exhibit short-range areas of order, but they usually don't last very long. This random orientation allows the liquid to change shape to match that of the container, and because the particles are in contact, a liquid isn't very compressible. In this section, I show you a few important macroscopic properties of liquids. The strength of the intermolecular forces is the key to these different properties.

Heat capacity

Heat capacity is the amount of energy needed to raise the temperature of a substance 1 K. The stronger the intermolecular forces between the molecules of a liquid are, the more energy that's required to break the forces and the greater the heat capacity. That also explains why liquids that have strong intermolecular forces have higher boiling points and vapor pressures than those that do not.

Capillary action

Another property of liquids that's related to intermolecular forces is capillary action. *Capillary action* is the rising of a liquid through a narrow tube against the force of gravity. It's a result of the competition of intermolecular forces within the liquid and the attractive forces between the liquid and the wall of the tube. The stronger the attraction between the liquid and the wall, the higher the level rises.

Water has a strong attraction to the walls of a glass tube and therefore has high capillary action. Mercury has a weak attraction to the walls of a glass tube, and so has low capillary action. This capillary action explains why you observe a meniscus with water contained in a thin tube. A *meniscus* is a concave water surface due to the attraction of the water molecules adjacent to the glass walls. Because of mercury's weak attraction to the glass walls, no meniscus is present. However, if you replace the glass tube with a plastic one, water behaves much more like the mercury did in the glass tube because very little attraction exists between the polar water molecules and the nonpolar plastic.

Only the liquid near the walls of the tube is attracted to the walls. The particles farther away from the walls pull the other molecules back. The narrower the tube, the fewer central molecules and the higher the liquid raises in the tube. This property is one of the ways in which water reaches the top of a tall tree.

Viscosity

Viscosity is the resistance to flow. Suppose you have a glass of gasoline (nonpolar molecules) and a glass of molasses (polar molecules). Try pouring each into another container. The molasses pours much more slowly that the gasoline. The two important factors influencing the viscosity of a liquid are as follows:

- **Intermolecular forces:** The stronger the intermolecular force is, the greater the viscosity. This is the factor in the gasoline and molasses situation. The mixture of polar molecules that make up the molasses attracts each other, and the result is a resistance to flow. The size of the molecules also affects intermolecular forces (and molasses). Large and complex molecules have difficulty moving past one another, so the viscosity is high.
- **Temperature:** Heat and cold also affect viscosity. If you heat the glass of molasses, it pours more easily because you reduce its viscosity by increasing the kinetic energy of the particles. The higher kinetic energy overcomes the intermolecular attractive forces, causing a lower viscosity. Putting the molasses in the refrigerator causes the opposite effect. The kinetic energy is reduced and the viscosity is increased; hence the phrase, "Slower than molasses in wintertime."

Some liquids have very high viscosities. If the viscosity is high enough, the liquid may not appear to flow at all and may be mistaken for a solid. A high viscosity liquid that appears to be a solid is commonly referred to as an *amorphous* solid. Sometimes scientists refer to these amorphous solids as *glasses* because glass is the most common example. Rubber and charcoal are other examples of amorphous solids.

Surface tension

Within the body of a liquid, intermolecular forces pull the molecules in all directions. However, at the liquid's surface, the molecules are pulled down into the body of the liquid and from the sides. As a result, no molecules and no attractive force above the surface pull in that direction. The effect of these unbalanced attractive forces is that the liquid tries to minimize its surface area. The *surface tension* is the resistance of a liquid to an increase in its surface area.

The minimum surface area for a given quantity of matter is a sphere. You may have seen video or pictures of a liquid, usually water, being released in zero gravity. The droplets form little spheres. In a large pool of liquid, where sphere formation is not possible, the surface behaves as if it had a thin, stretched elastic membrane or "skin" over it. Surface tension requires force to break these attractive forces at the surface. The greater the intermolecular force, the greater the surface tension. Polar liquids, especially those that utilize hydrogen bonding, have a much higher surface tension than nonpolar liquids.

Surface tension is what allows a bug to walk over water. It also allows you to add more water to a glass than its volume. Try it. Carefully add water to a glass and see the water dome at the top. Now touch a toothpick that has been dipped into dishwashing liquid to that dome and see it break. That happens because the dishwashing liquid is a *surfactant,* which disrupts surface tension.

Chapter 6

Not Full of Hot Air: Gases and Gas Laws

In This Chapter

- Recapping the kinetic molecular theory of gases
- Understanding gas laws
- Combining gases with reaction stoichiometry

The solid and liquid states are readily visible; you consider them without thinking. But the gaseous state is for the most part invisible. Because you don't directly see gas particles, you may tend to forget they're there until you check the pressure in your automobile's tires, blow up a balloon, or check the barometric pressure to see if it's falling or rising. And who could forget the propane in gas grills. Gases and their properties are part of our everyday lives. Many of the properties are interrelated, such as temperature and pressure. For instance, why is my tire pressure higher after I've driven a while? I explain the answer to this question and others in this chapter.

In this chapter, I introduce you to the properties of gases at both the microscopic and macroscopic levels. At the microscopic level I show you one of science's most successful theories — the kinetic molecular theory of gases. Back in the macroscopic level I explain the macroscopic properties of gases and show you the important interrelationships among them. Then I relate the properties of gases to reaction stoichiometry. This chapter may leave you gasping for gas (air)!

Reviewing the Kinetic Molecular Theory

Scientists find theories useful if they describe the physical system the scientists are examining and allow them to make a prediction of what will happen if the scientists change some variable. The *kinetic molecular theory of gases* has limitations — all theories do — but it's one of the most useful theories in chemistry. This section describes the theory's basic *postulates* — assumptions, hypotheses, axioms (pick your favorite word) you can accept as being true.

- **Gases are composed of tiny particles, either atoms or molecules.**

  The particles referred to as gases tend to be relatively small with relatively low atomic and molecular weights.

- **The gas particles are so small when compared to the distances between them that the volume the gas particles themselves take up is negligible and is assumed to be zero.**

  These gas particles do take up some volume — that's one of the properties of matter. But the gas particles are small, so if a container doesn't hold many of them, you say that their volume is negligible when compared to the volume of the container or the space between the gas particles. Because of all that space between the gas particles, they can be squeezed together to compress the gas. Solids and liquids can't be squeezed, because their particles are *much* closer together.

- **The gas particles are in constant random motion, moving in straight lines and colliding with the container's inside walls.**

  The gas particles are always moving in a straight-line motion. Gases have a higher *kinetic* energy — energy of motion — associated with them than solids or liquids do. They continue to move in these straight lines until they collide with something — either with each other or with the inside walls of the container. The particles also all move in different directions, so the collisions with the inside walls of the container tend to be uniform over the entire inside surface.

  The collision of the gas particles with the inside walls of the container is called *pressure.* The idea that the gas particles are in constant, random, straight-line motion explains why gases uniformly mix if put in the same container. It also explains why, when you drop a bottle of cheap perfume at one end of the room, the people at the other end of the room are able to smell it right away.

- **The gas particles are assumed to have negligible attractive or repulsive forces between each other.**

  In other words, the gas particles are assumed to be totally independent, neither attracting nor repelling each other. That said, it's hair-splitting time: This assumption is actually false. If it were true, chemists would never be able to liquefy a gas, which they can. But the reason you can accept this assumption as true (or at least useful) is that the attractive and repulsive forces are generally so small that they can safely be ignored. The assumption is most valid for nonpolar gases, such as hydrogen and nitrogen, because the attractive forces involved are London forces. However, if the gas molecules are polar, as in water and HCl, this assumption can become a problem.

- **The gas particles may collide with each other. These collisions are assumed to be elastic, with the total amount of kinetic energy of the two gas particles remaining the same.**

  Not only do the gas particles collide with the inside walls of the container, but they also collide with each other. If they hit each other, no kinetic energy is lost, but kinetic energy may be transferred from one gas particle to the other. These collisions of the gas particles with the inside walls of the container is what chemists call pressure, commonly measured in *atmospheres* or *torr* or *pascals*.

- **The Kelvin temperature is directly proportional to the *average* kinetic energy of the gas particles.**

  The gas particles aren't all moving with the same amount of kinetic energy. A few are moving relatively slow and a few are moving very fast, but most are somewhere in between these two extremes. Temperature, particularly as measured using the Kelvin temperature scale, is directly related to the *average* kinetic energy of the gas. If you heat the gas so that the Kelvin temperature (K) increases, the average kinetic of the gas also increases.

A gas that obeys all the postulates of the kinetic molecular theory is called an *ideal gas*. Obviously, no real gas obeys the assumptions made in the second and fourth postulates *exactly* (all gas particles actually do have small measures of volume and attractive or repulsive force). But a nonpolar gas at high temperatures and low pressure (concentration) approaches ideal gas behavior.

Obeying Laws: Gases, That Is

Various scientific laws describe the relationships among four of the important physical properties of gases:

- Volume
- Pressure
- Temperature
- Amount

This section covers those various laws. Boyle's, Charles's, and Gay-Lussac's laws each describe the relationship between two properties while keeping the other two properties constant. Another law — a combo of Boyle's, Charles's, and Gay-Lussac's individual laws — enables you to vary more than one property at a time. But that combo law doesn't let you vary the physical property of amount. Avogadro's law, however, does. And an ideal gas law even lets you take into account variations in all four physical properties.

When working gas law problems, you must express the temperature in Kelvin.

Boyle's law

Boyle's law, named after Robert Boyle, a 17th-century English scientist, describes the pressure-volume relationship of gases if the temperature and amount are kept constant. Figure 6-1 illustrates the pressure-volume relationship using the kinetic molecular theory.

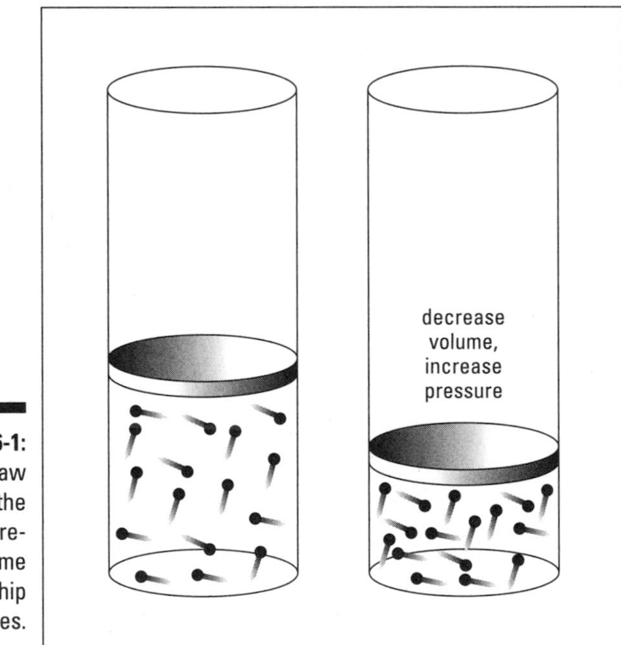

Figure 6-1: Boyle's law looks at the pressure-volume relationship of gases.

The left-hand cylinder in the figure contains a certain volume of gas at a certain pressure. When the volume is decreased, the same number of gas particles is now contained in a much smaller volume and the number of collisions increases significantly. Therefore, the pressure is greater.

Boyle's law states that an inverse relationship exists between the volume and pressure. As the volume decreases, the pressure increases, and vice versa. Boyle determined that the product of the pressure and the volume is a constant (k):

$$PV = k$$

Now consider a case where you have a gas at a certain pressure (P_1) and volume (V_1). If you change the volume to some new value (V_2), the pressure also changes to a new value (P_2). You can use Boyle's law to describe both sets of conditions:

$$P_1V_1 = k$$

$$P_2V_2 = k$$

The constant, k, is the same in both cases. So you can say

$$P_1V_1 = P_2V_2 \quad \text{(with temperature and amount constant)}$$

This equation is another statement of Boyle's law — and it's really a more useful one, because you normally deal with changes in pressure and volume. If you know three of the preceding quantities, you can calculate the fourth one. For example:

> Suppose that you have 10.00 liters of a gas at 5.00 atm pressure, and then you decrease the volume to 2.00 liters. What's the new pressure?

Before you even start calculating, ask yourself what will be the result of decreasing the volume. The pressure should increase because you have those same gas molecules now confined in a smaller volume, and because the temperature is held constant, their kinetic energies will be the same. They will be hitting the inside walls of the container more often — thus a higher pressure. To solve this problem, stick to these steps:

1. **Use the following equation.**

   $$P_1V_1 = P_2V_2$$

2. **Substitute 5.00 atmospheres for P_1, 10.00 liters for V_1, and 2.00 liters for V_2 and you get**

   $$(5.00 \text{ atm})(10.00 \text{ L}) = P_2(2.00 \text{ L})$$

3. **Solve for P_2, which gives you $P_2 = 25.0$ atm.**

The answer makes sense, because you decreased the volume and the pressure increased, which is exactly what Boyle's law says.

Charles's law

Charles's law looks at the relationship between volume and temperature, keeping the pressure and amount constant. You may not even realize it, but you run across situations dealing with this relationship in everyday life, especially in terms of the heating and cooling of balloons. Figure 6-2 shows the temperature-volume relationship.

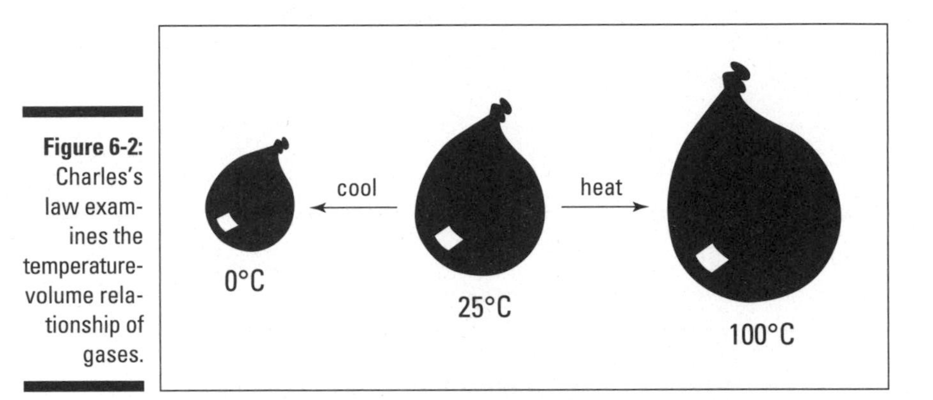

Figure 6-2: Charles's law examines the temperature-volume relationship of gases.

Look at the balloon in the middle of Figure 6-2. What would happen to the balloon if you heat the balloon? The balloon would expand and the volume would increase. On the other hand, if you placed it in the freezer or took it outside in subzero weather, what would happen? It'd get smaller. Inside the freezer or in arctic weather, the external pressure, or atmospheric pressure, is the same, but the gas particles inside the balloon aren't moving as fast, so the volume shrinks to keep the pressure constant. This correspondence is a *direct relationship* — as the temperature increases, the volume increases, and vice versa.

Jacques Charles, a 19th century French chemist, developed the mathematical relationship between temperature and volume. He also discovered that you must use the Kelvin (K) temperature when working with gas law expressions and calculations. This law says that the volume is directly proportional to the Kelvin temperature. Mathematically, the law looks like this:

$$V = \mathrm{b}T \text{ or } \frac{V}{T} = \mathrm{b} \text{ (where b is a constant)}$$

If the temperature of a gas with a certain volume (V_1) and Kelvin temperature (T_1) is changed to a new Kelvin temperature (T_2), the volume also changes (V_2) and because the constant is the same:

$$\frac{V_1}{T_1} = \mathrm{b} = \frac{V_2}{T_2} \text{ or } \frac{V_1}{T_1} = \frac{V_2}{T_2}$$

If you have any three of the four quantities, you can calculate the fourth. For example:

> Suppose you are vacationing in Canada and are outside in the middle of a really severe winter, where the temperature is –33 degrees Celsius. You blow up a balloon so that it has a volume of 1.00 liter. You then take it inside your home, where the temperature is a toasty 27 degrees Celsius. What's the new volume of the balloon?

To solve this problem, stick to these steps:

1. **Express the temperature in Kelvin.**

   –33°C + 273 = 240 K (outside)

   27°C + 273 = 300 K (inside)

2. **Use Charles' Law to solve.**

   Because the two variables involved in this problem are temperature and volume, you can use Charles's Law:

   $$\frac{V_1}{T_1} = \frac{V_2}{T_2}$$

3. **Arrange the equation to solve for V_2.**

   $$V_2 = \frac{V_1 T_2}{T_1}$$

4. **Substitute the values to calculate the following answer.**

   $$\frac{(1.00\text{ L})(300.\text{K})}{240.\text{K}} = 1.25\text{ L}$$

The answer is reasonable, because, according to Charles's law, if you increase the temperature, the volume also increases.

Gay-Lussac's law

Gay-Lussac's law, named after the 19th-century French scientist Joseph-Louis Gay-Lussac, focuses on the relationship between the pressure and temperature of a gas if its volume and amount are held constant. Imagine, for example, that you have a metal scuba tank that holds a certain amount of gas. The tank has a certain volume, and the gas inside has a certain pressure. If you heat the tank, you increase the kinetic energy of the gas particles. So they're now moving much faster, and they're not only hitting the inside walls of the tank more often but also with more force. The pressure increases.

Gay-Lussac's law says that the pressure is directly proportional to the Kelvin temperature. Figure 6-3 shows this relationship.

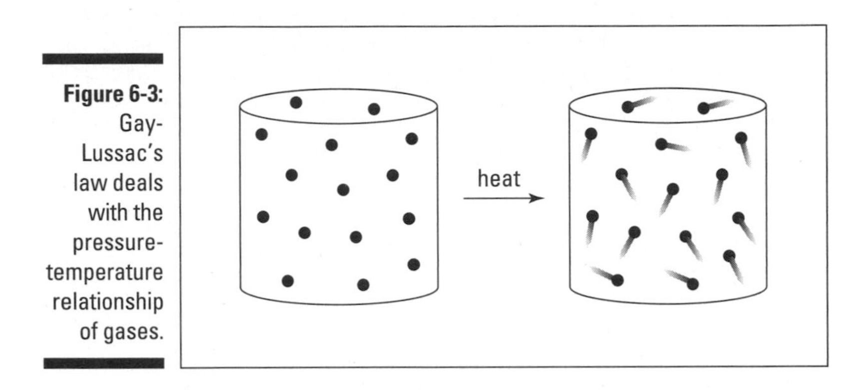

Figure 6-3: Gay-Lussac's law deals with the pressure-temperature relationship of gases.

Mathematically, Gay-Lussac's law is represented by the following equation, where P is pressure, T is temperature, and k is a constant:

$$P = \mathrm{k}T \text{ or } \frac{P}{T} = \mathrm{k} \text{ (at constant volume and amount)}$$

If you have a gas at a certain Kelvin temperature and pressure (T_1 and P_1), with the conditions being changed to a new temperature and pressure (T_2 and P_2):

$$\frac{P_1}{T_1} = \frac{P_2}{T_2}$$

If you have a tank of gas at 1,520 torr pressure and a temperature of 300. Kelvin, and it's heated to 500. Kelvin, what's the new pressure in atmospheres?

You can find the new pressure by following these steps:

1. **Convert the initial pressure to atmospheres at the beginning.**

   $$\frac{1520 \text{ torr}}{1} \times \frac{1 \text{ atm}}{760 \text{ torr}} = 2.00 \text{ atm}$$

2. **Solve for P_2 in Gay-Lussac's law.**

   $$P_2 = \frac{P_1 T_2}{T_1}$$

3. **Substitute the values.**

   $$P_2 = \frac{(2.00 \text{ atm})(500.\text{K})}{300. \text{ K}} = 3.33 \text{ atm}$$

This answer seems reasonable because if you heat the tank, the pressure should increase.

The combined gas law

All the examples in the preceding sections are set up so that two properties are held constant and one property is changed to see its effect on a fourth property. But life is rarely that simple. How do you handle situations in which two or even three properties change? You can treat each one separately, but wouldn't it be nice if you had a way to combine things so that wouldn't be necessary?

Actually, you can. You can combine the three preceding laws — Boyle's law, Charles's law, and Gay-Lussac's law — into one equation. I won't waste your time and show you exactly how it's done because it involves a lot of boring algebra, but the end result is called the *combined gas law,* and it looks like this:

$$\frac{P_1V_1}{T_1} = \frac{P_2V_2}{T_2}$$

P is the pressure of the gas (in atm, mm Hg, torr, and so on), V is the volume of the gas (in appropriate units), and T is the temperature (in Kelvin). The *1* and *2* stand for the initial and final conditions, respectively. The amount is still held constant: No gas is added, and no gas escapes. Six quantities are involved in this combined gas law, so knowing any five allows you to calculate the sixth.

Suppose that a weather balloon with a volume of 50.0 liters at 760. torr pressure and a temperature of 27 degrees Celsius (300. K) is allowed to rise to an altitude where the pressure is 380. torr and the temperature is –23 degrees Celsius (250. K). What's the balloon's new volume?

Before you work this problem, do a little reasoning. The temperature is decreasing, so that change should cause the volume to decrease (Charles's law). However, the pressure is also decreasing, which should cause the balloon to expand (Boyle's law). These two factors are competing, so at this point, you don't know which will win out.

You're looking for the new volume (V_2), so solve following these steps:

1. **Rearrange the combined gas equation.**

$$V_2 = \frac{P_1V_1T_2}{P_2T_1}$$

2. Substitute in your quantities.

$$V_2 = \frac{(760.\text{torr})(50.0\text{ L})(250.\text{K})}{(380.\text{torr})(300.\text{K})} = 83.3\text{ L}$$

Because the volume increased overall in this case, Boyle's law had a greater effect than Charles's law.

Avogadro's law

In order to account for amount of pressure, volume, and temperature, you need to know another law. Avogadro's law to the rescue!

Amedeo Avogadro determined, from his study of gases, that equal volumes of gases at the same temperature and pressure contain equal numbers of gas particles. So Avogadro's law says that the volume of a gas is directly proportional to the number of moles of gas (number of gas particles) at a constant temperature and pressure. Mathematically, Avogadro's law looks like this:

$V = kn$ (at constant temperature and pressure)

In this equation, k is a constant and n is the number of moles of gas. If you have a number of moles of gas (n_1) at one volume (V_1), and the moles change due to a reaction (n_2), the volume also changes (V_2), giving you the equation:

$$\frac{V_1}{n_1} = \frac{V_2}{n_2}$$

I won't work any problems with this law here, because it's basically the same idea as the other gas laws I have already covered.

Avogadro's law makes life easier by helping you be able to calculate the volume of a mole at any temperature and pressure. An extremely useful form to know when calculating the volume of a mole of gas is *1 mole of any gas at STP occupies 22.4 liters. STP* in this case is not an oil or gas additive. It stands for standard temperature and pressure.

- **Standard pressure:** 1.00 atm (760. torr or 760. mm Hg)
- **Standard temperature:** 273 K

This relationship between moles of gas and liters gives you a way to convert the gas from a mass to a volume. For example:

Suppose that you have 150.0 grams of nitrogen gas (N_2), and you want to know its volume at STP. Just convert from grams to moles to liters at STP:

$$\frac{150.0\text{ g N}_2}{1} \times \frac{1\text{ mole}}{28.02\text{ g}} \times \frac{22.4\text{ L}}{1\text{ mole}} = 120.\text{ L}$$

You now know that the 150.0 grams of nitrogen gas occupies a volume of 120. liters at STP. But what if the gas is not at STP?

What's the volume of 150.0 grams of nitrogen gas at 3.00 atm and 127.0 degrees Celsius (400.0 K)?

In the next section, I show you a simple way to do this problem, but right now, you can use the combined gas law, because you know the volume at STP:

P_1 = 1.00 atm, V_1 = 120.0 L, T_1 = 273 K

P_2 = 3.00 atm, T_2 = 400. K

$$\frac{P_1V_1}{T_1} = \frac{P_2V_2}{T_2}$$

$$V_2 = \frac{P_1V_1T_2}{P_2T_1}$$

$$V_2 = \frac{(1\text{ atm})(120.0\text{ L})(400.\text{K})}{(3.00\text{ atm})(273.\text{K})} = 58.6\text{ L}$$

The ideal gas equation

If you take Boyle's law, Charles's law, Gay-Lussac's law, and Avogadro's law and throw them into a blender on high for a minute, you get the *ideal gas equation* — a way of working in volume, temperature, pressure, *and* amount. The ideal gas equation has the following form:

$$PV = nRT$$

The P represents pressure in atmospheres (atm), the V represents volume in liters (L), the n represents moles of gas, the T represents the temperature in Kelvin (K), and the R represents the ideal gas constant, which is 0.0821 liters atm/K mol.

Using this value of the ideal gas constant, the pressure must be expressed in atm, and the volume must be expressed in liters. You can calculate other ideal gas constants if you really want to use torr and milliliters, for example, but why bother? Memorizing one value for R and then remembering to express the pressure and volume in the appropriate units is much easier. Naturally, you'll *always* express the temperature in Kelvin when working with any kind of gas law. Here I demonstrate how you can convert a gas from a mass to a volume if the gas is not at STP:

What's the volume of 150.0 grams of nitrogen gas at 3.00 atm and 127.0 degrees Celsius (400.0 K)?

To figure out this problem, stick to these steps:

1. **Convert from grams of nitrogen gas to moles of nitrogen gas.**

$$\frac{150.0 \text{ g}}{1} \times \frac{1 \text{ mole}}{28.02 \text{ g}} = 5.353 \text{ moles}$$

2. **Rearrange the equation to solve for volume and substitute in the information from the problem.**

$$V = \frac{nRT}{P} = \frac{(5.353 \text{ moles})\left(0.0821 \frac{\text{L atm}}{\text{K mol}}\right)(400.0 \text{ K})}{3.00 \text{ atm}} = 58.6 \text{ L}$$

This answer is exactly the same as what you get in the preceding section, but you calculate it in a much more straightforward way.

Dalton's and Graham's laws

As you may remember from your Chem I class, chemistry also has some miscellaneous laws you should have a basic knowledge of. The following sections give a quick overview to them. One relates to partial pressures and the other to gaseous effusion/diffusion. Like the other sections, this one is a real gas.

Dalton's law

Dalton's law of partial pressures says that in a mixture of gases, the total pressure is the sum of the partial pressures of each individual gas.

If you have a mixture of gases — gas A, gas B, gas C, and so on — then the total pressure of the system is simply the sum of the pressures of the individual gases. Mathematically, the relationship can be expressed like this:

$$P_{\text{total}} = P_A + P_B + P_C + \ldots$$

In the problems you encounter in Chem II, many times the problem will give the number of grams or moles of the gas along with the total pressure. The problem will ask you for the individual gas pressures in the mixture (called *partial pressures*). In order to accomplish this task, you can use the mole fraction(X) of the gas. The mole fraction of gas A will be the moles of gas A divided by the total number of moles of gas in the system:

$$X_A = \frac{\text{moles A}}{\text{moles A} + \text{moles B} + \text{moles C} + \ldots}$$

The pressure due to gas A will be the mole fraction times the total pressure:

$$P_A = X_A \times P_{total}$$

In a gas cylinder there are 0.20 moles of O_2, 0.80 moles of N_2, and 0.50 moles of Ne. The total pressure is 150.0 kPa. Calculate the partial pressures of each gas. A total of 1.50 moles of gas is in the cylinder.

To calculate the partial pressure, do the following:

1. **Calculate the mole fraction of each gas.**

$$X_{O_2} = \frac{0.20 \text{ moles}}{1.50 \text{ moles}} = 0.133$$

$$X_{N_2} = \frac{0.80 \text{ moles}}{1.50 \text{ moles}} = 0.533$$

$$X_{N_e} = \frac{0.50 \text{ moles}}{1.50 \text{ moles}} = 0.333$$

The sum of all the mole fractions should equal 1.

2. **Calculate the partial pressure of each gas by multiplying its mole fraction by the total pressure of 150.0 kPa:**

For O_2: 0.133 × 150.0 kPa = 20.0 kPa

For N_2: 0.533 × 150.0 kPa = 80.0 kPa

For Ne: 0.333 × 150.0 kPa = 50.0 kPa

Make sure the sum of all the partial pressures equals the total pressure.

Dalton's law is really handy when collecting a gas by displacement of water. The gas that is collected will be a mixture of the desired gas and water vapor. The vapor pressure of water at various temperatures is well known and tabulated in many sources such as the *CRC Handbook,* the *Merck Index,* and probably even in your textbook. By using the tabulated values and Dalton's law, you can calculate the pressure of just the generated gas. You can mathematically "dry out" the gas being collected.

Suppose, for example, that a sample of oxygen is collected over water at a total pressure of 755 torr at 20 degrees Celsius. Your task is to calculate the pressure of the oxygen.

You know that the total pressure is 755 torr. Your first task is to reference a table of vapor pressures of water versus temperature. (You can find such a table in a variety of places, such as the *Chemical Rubber Company [CRC] Handbook*.) After looking at the table, you determine that the partial pressure of water at 20 degrees Celsius is 17.5 torr. Now you're ready to calculate the pressure of the oxygen:

$$P_{\text{total}} = P_{\text{oxygen}} + P_{\text{water vapor}}$$

$$755\text{ torr} = P_{\text{oxygen}} + 17.5\text{ torr}$$

$$P_{\text{oxygen}} = 755\text{ torr} - 17.5\text{ torr} = 737.5\text{ torr}$$

Knowing the partial pressure of gases like oxygen is important in deep sea diving and the operation of respirators in hospitals.

Graham's law

Place a few drops of a strong perfume on a table at one end of a room, and soon people at the other end of the room can smell it. This process is called *gaseous diffusion,* the mixing of gases due to their molecular motion.

Place a few drops of that same perfume inside an ordinary rubber balloon and blow it up. Very soon you'll be able to smell the perfume outside of the balloon as it makes its way through the microscopic pores of the rubber. This process is called *gaseous effusion,* the movement of a gas through a tiny opening. The same process of effusion is responsible for the helium being quickly lost from rubber balloons.

Thomas Graham determined that the rates or velocities *(v)* of diffusion and effusion of gases are inversely proportional to the square roots of their molecular or atomic weights *(M)*. This relationship is called Graham's law. In general, it says that the lighter the gas, the faster it will effuse (or diffuse). Mathematically, Graham's law looks like this:

$$\frac{v_1}{v_2} = \sqrt{\frac{M_2}{M_1}}$$

Suppose that you fill two rubber balloons to the same size, one with helium (He) and the other with krypton (Kr). The helium, having a smaller molar mass, should effuse through the balloon pores faster. But how much faster? Using Graham's law, you can determine the answer:

1. **Write Graham's Law with the gases to be used.**

$$\frac{v_{He}}{v_{Kr}} = \sqrt{\frac{M_{Kr}}{M_{He}}}$$

2. **Substitute the atomic mass of each gas.**

$$\frac{v_{He}}{v_{Kr}} = \sqrt{\frac{83.8 \frac{\text{g}}{\text{mole}}}{4.00 \frac{\text{g}}{\text{mole}}}}$$

3. **Calculate the helium to krypton ratio.**

$$\frac{v_{He}}{v_{Kr}} = \sqrt{20.95} = 4.58$$

The helium should effuse out a little more than four and a half times as fast as the krypton.

Putting Gases Together with Reaction Stoichiometry

The ideal gas equation (and even the combined gas equation) allows chemists to work stoichiometry problems involving gases. In this section, you use the ideal gas equation to do such a problem, using a classic chemistry experiment — the decomposition of potassium chlorate to potassium chloride and oxygen by heating:

$$2\ KClO_3(s) \rightarrow 2\ KCl(s) + 3\ O_2(g)$$

Figure out the volume of oxygen gas produced at 0.950 atm and 27.0 degrees Celsius (300.0 K) from the decomposition of 5.00 grams of $KClO_3$.

To solve this problem, follow these steps:

1. **Convert the grams of $KClO_3$ to moles and then to moles of oxygen gas.**

$$\frac{5.00\ \text{g}\ KClO_3}{1} \times \frac{1\ \text{mole}}{122.55\ \text{g}\ KClO_3} \times \frac{3\ \text{mole}\ O_2}{2\ \text{mole}\ KClO_3} = 0.0612\ \text{moles}\ O_2$$

2. **Now you have the moles, pressure, and temperature, so use the ideal gas equation and solve for the volume.**

$$V = \frac{nRT}{P} = \frac{(0.0612\ \text{mole})\left(0.0821\ \text{L atm}/_{\text{K mol}}\right)(300.0\ \text{K})}{0.950\ \text{atm}} = 1.58\ \text{L}$$

Here is another example problem.

A sample weighing 5.500 grams was composed of a mixture of $KClO_3$ and KCl. It was heated and 0.525L of oxygen was collected over water at 20.0 degrees Celsius and the pressure was determined to be 757.5 torr. Calculate the mass of the both the $KClO_3$ and KCl in the sample.

Even though this problem looks complicated, you can do it. Just slow down and break it down in parts. The sample initially had both $KClO_3$ and KCl. When heated, the $KClO_3$ decomposed liberating oxygen gas, which was collected over water (Dalton's law). You know the volume, temperature, and pressure of the gas mixture. From a previous problem, I told you that the vapor pressure of water at 20 degrees Celsius was 17.5 torr, so you can solve this problem.

1. **Figure the pressure of just the oxygen gas.**

$$P_{\text{Total}} = P_{\text{oxygen gas}} + P_{\text{water vapor}}$$

$$757.5 \text{ torr} = P_{\text{oxygen gas}} + 17.5$$

$$P_{\text{oxygen gas}} = 740.0 \text{ torr} = 0.974 \text{ atm}$$

2. **Apply the ideal gas equation and solve for the number of moles of oxygen gas.**

$$n = \frac{PV}{RT} = \frac{(0.974 \text{ atm})(0.525 \text{ L})}{\left(0.0821 \text{L atm}/\text{K mol}\right)(293.0 \text{ K})} = 0.0213 \text{ moles oxygen}$$

3. **Calculate the number of grams of potassium chlorate using the reaction stoichiometry.**

The reaction is: $2\ KClO_3(s) \rightarrow 2\ KCl(s) + 3\ O_2(g)$

$$\frac{0.0213 \text{ moles } O_2}{1} \times \frac{2 \text{ mole } KClO_3}{3 \text{ mole } O_2} \times \frac{122.55 \text{ g}}{1 \text{ mole } KClO_3} = 1.740 \text{ g } KClO_3$$

Of the original 5.500 grams of mixture, 1.740 grams was $KClO_3$. The rest must have been KCl. The grams of KCl is

$$5.500 - 1.740 \text{ g} = 3.760 \text{ g KCl}$$

Part II
Diving Into Kinetics and Equilibrium

In this part . . .

In this part, you get into the real meat of Chemistry II. In Chapter 7 on kinetics, I discuss the factors associated with the speed of a reaction. I show you how to determine the rate law for a reaction. A rate law relates the changes in concentrations of reactants to the overall speed of a reaction. I also discuss the current model on how reactions occur and end up with a discussion of catalysts.

The rest of this part — three chapters worth — focuses on equilibrium. The study of equilibrium is probably the most important topic in a Chemistry II course. In Chapter 8 I introduce you to the basic concepts of equilibriums and then apply these basic concepts to homogeneous equilibrium systems, which are equilibrium systems of gas or those in solution. In Chapter 9 I discuss weak acid-base equilibrium systems, including buffers and titrations. Finally, in Chapter 10 I show you how to handle heterogeneous and complex-ion equilibriums. Heterogeneous systems are represented by the minute dissolving of insoluble salts, which establish what is called a solubility equilibrium.

You need to remember one important thing about the various types of equilibriums that you study in Chem II. They are all the same. Some of my students try to simply memorize the individual types and get really confused. The smart ones grasp the basic concepts presented in Chapter 8 and then simply apply those concepts to the various types of equilibriums. It's so much easier that way. Be smart!

Chapter 7

The Lowdown on Kinetics: Tortoise or the Hare?

In This Chapter

- Examining reaction rates
- Perusing rate laws
- Eyeing integrated rate laws
- Facing off with the collision theory
- Looking at activation energy
- Identifying mechanisms
- Peering closer at catalysts

Many times thermodynamics (Chapter 11) tells you whether or not a particular chemical reaction will occur. However, it can't answer the question as to how fast it may occur. Reactions may take years to complete, or they may occur in milliseconds. Chemists may get some idea on the speed of a proposed reaction by looking at things like the size and complexity of the particles, the temperature, and so on, but the only way they can really determine how fast a reaction will occur is to go into the laboratory and perform the reaction and measure its speed. In other words, *kinetics* (the study of the speed of reactions) is an experimental science.

In this chapter I show you some of the factors that affect the speed of reaction and then I dive into rate laws and mechanisms. Hopefully I don't go too *fast* for you (pun intended).

Comprehending Reaction Rates

In order for a reaction to occur, a collision must happen between particles at the right place transferring enough energy to break old bonds and reform new ones. (See the section on "Tackling the Collision Theory" later in this chapter for more information about this.) Five general factors can affect the rate of reaction:

- **Nature of the reactants:** Large complex molecules tend to react slower than smaller, simpler molecules. The larger ones tend to move slower and have less kinetic energy to transfer. Because they're moving slower, fewer collisions happen, thus making less of a chance for reaction. Because these large molecules are complex in structure, they have many more sites for collisions than simply the reactive site; therefore, their reactions will be slower.
- **The particle size:** In general, the larger the surface area of the particles is, the faster the rate. Striking a match next to a chunk of coal will cause no reaction to occur, but grind that chunk of coal into a very fine powder and strike that same match and — *explosion!* This reaction happens because the powder has a great deal more surface area than the chunk. If there is more surface area, then there are many more collisions, and the the reaction rate becomes much faster. For the same reasons, gases and liquids tend to react faster than solids.
- **The reactant concentration:** In most cases the greater the reactant concentration means the greater the statistical chance of collisions and the faster the rate of reaction. If gases are involved, this factor applies to an increase in pressure. However, if the *mechanism* (the series of steps in the reaction) is complex, such as in the reactions producing photochemical smog, then the effect of reactant concentrations may not be a simple one.
- **The temperature:** Increasing the temperature generally increases the reaction rate because the particles will have a greater amount of kinetic energy and the number of collisions is increased. This increase in kinetic energy means that a greater chance of transferring enough energy can initiate the reaction (to provide the activation energy, that minimum amount of energy needed to start the reaction; see Figure 7-1). In fact, in organic chemistry a general rule states that for every 10°C increase in temperature, the reaction rate doubles.

  Notice that at a given temperature not all of the molecules are moving with the same kinetic energy. A small number of molecules are moving very slow (low kinetic energy), while a few are moving very fast (high kinetic energy). A vast majority of the molecules are somewhere in between these two extremes. In fact, temperature is a measure of the average kinetic energy of the molecules. As you can see in Figure 7-1, increasing the temperature increases the average kinetic energy of the reactants, essentially shifting the curve to the right toward higher kinetic energies. But also notice that I mark the minimum amount of kinetic energy needed by the reactants in order to provide the activation energy during collision. The reactants have to collide at the reactive site, but they also have to transfer enough energy to break bonds so that new bonds can be formed. This process refers to the activation energy of the reaction. If the reactants don't have enough energy, a reaction won't occur even if the reactants do collide at the reactive site.

Notice that at the lower temperature, very few of the reactant molecules have the minimum amount of kinetic energy available to provide the activation energy. At the higher temperature, many more molecules possess the minimum amount of kinetic energy, which means that a lot more collisions will be energetic enough to lead to reaction.

Increasing the temperature not only increases the number of collisions but also increases the number of collisions that will be *effective,* or transfer enough energy to cause a reaction to take place.

- **Catalysts:** Catalysts speed up the reaction rate. I go into much greater detail about catalysts later in this chapter in "Keeping Tabs on Catalysts."

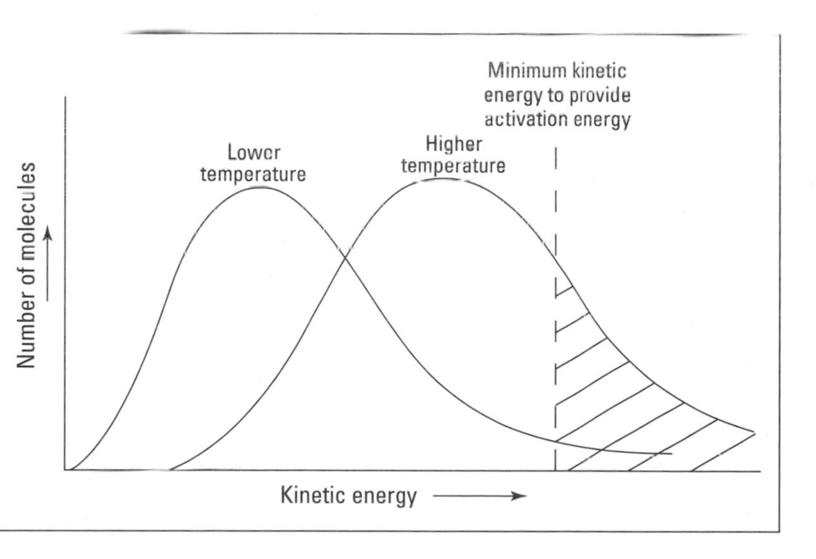

Figure 7-1: The effect of temperature on the kinetic energy of the reactants.

Getting to Know Rate Laws

The *reaction rate* (the speed of reaction) is the change in the concentration of a reactant or product per the change in time. You can write it as:

$$\text{Reaction rate} = \frac{\text{Change in concentration}}{\text{Change in time}}$$

Chemists normally measure concentration in terms of molarity, M, and time is usually expressed in seconds, s, which means that the units of the reaction rate are M/s. You can express the number of units in other ways such as:

$$\frac{M}{s} = Ms^{-1} = \frac{mole}{L \cdot s} = mole\ L^{-1}s^{-1}$$

The study of kinetics, the speed of chemical reactions, is essential to the study of chemistry and is a major topic in any Chem II class. Knowing the concepts of kinetics can help your understanding of why some reactions are fast and others slow and why some simple reactions are slow and other, more complex reactions are fast. Some instructors relate kinetics to equilibriums (although I'm not one of those). To help you get a firmer grasp of rate laws, the following sections explain this concept in greater depth and provide some example problems.

Slowing down the reaction

The rate of a reaction isn't constant. Reactions gradually slow down as they proceed. This decrease in the speed of reaction is due to the fact that as the reaction progresses, there are fewer reactant molecules present.

Consider the following general reaction:

$A_2(g) + B_2(g) \rightarrow 2\ AB(g)$

1. **Begin by measuring the rate of reaction for the disappearance of A_2.**

   The rate of disappearance of A_2 is

   $$-\frac{\Delta[A_2]}{\Delta t}$$

   Use the capital Greek letter delta, Δ, to represent a change and the square brackets, [ ], to indicate the concentration (molarity) of A_2. The *t* refers to the time. Finally, the concentration of A_2 decreases with time, so that the expression has a negative sign

2. **Determine the other rates without further experimentation.**

   The other rates come from a combination of A_2 rate with information from the balanced chemical equation.

   According to the balanced chemical equation, every time one A_2 molecule reacts, one B_2 molecule also reacts. Thus, the rate of B_2 is the same as that of A_2:

   $$-\frac{\Delta[A_2]}{\Delta t} = -\frac{\Delta[B_2]}{\Delta t}$$

   Note that the B_2 concentration is decreasing because it's also a reactant, so that it also has a negative sign.

   If you again refer to the balanced chemical equation, you see that every time one A_2 (or B_2) molecule reacts, two AB molecules form. Thus, the rate of formation for AB is twice the rate of disappearance of A_2.

3. **Divide the rate for AB by two to equal the rate for A_2.**

   $$-\frac{\Delta[A_2]}{\Delta t} = -\frac{\Delta[B_2]}{\Delta t} = \frac{1}{2}\frac{\Delta[AB]}{\Delta t}$$

The value for AB is positive because it's a product (whose concentration increases with time) and not a reactant (concentration decreasing with time).

Identifying orders of reaction

Many times when working reaction rate problems you often have to find the rate law for the reaction. The *rate law* is a mathematical relationship that shows what effect changing the reactant concentrations has on the reaction rate. This relationship involves a *rate constant* (a proportionality constant that allows you to work with an equality instead of a proportionality) and exponents known as *orders*. The general form of a rate law is:

$$\text{Rate} = k[\ ]^m[\ ]^n$$

In this equation, the rate is an experimentally determined value. The k is a *rate constant*, which you need to determine. The [ ] terms refer to the concentrations of substances involved in the reaction. The exponents, m and n, are the *orders of reaction*. The order(s) of reaction give you information about what will happen to the reaction rate if the concentration of that particular reactant changes. For example, if the order of a particular reactant is 2, it means that any change in the concentration of that reactant will lead to that change being squared. If the concentration of that reactant was tripled, the reaction rate would increase as $(3)^2 = 9$, or nine-fold. You must find each order in order to complete the rate law. If changing the concentration of a reactant has no effect on the speed of the reaction, then its order is zero (0).

You should be very careful with the k values. In this chapter, k, the rate constant, *must* be a small letter. In the equilibrium problems, which appear in other chapters in this book, you use a capital K. The terms k and K mean very different things.

Determining the rate law always begins with experimental data (the reaction rate). This is one of the few times you won't be using the balanced chemical equation of the reaction.

The rate constant may have any value, depending on the reaction being studied. The units of the rate constant depend on the rest of the rate law. The orders are usually small whole numbers; however, zero, negative, and fractional values are possible. The sum of the separate orders is the *overall order* of the reaction.

The rate law (the rate, the rate constant, and the orders of reaction) is determined experimentally. After the rate has been determined, the orders of reaction can be determined by conducting a series of reactions in which you change the concentrations of the reactant species one at a time and measure the rate of reaction. Then you can mathematically determine the effect on the reaction rate. After you have determined the orders of reaction, you can calculate the rate constant.

Determine the rate law for the following reaction:

$$NO(g) + O_3(g) \rightarrow NO_2(g) + O_2(g)$$

The following experimental data was collected:

Experiment	*[NO]*	*[O_3]*	*Rate (M/s)*
1	0.1	0.1	0.5
2	0.2	0.1	2.0
3	0.1	0.2	1.0

The general rate law for this reaction would be:

$$\text{Rate} = k[NO]^m[O_3]^n$$

You can use the preceding data to determine the rate equation in a couple different ways. If the concentrations and rates involved are simple numbers, then you can reason out the orders of reaction. In this example problem, you see that in going from experiment 1 to experiment 2, the [NO] doubles ([O_3] is held constant) and the rate increased four-fold, which means that the reaction is second order with respect to NO. If you compare experiments 1 and 3, you see that the $[O_3]$ doubles ([NO] is held constant) and the rate doubled. Therefore, the reaction is first order with respect to O_3 and the rate equation is:

$$\text{Rate} = k[\text{NO}]^2[\text{O}_3]$$

You can determine the rate constant by substituting the values of the concentrations of NO and O_3 from any of the experiments into the rate equation above and solving for *k*.

However, sometimes because the numbers involved are complex, you must determine the rate law by mathematically manipulating the equations. You use the ratio of the rate expressions of two experiments to determine the reaction orders. You choose equations in which the concentration of only one reactant has changed while the other(s) remain constant. In this example, you could use the ratio of experiments 1 and 2 to determine the effect of a change of the concentration of NO on the rate, because the concentration of ozone is the same (constant) in both experiments. Then you could use experiments 1 and 3 to determine the effect of O_3 on the reaction rate. You can't use experiments 2 and 3 because both NO and O_3 concentrations have changed.

In choosing experiments to compare, choose two in which the concentration of only one reactant changes while the others remain constant. To solve this problem, follow these steps:

1. **Mathematically compare Experiments 1 and 2.**

$$\frac{\text{Exp. 1}}{\text{Exp. 2}} = \frac{\text{Rate 1}}{\text{Rate 2}} = \frac{k[\text{NO}]^m[\text{O}_3]^n}{k[\text{NO}]^m[\text{O}_3]^n}$$

Comparing Experiments 1 to 2 (as I previously did) or vice versa doesn't make any difference.

2. **Put in the values from the data table.**

$$\frac{\text{Exp. 1}}{\text{Exp. 2}} = \frac{0.5\ ^{\text{M}}/_{\text{s}}}{2.0\ ^{\text{M}}/_{\text{s}}} = \frac{\cancel{k}[0.1]^m\cancel{[0.1]^n}}{\cancel{k}[0.2]^m\cancel{[0.1]^n}}$$

The *k*'s cancel as do both the $[0.1]^n$.

3. **Simplify.**

$$\frac{1}{4} = \left[\frac{1}{2}\right]^m$$

The value of m must be 2 — one-half squared equals one-fourth. You can easily see this with simple numbers, but experimental data is rarely this simple.

4. **Solve for *m*.**

   You can use logarithms. If you take the log of both sides, you get:

   $$\log 0.25 = m \log 0.50$$

   $$m = \frac{\log 0.25}{\log 0.50} = \frac{-0.6020}{-0.3010} = 2$$

5. **Do the same thing comparing Experiments 1 and 3, because the concentration of NO remains constant and the concentration of O_3 changes.**

   $$\frac{\text{Exp. 1}}{\text{Exp. 3}} = \frac{\text{Rate 1}}{\text{Rate 3}} = \frac{k\,[\text{NO}]^m[\text{O}_3]^n}{k\,[\text{NO}]^m[\text{O}_3]^n}$$

6. **Put in the values from the data table to get the following.**

   $$\frac{\text{Exp. 1}}{\text{Exp. 3}} = \frac{0.5\ \text{M/s}}{1.0\ \text{M/s}} = \frac{\cancel{k}\ \cancel{[0.1]^m}[0.1]^n}{\cancel{k}\ \cancel{[0.1]^m}[0.2]^n}$$

7. **Cancel and simplify.**

   $$\frac{1}{2} = \left(\frac{1}{2}\right)^n$$

   The value of *n* must equal 1.

8. **Write the rate law for this reaction.**

   Rate = $k[\text{NO}]^2[\text{O}_3]$

   The 1 order of reaction for the ozone is understood. This is the same rate law that you can arrive at by logic. In most situations, especially in the lab, you need to solve for the orders of reaction mathematically.

   The overall order of this reaction is the sum of the individual orders ($m + n$). In this case it would be $2 + 1 = 3$; the reaction is third order overall.

Finding the rate constant

After determining the orders of reaction, determining the rate constant is rather easy. You simply choose any one of the three experiments and substitute in the rate, the concentrations, and the orders of reaction. Then you solve for *k*.

The rate expression so far is:

Rate = $k[\text{NO}]^2[\text{O}_3]$

Choose Experiment 1 and substitute in all the values:

$$0.5 \text{ M/s} = k(0.1 \text{ M})^2(0.1\text{M})$$

$$0.5 \text{ M/s} = k\,(0.01\text{M}^2)(0.1 \text{ M})$$

$$0.5 \text{ M/s} = k\,(0.001 \text{ M}^3)$$

$$k = \frac{0.5 \text{ M/s}}{0.001 \text{ M}^3} = 500 \text{ M}^{-2}\text{s}^{-1}$$

Told you it was easy!

Grasping Integrated Rate Laws

The rate law that you determined in the previous section is just the first step in studying kinetics of a reaction. The next step is to make use of an integrated rate law. An *integrated rate law* describes how the concentrations relate to changes in time. The integrated rate law allows you to determine when a reaction is complete, or how much of a product can be produced per minute/hour/day. Using an integrated rate law allows a medical doctor to know when a patient needs a new dose of a medication. These are but a few of the applications of integrated rate laws.

Many types of integrated rate laws exist, but I limit my discussion to reactions in the following sections to ones such as this:

$$\text{A} \rightarrow \text{B}$$

with rate laws containing only one reactant (A) and an order of 1, 2, or 0.

That is, I only consider the following rate laws in these sections:

- **First order:** Rate = $k\,[\text{A}]^1$
- **Second order:** Rate = $k\,[\text{A}]^2$
- **Zero order:** Rate = $k\,[\text{A}]^0$

First order

I start with a first order rate law (Rate = $k\,[\text{A}]^1$). In this situation, one atom or molecule of A reacts to produce one or more products. One goal might be to determine how much A is left after a certain amount of time. (Doing so also tells you how much product has been formed.) To answer this equation, you have a couple methods to choose from.

- One method is to construct a graph with the horizontal axis being time and the vertical axis being ln[A]. Plotting the ln[A] versus time gives you a straight-line with the slope equal to k. Plotting ln[A] versus t only gives you a straight line for first order rate laws. Therefore, this method is also useful in determining if a reaction is first order.
- Another method to answer the question is to use an integrated rate law. For a first order reaction the integrated rate law is:

  $$\ln\frac{[A]_0}{[A]_t} = kt \quad \text{First order}$$

  In this equation, $[A]_0$ is the initial concentration of A, $[A]_t$ is the concentration of A after a certain amount of time, k is the rate constant, and t is the amount of time between $[A]_0$ and $[A]_t$.

Consider the following reaction:

$$2\,N_2O_5(g) \rightarrow 2\,N_2O_4(g) + O_2(g)$$

This reaction is first order and obeys the rate law:

Rate = $k\,[N_2O_5]$, where $k = 1.68 \times 10^{-2} s^{-1}$.

A sample of N_2O_5 with an initial concentration of 0.400 M begins reacting. What is the concentration of N_2O_5 after 2.00 minutes?

To solve this problem, stick to these steps:

1. **Apply the first order equation to this problem and solve for the $[N_2O_5]$ after t = 2.00 minutes (120 s).**

   $[A]_0 = [N_2O_5]_0 = 0.400$ M

   $k = 1.68 \times 10^{-2} s^{-1}$

   $t = 2.00$ min (120 s)

   $[A]_t = [N_2O_5]_t = ?$ M

2. **Substitute these values into the integrated rate law.**

   $$\ln\frac{[0.400]_0}{[A]_t} = \left(1.68 \times 10^{-2} s^{-1}\right)(120\ s)$$

3. **Carry out the mathematical operations on the right-hand side.**

   $$\ln\frac{[0.400]_0}{[A]_t} = 2.016$$

4. **Take the inverse natural logarithm (e^x) of each side.**

$$\frac{[0.400]_0}{[A]_t} = 7.508$$

5. **Rearrange.**

$$[A]_t = [0.400]_0 \div 7.508 = 0.0533 \text{ M } N_2O_5$$

Second order

If the reaction were second order in the reactant (rate = $k\,[A]^2$), you would plot $1/[A]$ versus t. This graph gives a straight line only if the reaction is second order. In all other cases, you get a curve. As I stated before in the previous section, the slope of this line is the rate constant. The integrated rate law for a second order reaction is:

$$\frac{1}{[A]_t} - \frac{1}{[A]_0} = kt \quad \text{Second order}$$

Zero order

For a zero order reaction where the rate law is rate = $k\,[A]^0 = k$, a plot of $[A]$ versus t gives you a straight line with the slope equaling k. The zero order integrated rate law is:

$$[A]_0 - [A]_t = kt \quad \text{Zero order}$$

Other integrated rate laws exist for additional reaction orders and other integrated rate laws exist for situations in which more than one substance is in the rate law. When more than one reactant appears in the rate law, determining the integrated rate law isn't always easy. To avoid this problem, the reaction is performed with all concentrations, except for one, very high. The concentration of the one reactant with a low concentration changes drastically during the reaction. The concentrations of the reactants with high concentrations change only slightly and are effectively constant during the reaction. Therefore, monitoring the effect of only one reactant on the reaction is possible. When doing this, the reaction conditions are *pseudo-first order.*

Half-life ($t_{1/2}$)

The *half-life* is the time required for one-half of the reactant to react. Determining the value of the half-life is possible by setting $[A]_t = (\frac{1}{2})\,[A]_0$, and using the appropriate integrated rate law. However, some equations are designed to determine the half-life directly and often much quicker.

For first order reactions, you can calculate the half-life by using this equation:

$$t_{1/2} = \frac{0.693}{k} \quad \text{First order}$$

In this case, the half-life doesn't depend on the concentration of the reactant, but only depends on the value of k. (The value 0.693 is ln 2.)

The half-life of a second order reaction is:

$$t_{1/2} = \frac{1}{k[\mathrm{A}]_0} \quad \text{Second order}$$

For second order reactions, the half-life depends upon k and the initial concentration of the reactant.

Finally, if the reaction is zero order, you can use:

$$t_{1/2} = \frac{[\mathrm{A}]_0}{2k} \quad \text{Zero order}$$

Like second order reactions, the half-life also depends upon both k and the initial concentration.

Table 7-1 summarizes the information on first, second, and zero order reactions. The straight-line plot tells what information is necessary for a graph to give a straight-line graph. It also shows the integrated rate law and the half-life relationships.

Table 7-1 Reviewing the First, Second, and Zero Order Kinetics

Straight-Line Plot	*Integrated Rate Law*	*Half-Life*
First order: Plot ln [A] versus t (Rate = k $[A]^1$)	$\ln\frac{[A]_0}{[A]_t} = kt$	$t_{1/2} = \frac{0.693}{k}$
Second order: Plot $1/[A]_t$ versus t (Rate = k $[A]^2$)	$\frac{1}{[A]_t} - \frac{1}{[A]_0} = kt$	$t_{1/2} = \frac{1}{k[A]_0}$
Zero order: Plot [A] versus t (Rate = $k[A]^0 = k$)	$[A]_0 - [A]_t = kt$	$t_{1/2} = \frac{[A]_0}{2k}$

Tackling the Collision Theory

In order for a chemical reaction to take place, the reactants must collide. It's like playing pool. In order to drop the 8-ball into the corner pocket, you must hit it with the cue ball. This collision transfers *kinetic energy* (energy of motion) from the cue ball to the 8-ball, sending the 8-ball (hopefully) toward the pocket. The collision between the molecules provides the energy needed to break the necessary bonds so that new bonds can be formed. This model is called *collision theory.*

However, when you play pool, not every shot you make causes a ball to go into the pocket. Sometimes you don't hit the ball hard enough and you don't transfer enough energy to get the ball to the pocket. This is also true with molecular collisions and reactions. Sometimes, even if a collision occurs, not enough kinetic energy is available to be transferred — the molecules aren't moving fast enough. You can help the situation somewhat by heating the mixture of reactants. The temperature is a measure of the average kinetic energy of the molecules; raising the temperature increases the kinetic energy available to break bonds during collisions.

Sometimes, even if you hit the ball hard enough, it doesn't go into the pocket because you didn't hit it in the right spot. The same is true during a molecular collision. The molecules must collide in the right orientation, or hit at the right spot, in order for the reaction to occur.

Here's an example: Suppose you have an equation showing molecule *A-B* reacting with *C* to form *C-A* and *B*, like this:

$$A\text{-}B + C \rightarrow C\text{-}A + B$$

The way this equation is written, the reaction requires that reactant *C* collide with *A-B* on the *A* end of the molecule. (You know this because the product side shows *C* hooked up with *A* — *C-A*.) If it hits the *B* end, nothing will happen. The *A* end of this hypothetical molecule is called the *reactive site*, the place on the molecule that the collision must take place in order for the reaction to occur. If *C* collides at the *A* end of the molecule, then there's a chance that enough energy can be transferred to break the *A-B* bond. After the *A-B* bond is broken, the *C-A* bond can be formed. The equation for this reaction process can be shown in this way (I show the breaking of the *AB* bond and the forming of the *CA* bond as "squiggly" bonds):

$$C\sim A\sim B \rightarrow C\text{-}A + B$$

So in order for this reaction to occur, a collision must occur between *C* and *A-B* at the reactive site. The collision between *C* and *A-B* has to transfer enough energy to break the *A-B* bond, allowing the *C-A* bond to form.

If instead of having a simple *A-B* molecule, you have a large complex molecule, like a protein or a polymer, then the likelihood of C colliding at the reactive site is much smaller. You may have a lot of collisions, but few at the reactive site. This reaction will probably be much slower than the simple case.

Energy is required to break a bond between atoms. Energy is released when a bond between atoms is made.

Note that this example is a simple one. I've assumed that only one collision is needed, making this a one-step reaction. Many reactions are one-step, but many others require several steps in going from reactants to final products. In the process, several compounds may be formed that react with each other to give the final products. These compounds are called *intermediates*. They're shown in the reaction *mechanism*, the series of steps that the reaction goes through in going from reactants to products. But in this chapter, I keep it simple and pretty much limit my discussion to one-step reactions.

Exothermic reactions

Imagine that the hypothetical reaction A-B + C → C-A + B is *exothermic* — a reaction in which heat is given off (released) when going from reactants to products. The reactants start off at a higher energy state than the products, so energy is released in going from reactants to products. Figure 7-2 shows an energy diagram of this reaction.

In Figure 7-2, E_a is the activation energy for the reaction — the energy that you have to put in to get the reaction going. I show the collision of *C* and *A-B* with the breaking of the *A-B* bond and the forming of the *C-A* bond at the top of an activation energy hill. This grouping of reactants at the top of the activation energy hill is sometimes called the *transition state* of the reaction. As I show in Figure 7-2, the difference in the energy level of the reactants and the energy level of the products is the amount of energy (heat) that is released in the reaction.

Some reactions may not give off heat, but they do give off energy. Light sticks are a good example. Mix two chemical solutions by flexing the light stick and it glows. It gives off light but not heat. Fireflies are another example. They do the same thing; they mix two chemicals in their bodies and give off light. I remember many evenings in North Carolina catching fireflies in a jar for a nightlight. Ah, the good old days! These are examples of *exergonic* reactions, reactions that gives off energy. If that energy is in the form of heat, the reaction is subclassified as an exothermic reaction.

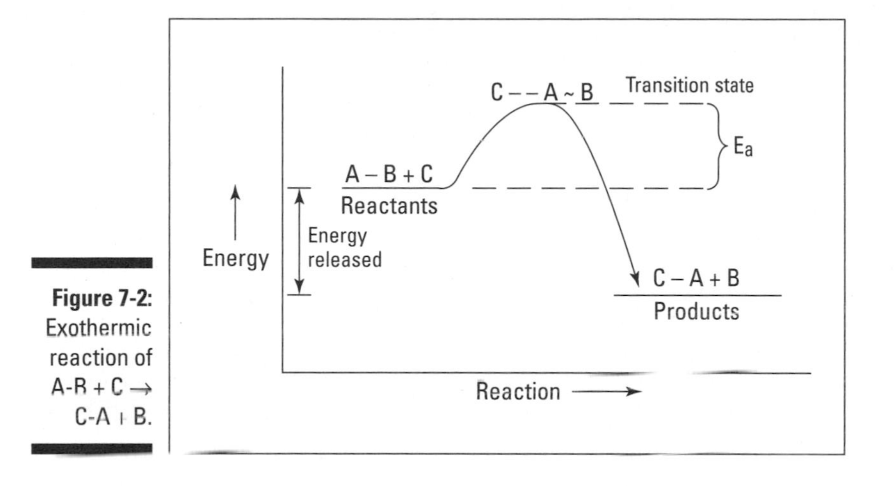

Figure 7-2: Exothermic reaction of A-B + C → C-A + B.

Endothermic reactions

Suppose that the hypothetical reaction A-B + C → C-A + B is *endothermic* — a reaction in which heat is absorbed in going from reactants to products — so the reactants are at a lower energy state than the products. Figure 7-3 shows an energy diagram of this reaction.

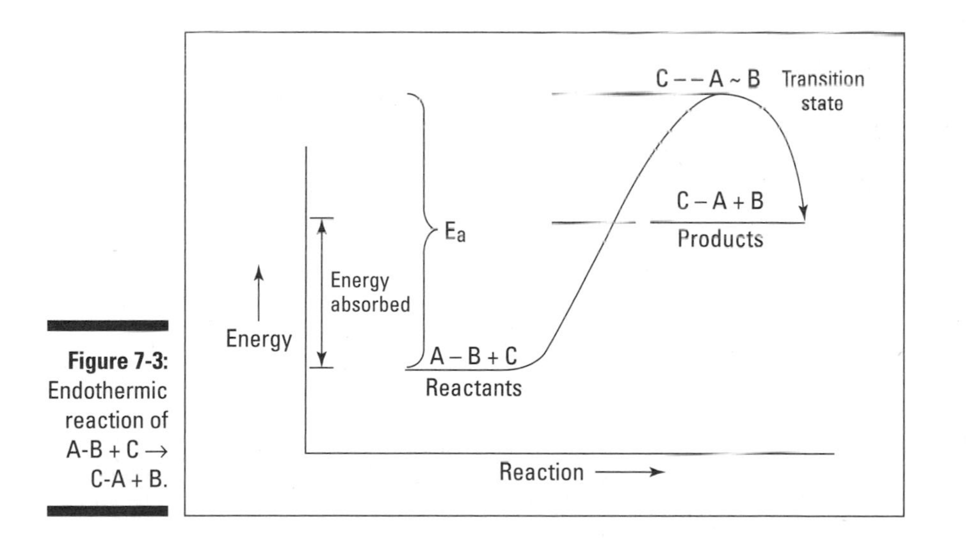

Figure 7-3: Endothermic reaction of A-B + C → C-A + B.

Just as with the exothermic-reaction energy diagram shown in Figure 7-2, this diagram shows that activation energy is associated with the reaction (represented by E_a). In going from reactants to products, you have to put in more energy initially to get the reaction started, and then you get some of that energy back out as the reaction proceeds. Notice that the transition state appears at the top of the activation energy hill — just like in the exothermic-reaction energy diagram. The difference is that, in going from reactants to products, energy (heat) must be absorbed in the endothermic example.

Cooking is a great example of an endothermic reaction. That ground beef isn't going to be a delicious hamburger unless you cook it. You have to continually supply energy in order for the chemical reactions called cooking to take place. Another example is the cold packs that athletic trainers use to treat injuries. Mix two solutions and the pack turns cold because it's absorbing heat from the surroundings.

Other reactions may absorb energy, but not necessarily heat. For example, some reactions absorb light energy in order to react. The general term that chemists use to describe reactions that absorb energy (heat or otherwise) is *endergonic*. Endothermic reactions are a proper subset of endergonic reactions (sorry, just had to get in that one piece of math).

Understanding Activation Energy

When developing an understanding of chemical kinetics, you also have to examine the activation energy. *Activation energy* simply is the energy that must be supplied initially to get the reaction going. You can get some idea about the magnitude of this energy from the value of the rate constant. Unfortunately, this may only reflect information about one step in a multi-step reaction. However, you can determine the activation energy from the Arrhenius equation.

The *Arrhenius equation* provides a means of relating the rate constant to the activation energy and the frequency factor. The Arrhenius equation is:

$$k = Ae^{-E_a/RT}$$

In this equation, k is the rate constant at a certain Kelvin temperature, T, E_a is the activation energy, and R is the universal gas constant, 8.314 J/mol × K. The term A is the frequency factor. The frequency factor is related to how often an effective collision occurs. The frequency factor is difficult to determine directly, and, in addition, it's difficult to estimate.

The Arrhenius equation agrees with experimental observations. You can use it to predict how changes in the frequency factor, *A*, temperature, *T*, and the activation energy, *k*, affect the rate constant. The rate constant's value increases with either an increase in the frequency factor or an increase in the temperature. The rate constant's value decreases with increasing activation energy.

You can use the Arrhenius equation to calculate the activation energy of a reaction. One way to do so is to plot the ln of *k* versus $1/T$. Doing so gives a straight line whose slope is $-E_a/\mathrm{R}$. Knowing the value of R, you can then calculate the value of E_a.

The problem with using the Arrhenius equation directly is that you have one equation with two unknowns (the frequency factor and the activation energy). The rate constant and the temperature are experimental values, while R is a constant. One way to get around this problem is to perform the experiment twice. To do so, you need an experimental value of the rate constant at two different temperatures. You can then assume that the frequency factor is the same at these two temperatures and bypass the frequency factor. (This assumption is reasonable if the reaction has no change.) You now have a new equation derived from the Arrhenius equation that allows you to find the activation energy. This equation is:

$$\ln\frac{k_1}{k_2} = \frac{E_a}{\mathrm{R}}\left[\frac{1}{T_2} - \frac{1}{T_1}\right]$$

The two rate constant values, k_1 and k_2, are determined at two different temperatures, T_1 and T_2. The temperatures must be in Kelvin units. The units on the rate constants cancel, leaving a unitless ratio. R is 8.314 J/mol × K. This leaves the activation energy in joules/mole.

Consider the reaction:

$$2\ \mathrm{HI}(g) \rightarrow \mathrm{H_2}(g) + \mathrm{I_2}(g)$$

The rate constant was measured at various temperatures, as the following shows. Using this information, calculate the activation energy for the reaction.

***T* (K)**	***k* (L/mole × s)**
555	3.52×10^{-7}
575	1.22×10^{-6}
645	8.59×10^{-5}
700	1.16×10^{-5}
781	3.95×10^{-2}

To solve this problem, follow these steps:

1. **Start with the preceding equation.**

   You could plot this data and determine the value of the activation energy from the slope of the line, but for working out this problem, use the equation instead.

   $$\ln\frac{k_1}{k_2}=\frac{E_a}{\mathrm{R}}\left[\frac{1}{T_2}-\frac{1}{T_1}\right]$$

2. **Assign the values from the table into the equation.**

   This equation requires only two lines from the table of data and it is probably best to pick the two lines that are furthest apart (again assuming no change in the reaction). Either of these lines may be assigned T_1 and k_1, while the other line will be T_2 and k_2. Assign the last line in the table T_1 and k_1 and the first line T_2 and k_2. This gives

   T_1 = 781 K $\qquad k_1$ = 3.95 – 10^{-2} L/mol × s

   T_2 = 555 K $\qquad k_2$ = 3.52 – 10^{-7} L/mol × s

   R = 8.314 J/mol × K $\qquad E_a$ = ?

3. **Simplify and round.**

   Calculations involving reciprocals and/or logarithms are very sensitive to rounding. Only round the final answer.

   $$\ln\frac{3.95\times10^{-2}\ \mathrm{L}/\mathrm{mole}\times\mathrm{s}}{3.52\times10^{-7}\ \mathrm{L}/\mathrm{mole}\times\mathrm{s}}=\frac{E_a}{8.314\ \mathrm{J}/\mathrm{mole}\cdot K}\left[\frac{1}{555\ \mathrm{K}}-\frac{1}{781\ \mathrm{K}}\right]$$

   $$\ln\left(1.122159\times10^{5}\right)=\frac{E_a}{8.314\ \mathrm{J}/\mathrm{mole}\cdot K}\left[\left(1.80180\times10^{-3}-1.2804097\times10^{-3}\right)\left(\frac{1}{K}\right)\right]$$

   $$11.62818=\frac{E_a}{8.314\ \mathrm{J}/\mathrm{mole}}\left(5.21392\times10^{-4}\right)$$

   $$E_a=\frac{(11.62818)8.314\mathrm{J}/\mathrm{mole}}{5.21392\times10^{-4}}=1.85420\times10^{5}=1.85\times10^{5}\ \mathrm{J}/\mathrm{mole}$$

Activation energies are normally large and positive. Because this is the case with the answer, it means that your answer is reasonable.

Recognizing Mechanisms

According to the collision theory, an effective collision must occur in order for a reaction to happen. An effective collision requires sufficient energy transfer, and the molecules must collide with the proper orientation. The orientation requirement limits reactions to a maximum of two molecules colliding because the probability of more than two molecules colliding simultaneously with sufficient energy and in the proper orientation is so extremely low as to make the reaction rate exceptionally slow. So slow, in fact, that this situation is normally disregarded.

If you disregard reactions involving three or more molecules, how can you explain the following reaction that is quite rapid?

$$2\ NO(g) + O_2(g) \rightarrow 2\ NO_2(g)$$

The reaction appears to involve the collision of three gas molecules (two molecules of NO and one molecule of O_2). Does this mean that the assumption to disregard reactions involving three molecules is wrong? Not necessarily. It only means that you must find another way to answer the problem without involving a three body collision. What you need is a reaction mechanism, which the following sections discuss.

Getting elementary with reactions

A *reaction mechanism* is a series of elementary steps which, when combined, produce the overall reaction. Reaction mechanisms use elementary steps to bypass the need for more than two molecules to participate in an effective collision. An *elementary step (reaction)* is a simple step in a mechanism that involves one or two molecules.

- If the elementary step involves one molecule, it is a *unimolecular step.*
- If an elementary step involves two molecules, it is a *bimolecular step.*

Reaction mechanisms use only unimolecular and/or bimolecular steps to explain a reaction. At no time should a mechanism step contain three or more molecules. Each step in the mechanism has its own rate law and activation energy. However, one-step will be slower than the rest. The slowest step in a mechanism is the *rate-determining step.*

When you determine rate laws (refer to the "Identifying orders of reaction" earlier in this chapter), you use experimental data to determine the relationships, not the balanced chemical equation. The balanced chemical equation is the sum of all the steps in a mechanism. If the mechanism contains only one step, the rate law matches the overall balanced chemical equation, but this occurrence is rare. However, the rate law will match the rate-determining step. Thus, you may go directly from the rate-determining step to the rate law and vice versa.

Postulating a mechanism for any reaction is possible. You need to realize that until chemists can actually observe individual molecules reacting, scientists may only postulate, not verify, a mechanism. When they postulate a mechanism, they seek additional experimental data to strengthen their belief in the correctness of the mechanism or to refute the mechanism.

A mechanism must follow a few simple rules:

- The sum of all the steps in the mechanism must yield the overall reaction.
- Only unimolecular and bimolecular steps may be present.
- One step will be the rate-determining step. This step must be consistent with the rate law of the reaction.

Examine the steps in creating a mechanism using the reaction I introduced previously. Recall that this reaction was:

$$2\ NO(g) + O_2(g) \rightarrow 2\ NO_2(g)$$

The experimental rate law for this reaction is Rate = k [NO] [O_2].

Before you begin to solve this problem, make sure you remember an important point: You can't have a mechanism involving all three molecules colliding simultaneously because a three-body collision is very unlikely and thus disallowed in a mechanism step. In addition, if it did occur, the rate law would be Rate = k $[NO]^2[O_2]$, which doesn't match the observed rate law. To solve this problem, stick to these steps:

1. **Pick any two of the three molecules, even two oxygen molecules.**

   For this example, I choose one NO and one O_2:

   $$NO(g) + O_2(g) \rightarrow$$

2. **Assume that a product molecule of NO_2(g) will be formed.**

   I'm not quite certain what must lie on the product side, but you can assume (yes, I know all about assumption).

   $$NO(g) + O_2(g) \rightarrow NO_2(g) \quad \text{(unbalanced)}$$

3. **Balance this equation.**

   In order to do so, you need an oxygen atom.

   $$NO(g) + O_2(g) \rightarrow NO_2(g) + O(g)$$

   The presence of oxygen atoms would give me something to verify in additional experimentation. Finding monoatomic oxygen atoms would lend strength to my belief that this is a correct step in the mechanism.

4. **Focus on the oxygen atom.**

   The oxygen atom doesn't appear in the overall reaction so you must use it in some manner. Begin the second equation in the proposed mechanism with the oxygen atom:

   $$NO(g) + O_2(g) \rightarrow NO_2(g) + O(g)$$

   $$O(g) \rightarrow$$

   The overall reaction requires two NO reactant molecules, and you've only used one, so use a second NO here:

   $$NO(g) + O_2(g) \rightarrow NO_2(g) + O(g)$$

   $$O(g) + NO(g) \rightarrow$$

5. **Combine the reactant species to give another molecule of NO_2.**

   $$NO(g) + O_2(g) \rightarrow NO_2(g) + O(g)$$

   $$O(g) + NO(g) \rightarrow NO_2(g)$$

6. **Add together these two steps to get.**

   $$NO(g) + O_2(g) \rightarrow NO_2(g) + O(g)$$

   $$O(g) + NO(g) \rightarrow NO_2(g)$$

   Sum: $2\ NO(g) + O_2(g) + O(g) \rightarrow 2\ NO_2(g) + O(g)$

7. **Cancel the oxygen atoms because they appear on both sides.**

   $$2\ NO(g) + O_2(g) + \cancel{O(g)} \rightarrow 2\ NO_2(g) + \cancel{O(g)}$$

   which leaves:

   $$2\ NO(g) + O_2(g) \rightarrow 2\ NO_2(g)$$

You began with this same overall reaction. Therefore, the proposed mechanism is consistent with the observed reaction. The proposed mechanism is:

$$NO(g) + O_2(g) \rightarrow NO_2(g) + O(g)$$

$$O(g) + NO(g) \rightarrow NO_2(g)$$

Overall: $2\ NO(g) + O_2(g) \rightarrow 2\ NO_2(g)$

This mechanism conforms to two of the three rules concerning mechanisms. (The sum of all the steps in the mechanism must yield the overall reaction and only unimolecular and bimolecular steps may be present.) The next section focuses on whether the mechanism conforms to the third rule, the one concerning the rate-determining step.

Conforming the rate-determining step

The rate-determining step must conform to the rate law. If the first step in the proposed mechanism is the rate-determining step, you would get Rate = k [NO][O_2]. Unlike overall reactions, you only need to examine the rate-determining step, and because only one NO and one O_2 are present, there must be one (indicated by the order) of each of these in the rate law. The rate law for the first step matches the rate law given, thus you can label the first step as the rate-determining step. You can now finish the mechanism:

$NO(g) + O_2(g) \rightarrow NO_2(g) + O(g)$ Rate determining step (slow)

$O(g) + NO(g) \rightarrow NO_2(g)$ (fast)

$2\ NO(g) + O_2(g) \rightarrow 2\ NO_2(g)$ Overall reaction

The oxygen atoms formed in the first step are an example of an intermediate. An *intermediate* is a substance that shows up in one step and is used in another. It is neither an original reactant nor a final product. Trying to experimentally find intermediates is a useful technique for strengthening the belief that a mechanism is correct. If further experimentation showed either that oxygen atoms weren't an intermediate or that there was a different intermediate, you would need to discard this mechanism and start over developing a different mechanism.

You assumed that the first step in the mechanism was the rate-determining step because it matched the rate law for the reaction. What would you have gotten if the second step were the rate-determining step? Initially, you would have Rate = k [NO] [O]. However, O is an intermediate. The final rate law can't involve any intermediates. You must ask the question, how does the intermediate relate to the materials in the overall equation? In this case, O is one-half of an O_2. We could then replace our intermediate with its equivalent (remembering the coefficient is the order). The new rate law would then be Rate = $k[NO][O_2]^{1/2}$. Because this isn't consistent with the observed rate law, the second step can't be the rate-determining step.

Keeping Tabs on Catalysts

Catalysts are substances that increase the reaction rate without themselves being changed at the end of the reaction. They increase the reaction rate by lowering the activation energy for the reaction. Catalysts won't increase the amount of product being formed, but they will allow you to make the same amount faster.

Look at Figure 7-1, for example. If you shift to the left that dotted line representing the minimum amount of kinetic energy needed to provide the activation energy, then many more molecules will have the minimum energy needed, and the reaction will be faster.

Catalysts lower the activation energy of a reaction in one of two ways:

- They provide a surface on which the reaction can take place and an orientation of one of the reactant molecules that increases the chance of another reactant hitting the reactive site.
- They provide an *alternative mechanism* (series of steps in the reaction) that has a lower activation energy.

Catalysts are very important in biological systems. An *enzyme* is a biological catalyst. The enzyme is normally a very large molecule that catalyzes a very specific reaction. Nearly every biologically important reaction involves one or more enzymes that catalyze the process. For example, the decomposition of sucrose in the body is catalyzed by the enzyme sucrase that is found in the small intestine.

The following sections give you an overview to the types of catalyst problems you may encounter in your Chem II course.

Heterogeneous catalysis

In the earlier section, "Tackling the Collision Theory," I describe how molecules react, using this generalized example:

$$C\text{~}A\text{-}B \rightarrow C\text{-}A + B$$

Reactant *C* must hit the reactive site on the *A* end of molecule *A-B* in order to break the *A-B* bond and form the *C-A* bond shown in the equation. The probability of the collision occurring in the proper orientation is pretty much determined by chance. The reactants are moving around, running into each other, and sooner or later the collision may occur at the reactive site. But what would happen if you could tie the *A-B* molecule down with the *A* end exposed? Hitting *A* with this scenario would be much easier and more probable.

This is what a heterogeneous catalyst accomplishes: It ties one molecule to a surface while providing proper orientation to make the reaction easier. Figure 7-4 shows the process of heterogeneous catalysis.

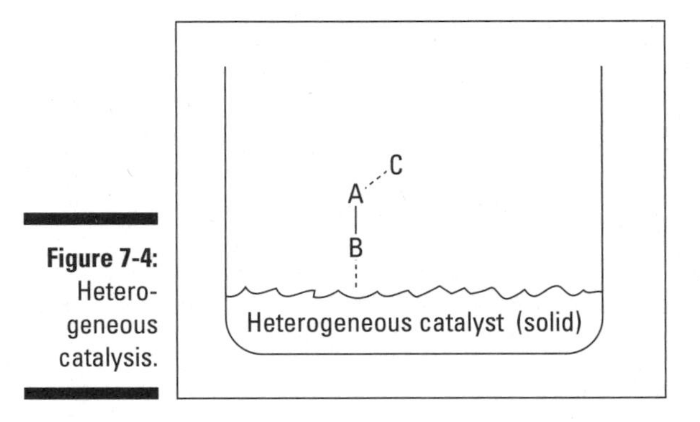

Figure 7-4: Heterogeneous catalysis.

The catalyst is called a *heterogeneous* catalyst because it's in a different phase than the reactants. This catalyst is commonly a finely divided solid metal or metal oxide, while the reactants are gases or in solution. This heterogeneous catalyst tends to attract one part of a reactant molecule due to rather complex interactions that aren't fully understood. After the reaction takes place, the forces that bound the *B* part of the molecule to the surface of the catalyst are no longer there, so *B* can drift off, and the catalyst can be ready to do it again.

Most people sit very close to a heterogeneous catalyst every day — the catalytic converter in your automobile. It contains finely divided platinum and/or palladium metal and speeds up the reaction that causes harmful gases from gasoline (such as carbon monoxide and unburned hydrocarbons) to decompose into mostly harmless products (such as water and carbon dioxide).

Homogeneous catalysis

The second type of catalyst is a *homogeneous catalyst* — one that's in the same phase as the reactants. It provides an alternative mechanism, or reaction pathway, that has a lower activation energy than the original reaction.

For example, check out the decomposition reaction of hydrogen peroxide:

$$2\ H_2O_2(l) \rightarrow 2\ H_2O(l) + O_2(g)$$

This reaction is slow, especially if it's kept cool in a dark bottle. It may take years for that bottle of hydrogen peroxide in your medicine cabinet to decompose. However, if you put a little bit of a solution containing the ferric ion in the bottle, the reaction will be much faster, even though it will decompose via a two-step mechanism instead of a one-step mechanism:

$$\text{(Step 1) } 2\ Fe^{3+}(aq) + H_2O_2(l) \rightarrow 2\ Fe^{2+}(aq) + O_2(g) + 2\ H^+(aq)$$

$$\text{(Step 2) } 2\ Fe^{2+}(aq) + H_2O_2(l) + 2\ H^+(aq) \rightarrow 2\ Fe^{3+}(aq) + 2\ H_2O(l)$$

If you add the two preceding reactions together and cancel the species that are identical on both sides, you get the original, uncatalyzed reaction:

$$\cancel{2\ Fe^{3+}(aq)} + H_2O_2(l) + \cancel{2\ Fe^{2+}(aq)} + H_2O_2(l) + \cancel{2\ H^+(aq)} \rightarrow$$

$$\cancel{2\ Fe^{2+}(aq)} + O_2(g) + \cancel{2\ H^+(aq)} + \cancel{2\ Fe^{3+}(aq)} + 2\ H_2O(l)$$

$$\text{Overall: } 2\ H_2O_2(l) \rightarrow 2\ H_2O(l) + O_2(g)$$

The ferric ion catalyst was changed in the first step and then changed back in the second step. This two-step catalyzed pathway has a lower activation energy and is thus faster.

Chapter 8

All Present in the Same State: Homogeneous Equilibrium

In This Chapter

- Examining chemical equilibrium
- Working equilibrium problems
- Deciphering the LeChatelier's Principle
- Grasping the Haber Process

In your Chem I class, you probably mastered how to calculate the amounts of reactants and products consumed or formed in a chemical reaction — the so-called *reaction stoichiometry*. (For a quick review, see Chapter 4.) But many times you produce less amount of product than expected; hence, the percent yield calculation (also reviewed in Chapter 4). You actually form less product than expected because of several reasons, but the major reason is that most reactions don't go to completion — they establish a chemical equilibrium. This chapter gives you the lowdown on chemical equilibriums and provides plenty of example problems so you can tackle them head on when you encounter them in your Chem II class.

Considering Chemical Equilibrium

In a chemical equilibrium you actually have two reactions taking place at the same time — the forward reaction, left to right, reactants to products (the reaction you are familiar with writing) and the reverse reaction, right to left, products back to reactants. When a system reaches equilibrium, the rate (speed) of both of the reactions are the same and the concentrations of the reactants and products become constant.

Almost half of Chem II is related to chemical equilibrium in some way. There are equilibrium systems related to weak acids and bases, the dissolving of sparingly soluble salts, and the formation of complex ions. Knowledge of these equilibrium systems goes a long way to ensuring that you do well in Chem II.

A perfect example of a chemical equilibrium occurs during the synthesis of ammonia from nitrogen and hydrogen gases, which happens to be my favorite reaction. (This reaction is the Haber process; check out "Putting It All Together: The Haber-Bosch Process" later in this chapter for more information.) After balancing the reaction you end up with:

$$N_2(g) + 3\ H_2(g) \rightarrow 2\ NH_3(g)$$

Written this way, the reaction says that hydrogen and nitrogen react to form ammonia, which keeps on happening until you use up one or both of the reactants. But this isn't quite true. (Yep. It's hair-splitting time again.)

If this reaction occurs in a closed container (which it has to, with everything being gases), then the nitrogen and hydrogen gases react and ammonia gas is formed, but then some of the ammonia soon starts to decompose to nitrogen and hydrogen, like this:

$$2\ NH_3(g) \rightarrow N_2(g) + 3\ H_2(g)$$

In the container, then, you actually have *two* exactly opposite reactions occurring — nitrogen and hydrogen combine to produce ammonia, and ammonia decomposes to give nitrogen and hydrogen. Chemists say that the reactions are *reversible.*

Instead of showing the two separate reactions, you can show one reaction and use a double arrow like this:

$$N_2(g) + 3\ H_2(g) \rightleftarrows 2\ NH_3(g)$$

You put the nitrogen and hydrogen gases on the left because that's what you initially put into the reaction container.

Now initially these two reactions occur at different speeds, but sooner or later, the two speeds become the same, and the relative amounts of nitrogen, hydrogen, and ammonia become constant (but not necessarily the same). This is an example of a dynamic chemical equilibrium. A *dynamic chemical equilibrium* is established when two exactly opposite chemical reactions are occurring at the same place, at the same time, with the same rates (speed) of reaction. I call this example a dynamic chemical equilibrium because when the reactions reach equilibrium, things don't just stop. At any given time, you have nitrogen and hydrogen reacting to form ammonia, and ammonia decomposing to form nitrogen and hydrogen. When the system reaches equilibrium, the amounts of all chemical species become constant but not necessarily the same.

Here's an example to help you understand what I mean by this dynamic stuff: I was raised on a farm in North Carolina, and my mother, Grace, *loved* small dogs. Sometimes we'd have close to a dozen dogs running around the house. When Mom opened the door to let them outside, they'd start running out. But some would change their minds after they got outside and would then start running back into the house. They'd then get caught up in the excitement of the other dogs and start running back outside again. There'd be a never-ending cycle of dogs running in and out of the house. Sometimes there'd only be two or three in the house, with the rest outside, or vice versa. The number of dogs inside and outside would be *constant* but not the same. And at any given point, there'd be dogs running out of the house and dogs running into the house. It was a dynamic equilibrium (and a noisy one).

One goal of this chapter is to show you how to predict how much of each chemical species on the right and left of the equilibrium arrow is present when the reaction comes to equilibrium. In this chapter I rarely talk about reactants and products, but instead I talk about left-hand and right-hand species. The reason I use these terms instead of reactants and products is that the reactants in the forward (left to right) reaction are the products in the reverse (right to left) reaction. It, therefore, is easier to identify them as left-hand and right-hand species.

In order to achieve my goal of showing you how to calculate these concentrations at equilibrium, the following sections introduce a couple of new terms — the equilibrium quotient and the equilibrium constant.

Looking at reaction (equilibrium) quotients: The Law of Mass Action

The *Law of Mass Action* states that during a reversible reaction a relationship occurs between the quantities of products and the quantities of reactants. For any reversible reaction, this law allows you to write a mass action expression, a *reaction quotient, Q,* that has the following general form:

$$Q = \frac{\text{Products}}{\text{Reactants}}$$

The *Products* term contains information concerning the products of the reaction, and the *Reactants* term contains information about the reactants of the reaction. This information may appear in various ways such as *concentrations* or *pressures* and includes other information related to the specific reaction.

Mass action expressions actually require the use of activities. The activity of a substance is related to the concentration or the pressure. Unlike concentration and pressure, an *activity* is a quantity that has no units, which means that the definition of a reaction quotient comes from unitless quantities; therefore, reaction quotients have no units. This even applies when concentrations or pressures take the place of the activity values in calculations.

The form of the mass action expression is related directly to the specific reaction. You use the following reaction to demonstrate how to write a mass action expression.

$$2\ NO(g) + O_2(g) \rightleftarrows 2\ NO_2(g)$$

Start by placing the concentrations of the products over the concentrations of the reactants.

$$Q = \frac{[NO_2]}{[NO][O_2]}$$

In this example you use concentration terms for the substances in the reaction. The concentrations are in terms of molarity, M. Instead of using the M abbreviation for molarity, use square brackets.

This form though isn't the final form for the mass action expression. The coefficient for each reactant or product in the balanced chemical equation becomes the exponent for that species in the mass action expression:

$$Q = \frac{[NO_2]^2}{[NO]^2[O_2]}$$

Now you have the complete mass action expression, the reaction quotient. When the reaction is at equilibrium, it is also the equilibrium constant expression.

Transforming into an equilibrium constant

The numerical value of the reaction quotient changes as the reaction proceeds until the system reaches a state of equilibrium. At this point, the reaction quotient becomes a constant, the *equilibrium constant*.

An equilibrium constant always represent a chemical equilibrium. The letter K represents this equilibrium constant and a lower case k represents a rate constant. Don't confuse the two. This equilibrium constant describes the relationship between the amounts of the reactants and the products at a certain temperature. The value of the equilibrium constant is only constant if the temperature is constant. In order to convey additional information, the K may also have a subscript such as K_a, K_b, K_c, K_d, K_{eq}, K_f, K_p, K_{sp}, K_w, and others. The presence of a subscript provides additional helpful information; it does not change the definition of K.

The subscript *eq* on the K of equilibrium constant reinforces the idea that the reaction is at equilibrium. However, because the value of K is calculated using concentrations, K_c is commonly used in place of the general K_{eq}.

$$K_c = \frac{[NO_2]^2}{[NO]^2[O_2]}$$

In order to calculate the numerical value of K_c, you need to know the equilibrium concentrations of the three substances.

K_c' and K_p

Before I start to snow you under with problems, I take a look at a couple more aspects of this equilibrium constant expression.

Suppose you were looking at an equilibrium system involving the decomposition of NO_2 gas into NO and O_2 gases, which is essentially the reverse of the reaction I show you in the previous section. You follow the same procedure to obtain the equilibrium constant expression.

$$2\ NO_2(g) \rightleftarrows 2\ NO(g) + O_2(g)$$

$$K_c' = \frac{[NO]^2[O_2]}{[NO_2]^2}$$

This expression is essentially the reciprocal of the original expression. Therefore, for these pairs of reversed equilibrium reactions:

$$K_c' = \frac{1}{K_c}$$

I use prime (K'_c) so you can distinguish between the two K_c values.

The relationship between any K and its reciprocal, K', is one that you can find useful in other places. If you simply reverse the equation, you invert the equilibrium constant expression to give an equilibrium constant that is the reciprocal of the first.

Using pressures in the place of concentrations in the equilibrium constant expression is possible. When using pressures, you use the partial pressure, *P*, of each gas in the reaction. In the mass action expression, you use *P* instead of the square brackets, [ ]. Using the first example from this section, you can derive the following equilibrium constant expression using gas pressures:

$$2\ \mathrm{NO}(g)+\mathrm{O_2}(g) \rightleftarrows 2\ \mathrm{NO_2}(g)$$

$$\mathrm{K}_p=\frac{P_{\mathrm{NO_2}}^2}{P_{\mathrm{NO}}^2 P_{\mathrm{O_2}}}$$

In this case, use a *p* subscript on the equilibrium constant to emphasize the fact that this expression uses the equilibrium partial pressures and not the equilibrium concentrations of the three substances.

Most of the calculations in this book involve K values based upon the concentrations of the substances in the reactions. A few calculations do use a pressure-based K value. However, you may encounter cases where you need to convert a concentration-based K value, K_c, to a pressure-based K value, K_p. The following relationship relates the two K values, but in order to relate these two types of equilibrium constants, you must pay particular attention to the number of moles of gas involved in the reaction.

$$\mathrm{K}_p=\mathrm{K}_c(\mathrm{RT})^{\Delta n_g}$$

In this equation, K_p is the pressure-based equilibrium constant for a reaction, and K_c is the concentration-based equilibrium constant for the same reaction. R is the ideal gas constant and T is the temperature. Even though the K values have no units, you can't use any R or T unit. If you use atmosphere-based activities to calculate the K_p value, then the value of R must be 0.08206 L × atm/mole × K. The temperature must be in Kelvin units. The superscript, Δn_g, is the change in the number of moles of gas. The value of Δn_g is the number of moles of product gases minus the number of moles of reactant gases.

The number of moles of gas in a reaction is simply the coefficients of the gases. Consider the following reaction:

$$2\ \mathrm{NO}(g)+\mathrm{O_2}(g) \rightleftarrows 2\ \mathrm{NO_2}(g)$$

For this reaction, the calculation of Δn_g would be [2 moles product gas] – [(2 + 1) moles of reactant gas] = –1.

Working Equilibrium Problems

Before you start working problem, I encourage you to not simply try to memorize how to do a particular problem type, but concentrate on the reasoning behind how to solve the problem. After you have that reasoning, then you can find that all equilibrium problems are basically the same. Take a deep breath and off you go!

Nitrogen and hydrogen gas are placed into a container and heated to 1000K. An equilibrium is established at 1000 K:

$$N_2(g) + 3\,H_2(g) \rightleftarrows 2\,NH_3(g)$$

If the equilibrium concentrations were $[N_2]$ = 1.03 M, $[H_2]$ = 1.62 M, and $[NH_3]$ = 0.120 M, what is the value of the equilibrium constant?

To solve this type of problem, follow these steps:

1. **Write the equilibrium constant expression.**

   Writing the expression is generally the first step in all equilibrium problems. Because concentrations are given, you use a K_c, not a K_p. The K_c expression is:

   $$K_c = \frac{[NH_3]^2}{[N_2][H_2]^3}$$

2. **Substitute the values into the appropriate position in the equilibrium expression.**

   From the problem statement you know the following equilibrium concentrations: $[N_2]$ = 1.03 M, $[H_2]$ = 1.62 M, and $[NH_3]$ = 0.120 M.

3. **Round the final answer to the correct number of significant figures.**

   Because it's an equilibrium constant, it doesn't require any units.

   $$K_c = \frac{[NH_3]^2}{[N_2][H_2]^3} = \frac{[0.120]^2}{[1.03][1.62]^3} = 3.29 \times 10^{-3}$$

Here is another problem. Consider the following reaction:

$$N_2(g) + 3\,H_2(g) \rightleftarrows 2\,NH_3(g)$$

What will be the $[NH_3]$ if $[N_2]$ = 2.00 M, $[H_2]$ = 3.00 M, and K = 2.37×10^{-3}.

Even though this equilibrium is the same, the value of the equilibrium constant is different. The different equilibrium constant tells you that the equilibrium in this problem is at a different temperature from that in the preceding example.

To solve this problem, follow these steps:

1. **Write the equilibrium expression.**

$$K_c = \frac{[NH_3]^2}{[N_2][H_2]^3}$$

2. **Enter the two given concentrations along with the given value of the equilibrium constant.**

$$K_c = \frac{[NH_3]^2}{[N_2][H_2]^3} = \frac{[NH_3]^2}{[2.00][3.00]^3} = 2.37 \times 10^{-3}$$

3. **Rearrange the equation.**

$$[NH_3]^2 = [2.00]\,[3.00]^3\,(2.37 \times 10^{-3})$$

$$[NH_3]^2 = 0.12798$$

$$[NH_3] = 0.358\ M$$

Now let me show you a K_p example.

Ozone is introduced into a container and then heated to 2000.0 K. The following equilibrium was established:

$$2\,O_3(g) \rightleftarrows 3\,O_2(g)$$

At equilibrium, the total pressure was 7.33 atm. For this equilibrium K_p = 4.17×10^{14}. Calculate the equilibrium partial pressures of each gas.

To solve this problem, work through these steps:

1. **Write the equilibrium constant expression using pressures.**

$$K_p = \frac{p_{O_2}^3}{p_{O_3}^2} = 4.17 \times 10^{14}$$

2. **Use Dalton's Law.**

This law states that the total pressure is simply the sum of the partial pressures. Refer to Chapter 6 for more information about this law.

$$P_{total} = P_{O_2} + P_{O_3} = 7.33\ \text{atm}$$

3. **Simplify this problem significantly by using a reasonable approximation.**

   This equilibrium expression could be rather difficult to solve. Fortunately, the equilibrium constant is very large, so make an approximation.

   The very large positive value of K_p indicates that either the numerator is very large, the denominator is very small, or both, which means that the equilibrium lies far to the right. Thus, most of the oxygen is present as O_2 and only a very small amount of oxygen is present as O_3. Therefore, you can assume that the partial pressure of ozone is very nearly zero. Therefore,

   $$P_{O_2} \cong 7.33 \text{ atm}$$

   You can temporarily make this assumption in order to simplify this problem; however, I show you how to check the validity of this assumption later. If you make an assumption, you must check its validity at the end of the calculation.

4. **Substitute the partial pressure of O_2 into the equilibrium expression.**

   $$K_p = 4.17 \times 10^{14} = \frac{(7.33 \text{ atm})^3}{P_{O_3}^2}$$

5. **Solve for the partial pressure of O_3.**

   $$P_{O_3} = 9.72 \times 10^{-7} \text{ atm}$$

6. **Enter this value into the Dalton's law expression.**

   $$P_{total} = P_{O_2} + P_{O_3} = 7.33 \text{ atm}$$

   $$7.33 \text{ atm} = P_{O_2} + 9.72 \times 10^{-7} \text{ atm}$$

   $$P_{O_2} = 7.33 \text{ atm} - 9.72 \times 10^{-7} \cong 7.33 \text{ atm}$$

   The approximation that you made is valid. If you subtract 10^{-7} from 7.33, for all practical purposes you have 7.33.

You can check the validity of this type of assumption following various rules. In this book, I use the simple rule that an assumption is justified if the difference affects no more than the last significant decimal place of the original value.

Here is another problem for you to try.

For the reaction:

$$I_2(g) \rightleftarrows 2\,I(g)$$

at 1000.0 K, the value of K is 3.76×10^{-5}. If 1.00 mole of I_2 is placed in a 2.00 L container at 1000.0 K, what will be the equilibrium concentrations of $I_2(g)$ and $I(g)$?

Here are the steps to follow to solve this problem.

1. **Write the equilibrium expression.**

$$K_c = \frac{[I]^2}{[I_2]} = 3.76 \times 10^{-5}$$

2. **Calculate the initial concentration of diatomic iodine.**

$$[I_2] = \frac{1.00 \text{ mole}}{2.00 \text{ liters}} = 0.500 \text{ M}$$

The initial concentration of monoatomic iodine, I, is zero because none was placed into the container initially.

3. **Create a table in order to help keep track of the changes in concentration that take place in this system.**

The table has columns for each reactant and product. The first row in the table contains the initial concentrations.

	[I₂]	*[I]*
Initial	0.500	0

As the system proceeds to equilibrium, some of the iodine molecules, I_2, become iodine atoms, I. Thus, the value of $[I_2]$ decreases and the value of [I] increases. The second line in the table accounts for this change.

The balanced chemical reaction shows the relationship between these changes. The equation indicates that for every I_2 that decomposes, two iodine atoms will form. If the change in $[I_2]$ were $-x$, then the change in [I] would be $+2x$ because the coefficient of I in the balanced equation is a 2.

This stoichiometric relationship allows you to finish the second line in the table:

	$[I_2]$	*[I]*
Initial	0.500	0
Change	$-x$	$+2x$

The equilibrium concentrations, or partial pressures, can never be zero or a negative number. If you get such a value, you have made an error.

The last row in each column represents the concentration at equilibrium. The values in this row are the sum of the values in each column. For I_2, you know that you started with a concentration of 0.500 M, but x amount was consumed as the system established equilibrium. Therefore the amount remaining at equilibrium is $0.500 - x$.

	[I_2]	*[I]*
Initial	0.500	0
Change	$-x$	$+2x$
Equilibrium	$0.500 - x$	$+2x$

This table is commonly referred to as an ICE table for **I**nitial, **C**hange, **E**quilibrium table. Using such a table is a very logical approach in solving many types of equilibrium problems.

4. **Use the values in the last row of the table in your equilibrium expression.**

$$K_c = \frac{[2x]^2}{[0.500 - x]} = 3.76 \times 10^{-5}$$

5. **Choose one of two methods to solve.**

   At this point you have two ways to proceed. This first method of using the quadratic equation works all the time. The second method involves making an assumption. I show you the first method first and then follow with the second to compare them.

6. **Simplify.**

$$4x^2 - (0.500 - x)(3.76 \times 10^{-5})$$

$$4x^2 + 3.76 \times 10^{-5} x - 1.88 \times 10^{-5} = 0$$

7. **Apply the quadratic equation.**

$$x = \frac{-b + \sqrt{b^2 - 4ac}}{2a}$$

$$x = \frac{-(3.76 \times 10^{-5}) \pm \sqrt{(3.76 \times 10^{-5})^2 - 4(4)(-1.88 \times 10^{-5})}}{2(4)}$$

8. **Solve the equation.**

When solving the quadratic equation, you get two answers, but one is impossible — like a negative concentration. You can have a concentration less than one, but you can never, never have a concentration less than zero.

$$x = 2.16 \times 10^{-3}\ M$$

9. **Use this x value to complete the bottom line of the table.**

   Doing so gives the equilibrium concentrations of the two substances:

$$[I_2] = 0.500 - 2.16 \times 10^{-3} = 0.498\ M$$

$$[I] = 2(2.16 \times 10^{-3}) = 4.32 \times 10^{-3}\ M$$

This is the final answer. Now I take you back to Step 5 and use the assumption method to show you how the two methods compare to each other. (Everyone knows about assumptions, but good ones can simplify the calculations quite a bit.)

In this method, go ahead and create the ICE table and put the values into the equilibrium constant expression:

$$K_c = \frac{[2x]^2}{[0.500 - x]} = 3.76 \times 10^{-5}$$

If x is a small number (and in this case you know that it is because you solved the quadratic equation and found it was 2.16×10^{-3} M), then in the denominator you can approximate $0.500 - x$ as 0.500. You can't do the same with the numerator because x isn't being compared to another number.

Think of it this way: You're walking down the street after hitting the lottery with a net worth of $10 million. You give a homeless person a $1. Your net worth is still approximately $10 million. The amount you gave away is insignificant in comparison to your net worth. If however, you're walking down the street with a net worth of $10 and you give a homeless person a $1, you have changed your net worth significantly because the amount you gave away was much closer in size to your net worth.

The same thing is true in this case. If x is small compared to 0.500, then your assumption is valid; if it introduces too much error, you must resort to the quadratic equation to solve for x.

If you're going to assume, stick to these steps when you get to Step 5 in the preceding list.

1. **Simply the K_c expression if x is small enough to.**

   $$K_c = \frac{[2x]^2}{[0.500]} = 3.76 \times 10^{-5}$$

2. **Solve for x.**

   $4x^2 = (3.76 \times 10^{-5})(0.500)$

   $x = 2.17 \times 10^{-3}$

3. **Solve for the concentrations.**

   $[I_2] = 0.500 - 2.17 \text{ x } 10^{-3} = 0.498$ M

   $[I] = 2(2.17 \times 10^{-3}) = 4.34 \times 10^{-3}$ M

These values are essentially the same that you got when you used the quadratic equation earlier in this section and with a lot less mathematical manipulations.

4. **Check the assumption to see whether it was valid without resorting to the quadratic equation.**

   The assumption was:

   $$0.500 - x \approx 0.500$$

   $$0.500 - 92.17 \times 10^{-3} \approx 0.500$$

   So, is 0.4998 close enough to 0.500? Different teachers use different rules. Many say if the two values are within 5% of each other, then the approximation is valid. I'm partial to the rule that I introduced in the K_p calculations previously in this chapter that say the assumption is justified if the difference affects no more than the last significant decimal place of the original value. And that certainly is the case here.

Making Sense of LeChatelier's Principle

The equilibrium constant expression can tell you how much product can be formed at a certain temperature, but what if that is too small? A chemist wants lots of product. Is there a way to manipulate the equilibrium in order to affect the amount of chemical species? Yes, you can with the LeChatelier's Principle.

A French chemist, Henri LeChatelier, discovered that if you apply a change of condition (called *stress*) to a chemical system that's at equilibrium, the reaction will return to equilibrium by shifting in such a way as to counteract the change (the stress). This is called *LeChatelier's Principle.*

You can stress an equilibrium system in three ways:

- Change the concentration of a reactant or product.
- Change the temperature.
- Change the pressure on a system that contains gases.

Now if you're a chemist who's looking for a way to make as much ammonia (money) as possible for a chemical company, you can use LeChatelier's Principle to help you along. The following sections explain how.

But first, I want to show you a quick, useful analogy. A reaction at equilibrium is like one of my favorite pieces of playground equipment, a teeter-totter. Everything is well balanced, as shown in Figure 8-1.

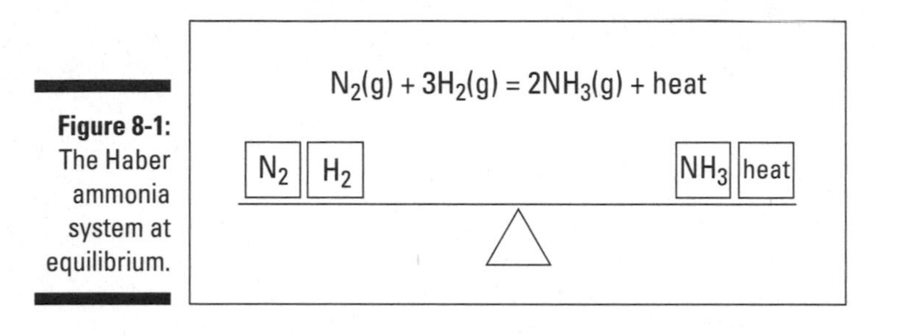

Figure 8-1: The Haber ammonia system at equilibrium.

The Haber process, the synthesis of ammonia from nitrogen and hydrogen gases, is exothermic: It gives off heat. I show that heat on the right-hand side of the teeter-totter in the figure.

Altering the concentration

If you add more of a reactant or product, the reaction will shift to the other side of the equilibrium to use it up. If you remove some reactant or product, the reaction shifts to that side in order to replace it. For example, suppose that you have the ammonia system at equilibrium (see Figure 8-1), and you then put in some more nitrogen gas. Figure 8-2 shows what happens to the teeter-totter when you add more nitrogen gas.

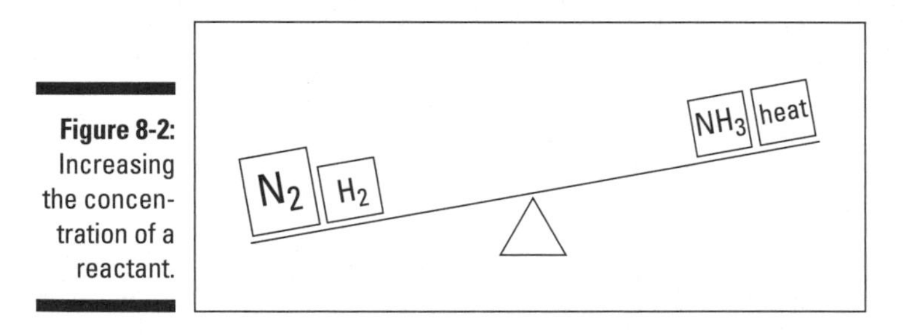

Figure 8-2: Increasing the concentration of a reactant.

In order to re-establish the balance (equilibrium), weight has to be shifted from the left to the right, using up some nitrogen and hydrogen and forming more ammonia and heat. Figure 8-3 shows this shifting of weight.

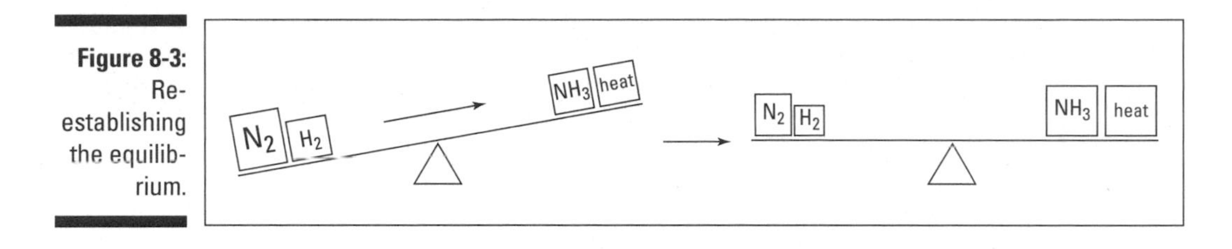

Figure 8-3: Re-establishing the equilibrium.

The equilibrium has been reestablished. There is less hydrogen and more nitrogen, ammonia, and heat than you had before you added the additional nitrogen. The same thing would happen if you had a way of removing ammonia as it was formed. The right-hand side of the teeter-totter would again be lighter, and weight would be shifted to the right in order to re-establish the equilibrium. Again, more ammonia would be formed.

Altering the temperature

In general, heating a reaction causes it to shift to the endothermic side. (If you have an exothermic reaction where heat is produced on the right side, then the left side is the endothermic side.) Cooling a reaction mixture causes the equilibrium to shift to the exothermic side.

Suppose that you heat the reaction mixture. You know that the reaction is exothermic (heat is given off, showing up on the right-hand side of the equation). So, if you heat the reaction mixture, the right side of the teeter-totter gets heavier, and weight must be shifted to the left in order to re-establish the equilibrium. This weight shift uses up ammonia and produces more nitrogen and hydrogen. And as the reaction shifts, the amount of heat also decreases, lowering the temperature of the reaction mixture. Figure 8-4 shows this shift in weight.

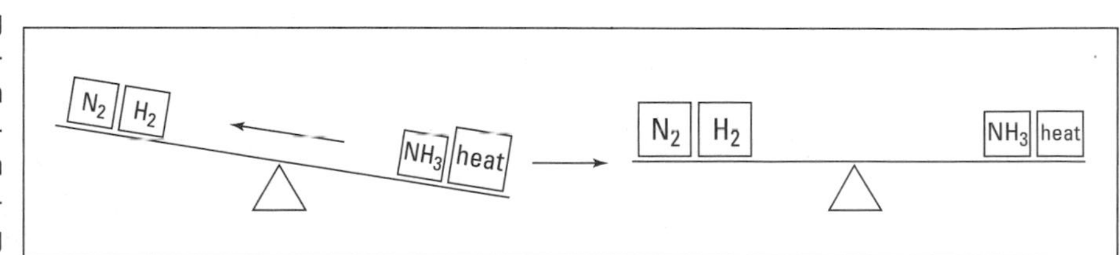

Figure 8-4: Increasing the temperature on an exothermic reaction and re-establishing the equilibrium.

That's not what you want! You want more ammonia, not more nitrogen and hydrogen. So you have to cool the reaction mixture, using up heat, and then the equilibrium shifts to the right in order to replace it. This process helps you make more ammonia and more profit. That's more like it.

Altering the pressure

In general, increasing the pressure on an equilibrium mixture causes the reaction to shift to the side containing the *fewest* number of gas molecules. Changing the pressure only affects the equilibrium if reactants and/or products are gases. In the Haber process (check the "Putting It All Together: The Haber-Bosch Process" section for more info) all species are gases, so there is a pressure effect.

My teeter-totter analogy of equilibrium systems begins to break down when I explain pressure effects, so I have to take another approach. Think about the sealed container where your ammonia reaction is occurring. (The reaction has to occur in a sealed container with everything being gases.) You have nitrogen, hydrogen, and ammonia gases inside. There is pressure in the sealed container, and that pressure is due to the gas molecules hitting the inside walls of the container.

Now suppose that the system is at equilibrium, and you want to increase the pressure. You can do so by making the container smaller (with a piston type of arrangement) or by putting in an unreactive gas, such as neon. You get more collisions on the inside walls of the container, and therefore you have more pressure. Increasing the pressure stresses the equilibrium; in order to remove that stress and re-establish the equilibrium, the pressure must be reduced.

Take another look at the Haber reaction and see if there may be some clues as to how this may happen.

$$N_2(g) + 3\,H_2(g) \rightleftarrows 2\,NH_3(g)$$

Every time the forward (left to right) reaction takes place, four molecules of gas (one nitrogen and three hydrogen) form two molecules of ammonia gas. This reaction reduces the number of molecules of gas in the container. The reverse reaction (right to left) takes two ammonia gas molecules and makes four gas molecules (nitrogen and hydrogen). This reaction increases the number of gas molecules in the container.

The equilibrium has been stressed by an increase in pressure; reducing the pressure will relieve the stress. Reducing the number of gas molecules in the container reduces the pressure (fewer collisions on the inside walls of the container), so the forward (left to right) reaction is favored because four gas molecules are consumed and only two are formed. As a result of the forward reaction, more ammonia is produced!

Looking at the effect of a catalyst

The presence of a catalyst doesn't change the equilibrium amounts. Catalysts affect the reaction kinetics, allowing the reaction mixture to get to equilibrium faster.

Here I put it together and have you work a LeChatelier Principle problem.

Consider the following equilibrium:

$$4\ NO(g) + 3\ O_2(g) \rightleftharpoons 2\ N_2O_5(g) \quad \Delta H = -447\ kJ$$

Predict what changes, if any, would occur if you applied the following stresses after equilibrium was established?

Add NO: Adding NO gas increases the NO concentration on the left. The reaction shifts to the right in order to use up the extra NO, thus increasing the N_2O_5 concentration and decreasing the O_2 concentration.

Remove O_2: Removing some O_2 gas decreases its concentration on the left-hand side. The reaction shifts to the left to replace the O_2, increasing the NO concentration and decreasing the N_2O_5 concentration.

Increase T: The reaction is exothermic (ΔH is on the right and is negative) so increasing the temperature causes the heat to increase. The reaction shifts to the left, increasing the NO and O_2 concentrations and decreasing the N_2O_5 concentration.

Decrease P: Decreasing the pressure causes the reaction to shift in order to increase the pressure. It therefore shifts to the side containing the greater number of moles of gas. In this example there are 7 moles of gas on the left and 2 on the right. The reaction shifts to the left, increasing the NO and O_2 concentrations and decreasing the N_2O_5 concentration.

Putting It All Together: The Haber-Bosch Process

The Haber-Bosch Process, often just referred to as the Haber Process, is basically the process by which ammonia gas is produced from hydrogen and nitrogen gases. This chemical process is one of the most important in the world. It is the first step is in the production of synthetic fertilizers. Without this process, famine would spread across the globe.

In the modern Haber-Bosch process, atmospheric nitrogen is produced from liquefying air and the hydrogen gas is produced by reacting methane with steam:

$$CH_4(g) + H_2O(g) \rightleftarrows CO(g) + 3\ H_2(g) \quad K_p = 10$$

The nitrogen and hydrogen gases combine exothermically to produce ammonia:

$$N_2(g) + 3\ H_2(g) \rightleftarrows 2\ NH_3(g) \quad \Delta H = -92.4\ \text{kJ}$$

The equilibrium constant expression, K_c, for this reaction at 25°C is 4×10^8.

$$K_c = \frac{[NH_3]^2}{[N_2][H_2]^3}$$

This large value of the equilibrium constant indicates that a large quantity of ammonia should be produced, but this reaction is so slow that not much ammonia is produced in a reasonable period of time. If the temperature is increased in order to increase the speed of reaction (kinetics), the K_c decreases dramatically. In order for the reaction to be economically practical, LeChatelier's Principle must be applied.

LeChatelier's Principle allows chemists to adjust the concentrations of the reactants, the pressure, and the temperature in order to increase the amount of ammonia produced. This increased efficiency allows more fertilizer to be made at a lower price. This is certainly important to US farmers but is of critical importance to farmers in third world countries.

The history of the Haber process

In 1909, a German chemist, Fritz Haber, developed a method of combining nitrogen gas with hydrogen gas under easily achievable conditions to produce ammonia, NH_3. The Germans used the ammonia produced by the Haber process to manufacture both fertilizer and explosives, both needed by Germany during World War I. Today, the Haber process is still used and is one of the most important chemical processes in the world. Much of today's synthetic fertilizer, in addition to plastics, dyes, and explosives, can be traced back to the Haber process.

Concentration

The nitrogen and hydrogen gases are fed into the reaction vessel in a 1:3 proportion, the proportion dictated by the balanced reaction. Some plants use an excess of nitrogen to force the reaction to the right, but too high an excess would reduce the number of effective collisions. The concentrations are kept high in order to maximize collisions and reactions. As the gas mixture containing nitrogen, hydrogen, and ammonia exits the reactor, it is cooled and the ammonia liquefies, removing it from the gas mixture. The unreacted gases are recycled back into the reaction vessel to produce additional ammonia.

Temperature

Because the reaction is exothermic, cooling the reaction mixture should increase the yield of ammonia because the heat being produced in the reaction is being consumed and the reaction should then shift to the right to replace it. Thus, the reaction should produce more ammonia. However, the lower the temperature the slower is the reaction rate. An increased yield, which takes place over a period of months or years, isn't economically advantageous. Therefore, a compromise between Le Châtelier's Principle and the reaction kinetics is reached by using a temperature of about 400 to 450°C.

Pressure

During the reaction, 4 moles of reactant gas combine to form 2 moles of product gas. LeChatelier's Principle indicates that increasing the pressure should increase the production of ammonia. However, there are practical limits and safety considerations to working with high gas pressures. Engineering and building high-pressure systems are expensive, and these systems are expensive to maintain. Again, chemical engineers reach a compromise by using a pressure of approximately 200 atmospheres.

Catalyst

The use of a catalyst doesn't directly affect the equilibrium amount of ammonia produced. However, a catalyst affects the reaction kinetics, speeding up the reaction and allowing the reaction to reach equilibrium much faster. Chemists often use a treated iron catalyst in the Haber process to speed up the reaction.

Other considerations

Using the previous conditions, a typical ammonia plant has about a 15 to 20 percent yield on a single pass of reactants through the reaction vessel. However, the continual recycling of the reactant gases allows plants to achieve an overall 98 to 99 percent yield.

Chapter 9

Neutralizing Effects: Acid-Base Equilibrium

In This Chapter

- Zooming in on the macroscopic properties of acids and bases
- Naming the microscopic views of acids and bases
- Identifying strong and weak acids and bases
- Figuring out pH
- Dipping into water's auto-ionization
- Working through some equilibrium problems
- Understanding buffers
- Eyeing titration curves and indicators

You probably studied acids and bases in your Chem I course. Acids and bases are a basic and essential part to chemistry. That's why I want to give you a very quick and brief review of acids and bases in the first part of this chapter. For a more complete review refer my book *Chemistry For Dummies,* 2nd edition (John Wiley & Sons, Inc.). The rest of this chapter focuses on acid-base equilibriums and buffers.

Considering the Macroscopic Properties of Acids and Bases

Macroscopic properties are those that you can observe in the real world. For acids, these macroscopic properties include:

- Tastes sour (but remember, in the lab, you test, not taste)
- Produces a painful sensation on the skin
- Reacts with certain metals (magnesium, zinc, and iron) to produce hydrogen gas

- Reacts with limestone and baking soda to produce carbon dioxide
- Reacts with litmus paper and turns it red

For bases, the properties are

- Tastes bitter (again, in the lab, you test, not taste)
- Feels slippery on the skin
- Reacts with oils and greases
- Reacts with acids to produce a salt and water
- Reacts with litmus paper and turns it blue

You can experience the sour taste of acids by sucking on a lemon (citric acid). You can experience the bitter taste of a base by taking a sip of tonic water (contains quinine, an organic base).

Looking Closely at the Microscopic Properties of Acids and Bases

The *microscopic* properties of acids and bases focus on the theories (models) that chemists use to explain the behavior of these substances. Chemistry has three main theories, which I discuss in the following sections that are used when referring to acids and bases: the Arrhenius theory, the Bronsted-Lowry theory, and the Lewis theory.

Why so many? One reason is that these theories were developed over a period of time. A theory may become entrenched and scientists are reluctant to give it up. And some theories explain certain things in an easier, more complete fashion than others. Just as there are no perfect people, there are no perfect theories. So scientists use the theories that best describe what they're dealing with at that particular moment.

Dissolving in water: The Arrhenius theory

The Arrhenius theory, introduced in 1887 by the Swedish scientist Svante Arrhenius, was the first modern acid-base theory developed. In this theory, an acid is a substance that yields H^+ (hydrogen) ions when dissolved in water, and a base is a substance that yields OH^- (hydroxide) ions when dissolved in water. HCl*(g)* can be considered a typical Arrhenius acid, because when this gas dissolves in water, it ionizes (forms ions) to give the H^+ ion.

$$HCl(aq) \rightarrow H^+(aq) + Cl^-(aq)$$

According to this theory, sodium hydroxide is classified as a base, because when it dissolves, it yields the hydroxide ion:

$$NaOH(aq) \rightarrow Na^+(aq) + OH^-(aq)$$

Arrhenius also classified the reaction between an acid and base as a *neutralization* reaction, because if you mix an acidic solution with a basic solution, you end up with a neutral solution composed of water and a salt.

$$HCl(aq) + NaOH(aq) \rightarrow H_2O(l) + NaCl(aq)$$

Look at the ionic form of this equation (the form showing the reaction and production of ions) to see where the water comes from:

$$\mathbf{H^+}(aq) + Cl^-(aq) + Na^+(aq) + \mathbf{OH^-}(aq) \rightarrow \mathbf{H_2O}(l) + Na^+(aq) + Cl^-(aq)$$

As you can see, the water is formed from combining the hydrogen and hydroxide ions. In fact, the *net-ionic* equation (the equation showing only those chemical substances that are changed during the reaction) is the same for all Arrhenius acid-base reactions:

$$H^+(aq) + OH^-(aq) \rightarrow H_2O(l)$$

The Arrhenius theory is still used quite a bit. But, like all theories, it has some limitations. It specifies that the reactions must take place in water and that bases must contain hydroxide ions, but many reactions resemble acid-base reactions but don't meet these stipulations. For example, look at the gas phase reaction between ammonia and hydrogen chloride gases:

$$NH_3(g) + HCl(g) \rightarrow NH_4^+(g) + Cl^-(g) \rightarrow NH_4Cl(s)$$

The two clear, colorless gases mix, and a white solid of ammonium chloride forms. I show the intermediate formation of the ions in the equation so that you can better see what's actually happening. The HCl transfers an H^+ to the ammonia, which is basically the same thing that happens in the HCl/NaOH reaction, but the reaction involving the ammonia can't be classified as an acid-base reaction, because of two reasons:

- It doesn't occur in water.
- It doesn't involve the hydroxide ion.

But again, the same basic process is taking place in both cases. In order to account for these similarities, a new acid-base theory was developed, the Bronsted-Lowry theory, which I discuss in the next section.

Accepting hydrogen: The Bronsted-Lowry theory

The Bronsted-Lowry theory was introduced in 1923 by the Danish chemist Johannes Bronsted and the English chemist Thomas Lowry. This theory attempted to overcome the limitations of the Arrhenius theory by defining an acid as a proton (H^+) donor and a base as a proton (H^+) acceptor. The base accepts the H^+ by furnishing a lone pair of electrons for a *coordinate-covalent* bond, which is a covalent bond (shared pair of electrons) in which one atom furnishes both of the electrons for the bond. Normally, one atom furnishes one electron for the bond and the other atom furnishes the second electron (see Chapter 3 for a review of bonding). In the coordinate-covalent bond, one atom furnishes both bonding electrons.

Taking and giving electrons: The Lewis acid-base theory

The third acid-base theory that is commonly used (especially in organic chemistry) is the Lewis acid-base theory, first introduced in 1923 by the American chemist Gilbert Lewis. According to this theory, a Lewis acid accepts a pair of electrons and a Lewis base donates a pair of electrons. In order to more easily identify which species is donating electrons, you can use Lewis structures for the reactants and, if possible, for the products.

The following is an example of a Lewis acid-base reaction.

$$H^+(aq) + :NH_3(aq) \rightarrow H\text{–}NH_3^+(aq)$$

The hydrogen ion accepts the lone pair of electrons from the ammonia to form the ammonium ion. The hydrogen ion, because it accepts a pair of electrons, is the Lewis acid. The ammonia, because it donates a pair of electrons, is the Lewis base. This reaction is also a Bronsted-Lowry acid-base reaction. This fact illustrates that a substance may be an acid or a base by more than one definition. All Bronsted-Lowry acids are Lewis acids, and all Bronsted-Lowry bases are Lewis bases. However, the reverse is not necessarily true.

Examining Strong and Weak Acids and Bases

Acid-base strength is not the same as concentration. *Strength* refers to the amount of ionization or breaking apart that a particular acid or base undergoes. *Concentration* refers to the amount of acid or base that you initially have. You can have a concentrated solution of a weak acid, or a dilute solution of a strong acid, or a concentrated solution of a strong acid, or . . . well, I'm sure you get the idea. Let me point out the main differences between strong and weak acids and bases in the following sections.

Disclosing the truth about acid strength

The strength of an acid is related to the strength of the bond between the hydrogen that is leaving and the rest of the molecule. The weaker that bond is, the stronger the acid. These sections focus on the strength (and weakness) of acids.

Binary acids: Comprising hydrogen and one other element

A binary acid consists of hydrogen and one other element. Chemistry unfortunately doesn't have a simple way of predicting the strength of a binary acid. However, in general, within a column of the periodic table, the lower the nonhydrogen atom is on the periodic table the stronger the acid. Within a period, the further the nonhydrogen atom is to the right, the stronger the acid is. Both of these trends are in part due to a complex interaction concerning the bond energies and the bond polarities.

Oxyacids: Containing hydrogen, oxygen, and another element

Ternary acids involve three elements. The only ternary acids I discuss are the *oxyacids*, which are acids that contain hydrogen, oxygen, and one other element. The trends for oxyacids depend on the polarity of the O-H bond, not the strength of the bond, which is relatively constant. For oxyacids differing only in the identity of the central element, the stronger acid is the one containing the more electronegative element. For example, H_2SO_4 is stronger than H_2SeO_4. The pair H_3PO_4 and H_2SO_4 don't work because the number of hydrogen atoms isn't the same. For oxyacids with the same central element, the one with the most oxygen is the strongest. Thus, H_2SO_4 is stronger than H_2SO_3.

The oxyacids of chlorine are useful in illustrating why the number of oxygen atoms present affects an acid's strength. The chlorine oxyacids are hypochlorous acid, HOCl, chlorous acid, $HClO_2$, chloric acid, $HClO_3$, and perchloric acid, $HClO_4$. The strength of the acid is related to how easily the molecule can lose an H^+. The greater the partial positive charge on the hydrogen atom, the more easily it is lost. The simplified structures of these acids are:

H–O–Cl	hypochlorous acid
H–O–ClO	chlorous acid
H–O–ClO_2	chloric acid
H–O–ClO_3	perchloric acid

In all four cases, the H–O–Cl group is present. The addition of zero to three oxygen atoms to this group changes the acid strength. Oxygen is very electronegative. The addition of an oxygen atom results in a shift of electron density toward the oxygen. The least electronegative element present (hydrogen) is affected the most. The greater the number of oxygen atoms added, the greater the shift in electron density. As the number of oxygen atoms increases so does the positive charge on the hydrogen atom. Thus, the acid strength increases in the order:

$$HOCl < HClO_2 < HClO_3 < HClO_4$$

The change in strength is normally very large; often the acid strength increases by 10,000 to 1,000,000 times with the addition of each oxygen atom.

Adding an ion affects acid strength

The trends in acid strength are different if an ion is present. Anions have excess electrons; it would require more energy for the positive hydrogen ion to leave the negatively charged ion than a neutral species. Thus, anionic species are weaker acids than the corresponding neutral molecules. Cations, with their deficiency of electrons, behave in the opposite manner.

Certain metal cations also produce acidic aqueous solutions. Metal ions with a +3 or greater charge usually fall into this category. Water molecules in aqueous solution coordinate to (surround) these cations. Unlike cations with lower charges, the high metal charge pulls on the oxygen very strongly. The result of this strong pull is that one of the hydrogen-oxygen bonds in water breaks so that the oxygen can form a bond to the metal, which releases a hydrogen ion, making the solution acidic.

$$M^{3+} + H-O-H \rightleftarrows [M-O-H]^{2+} + H^{+}$$

Uncovering the basic truth about base strength

The strong bases are the water-soluble metal hydroxides. Many other insoluble metal hydroxides, such as $Mg(OH)_2$, readily react with acids as if they were strong bases. Many of the weak bases contain nitrogen. The simplest nitrogen containing weak base is ammonia. In aqueous solution, ammonia behaves in the following fashion:

$$NH_3(aq) + H_2O(l) \rightleftarrows NH_4^{+}(aq) + OH^{-}(aq)$$

The formation of the hydroxide ion by this equilibrium is why chemistry classifies ammonia as a base. The hydrogen ion from the water forms a coordinate covalent bond with the nonbonding electron pair on the nitrogen. Replacing one or more of the hydrogen atoms in the ammonia with other atoms alters the strength. The addition of an atom or group of atoms with a high electronegativity (relative to hydrogen) tends to produce a weaker base than ammonia, because doing so would withdraw electron density from the nitrogen lone-pair. The addition of an atom or group of atoms with a low electronegativity (relative to hydrogen) tends to produce a stronger base than ammonia, because the other groups attached to the nitrogen aren't withdrawing electron density.

Strong acids

Acids that ionize essentially 100 percent in water are called *strong acids*. Table 9-1 lists the most common strong acids you're likely to encounter.

Table 9-1 Common Strong Acids

Name	*Formula*
Hydrochloric acid	HCl
Hydrobromic acid	HBr
Hydroiodic acid	HI
Nitric acid	HNO_3
Perchloric acid	$HClO_4$
Chloric acid	$HClO_3$
Sulfuric acid (first ionization only)	H_2SO_4

Sulfuric acid is called a *diprotic* acid. It can donate 2 protons, but only the first ionization goes 100 percent. The other acids listed in Table 9-1 are *monoprotic* acids, because they donate only one proton.

Allow me to show you an example of what happens when working with a strong acid. For example, if you dissolve hydrogen chloride gas in water, the HCl reacts with the water molecules and donates a proton to them:

$$HCl(g) + H_2O(l) \rightarrow Cl^-(aq) + H_3O^+(aq)$$

The H_3O^+ ion is called the *hydronium ion*. You can think of the hydronium ion as a water molecule plus a hydrogen cation ($H_2O + H^+$).

You can show this reaction in another way:

$$HCl(g) \rightarrow Cl^- (aq) + H^+(aq)$$

Here I have simply omitted the water from both sides. This short-hand notation for the reaction shows the hydronium ion. Because I can be rather lazy, many times I use this short-hand notation. If you see a $H^+(aq)$, it really is an hydronium ion or actually a mixture of one, two, and even three water molecules surrounding a hydrogen ion.

This reaction goes essentially to *completion,* meaning the reactants keep creating the product until they're all used up. In this case, all the HCl ionizes to H_3O^+ and Cl^-; no more HCl is present. Note that water, in this case, acts as a base, accepting the proton from the hydrogen chloride.

Because strong acids ionize completely, calculating the concentration of the hydronium ion and chloride ion in solution is easy if you know the initial concentration of the strong acid. For example, suppose that you add 1.0 moles of nitric acid, HNO_3 to enough water to make a liter of solution. You can say that the initial concentration of HNO_3 is 1.0 M (1.0 mol/L). *M* stands for molarity, and *mol/L* stands for moles of solute per liter of solution. (For a review of molarity and other concentration units, see Chapter 5.)

You can represent this 1.0 M concentration for the HNO_3 in this fashion: $[HNO_3] = 1.0$. The brackets around the compound indicate molar concentration, mol/L. Because the HNO_3 completely ionizes, you see from the chemical formula that for every HNO_3 that ionizes, you get one hydronium ion and one nitrate ion. So the concentration of ions in that 1.0 M HNO_3 solution is

$$[H_3O^+] = 1.0 \text{ and } [NO_3^-] = 1.0$$

This idea is valuable when you calculate the pH of a solution, which I show you how to do in the "Determining pH" section later in this chapter.

Strong bases

A *strong base* is one that dissociates completely in water. You normally see only one strong base, and that's the hydroxide ion, OH^-. A strong base basically is a salt containing the hydroxide ion.

Calculating the hydroxide ion concentration is really straightforward. Suppose that you have a 2.5 M (2.5 mol/L) $Ca(OH)_2$ solution. The calcium hydroxide, a salt, completely *dissociates* (breaks apart) into ions:

$$Ca(OH)_2 \rightarrow Ca^{+2}(aq) + 2\ OH^-(aq)$$

In order to figure the concentration of ions, you must take into account that for every calcium hydroxide unit that dissolves you get one calcium cation and two hydroxide anions. If you start with 2.5 mol/L $Ca(OH)_2$, then the concentration of ions would be:

$$[Ca^{+2}] = 2.5 \text{ and } [OH^-] = 5.0$$

Weak acids

Acids that only partially ionize are called *weak acids.* One example is acetic acid (CH_3COOH). If you dissolve acetic acid in water, it reacts with the water molecules, donating a proton and forming hydronium ions. It also establishes an equilibrium in which you have a significant amount of unionized acetic acid. (In reactions that go to completion, the reactants are completely used up creating the products. But in equilibrium systems, two exactly opposite chemical reactions — one on each side of the reaction arrow — are occurring at the same place, at the same time, with the same speed of reaction. See Chapter 8 for a more detailed discussion of the equilibrium process.)

The acetic acid reaction with water looks like this:

$$CH_3COOH(aq)+H_2O(l)\rightleftarrows CH_3COO^-(aq)+H_3O^+(aq)$$

The acetic acid that you added to the water is only partially ionized, so it's a weak acid. In the case of acetic acid, about 5 percent ionizes, and 95 percent remains in the molecular form. The amount of hydronium ion that you get in solutions of acids that don't ionize completely is much less than it is with a strong acid.

Calculating the hydronium ion concentration in weak acid solutions isn't as straightforward as it is in strong solutions, because not all the weak acid that dissolves initially has ionized. In order to calculate the hydronium ion concentration, you must use the equilibrium constant expression for the weak acid. For weak acid solutions, you use a mathematical expression called the K_a — the *acid ionization constant.* Check out the "Tackling More Equilibrium Problems" later in this chapter for how to deal with these solutions.

If you want to see whether a person is a chemist, ask him to pronounce *unionized.* A chemist pronounces it *un-ionized,* meaning "not ionized." Everyone else pronounces it *union-ized,* meaning "being part of a union."

Weak bases

Weak bases also react with water to establish an equilibrium system. Ammonia is a typical weak base. It reacts with water to form the ammonium ion and the hydroxide ion:

$$NH_3(g) + H_2O(l) \rightleftarrows NH_4^+(aq) + OH^-(aq)$$

Like a weak acid, a weak base is only partially ionized. The modified equilibrium constant expression for weak bases is K_b. You use it exactly the same way you use any other equilibrium constant, except you are normally solving for the $[OH^-]$.

Acidic/basic oxides

Most *oxides* (compounds containing the oxide ion, O^{2-}) exhibit acidic or basic properties. These properties become apparent when the oxide reacts with an acid or a base, and, in some cases, when the oxide reacts with water.

Most nonmetal oxides are *acidic oxides*. An acid anhydride produces an acid when added to water. A few nonmetal oxides, such as CO and NO, don't react. Usually, when a nonmetal oxide reacts, it forms only a simple acid, and the nonmetal remains in the same oxidation state.

In terms of chemical reactions, metals are the opposite of nonmetals. Thus, metal oxides are the opposite of nonmetal oxides. Metal oxides are *basic oxides*. The addition of a base anhydride to water produces a base or, like all bases, reacts with an acid. These tend to be relatively straightforward reactions yielding only a base.

$$K_2O(s) + H_2O(l) \rightarrow 2\ KOH(aq)$$

Many metal oxides don't appear to be bases because of their limited solubility in water. However, when added to acids, their basic nature becomes obvious. For example:

$$CuO(s) + H_2O(l) \rightarrow \quad \text{No reaction}$$

$$CuO(s) + 2\ HCl(aq) \rightarrow CuCl_2(aq) + H_2O(l)$$

An exception occurs when the metal is in a high oxidation state (+5 or greater). Oxides containing metal ions in high oxidation states produce acids upon addition to water. An example is chromium(VI) oxide dissolving in water, as shown here:

$$CrO_3(s) + H_2O(l) \rightarrow H_2CrO_4(aq)$$

Bronsted-Lowry acid-base reactions

A Bronsted-Lowry acid-base reaction gives a salt, but it may or may not produce water. Figure 9-1 shows the NH_3/HCl reaction using the electron-dot structures of the reactants and products.

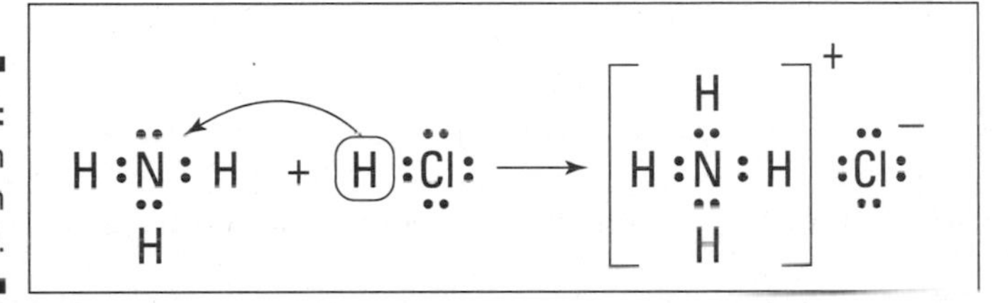

Figure 9-1: Reaction of NH_3 with HCl.

HCl is the acid, so it's the proton donor, and ammonia is the base, the proton acceptor. Ammonia has a lone pair of nonbonding electrons that it can furnish for the coordinate-covalent bond.

The reaction between hydrochloric acid and sodium hydroxide that I use as an example of an Arrhenius acid-base reaction earlier in this chapter is also a Bronsted-Lowry acid base reaction. However, I can only classify the ammonia/ hydrochloric acid reaction as an acid-base reaction in the Bronsted-Lowry theory because no water is produced.

One important consequence of defining acids and bases only in terms of hydrogen ions is the concept of conjugate acid-base pairs. A *conjugate acid-base pair* is two substances differing by a single hydrogen ion. For example, consider the ionization of nitrous acid:

$$HNO_2(aq) \rightleftarrows H^+(aq) + NO_2^-(aq)$$

This acid is present in solution in an equilibrium. The HNO_2 is capable of donating a hydrogen ion, thus it's a Brønsted-Lowry acid. The reverse arrow of the equilibrium symbol means the reaction can proceed to the left. In the left reaction, the nitrite ion behaves as a Bronsted-Lowry base because it's accepting a hydrogen ion. The nitrous acid and the nitrite ion form a conjugate acid-base pair. The nitrous acid, by virtue of having one more hydrogen ion, is the conjugate acid, and the nitrite ion, with one less hydrogen ion, is the conjugate base. In all cases, the *conjugate acid* is the member of the conjugate acid-base pair with *one* more hydrogen ion. The *conjugate base* is the member of the conjugate acid-base pair with *one* less hydrogen ion. Conjugate acid-base pairs differ by one and only one hydrogen ion: the conjugate base of H_2SO_4 is HSO_4^-, not SO_4^{2-}.

In the Arrhenius theory, acid-base reactions are neutralization reactions. In the Bronsted-Lowry theory, acid-base reactions are a competition for a proton. For example, take a look at the reaction of ammonia with water:

$$NH_3(g) + H_2O(l) \rightleftarrows NH_4^+(aq) + OH^-(aq)$$

Ammonia is a base (it accepts the proton), and water is an acid (it donates the proton) in the forward (left to right) reaction. But in the reverse reaction (right to left), the ammonium ion is an acid and the hydroxide ion is a base. If water donates a hydrogen ion more readily than the ammonium ion, then there will be a relatively large concentration of ammonium and hydroxide ions at equilibrium. If, however, the ammonium ion donates hydrogen ions more readily, much more ammonia than ammonium ion is present at equilibrium.

Bronsted and Lowry said that an acid reacts with a base to form conjugate acid-base pairs, which differ by a single H^+. NH_3 is a base, for example, and NH_4^+ is its conjugate acid. H_2O is an acid in the reaction between ammonia and water, and OH^- is its conjugate base. In this reaction, the hydroxide ion accepts hydrogen ions much more readily than ammonia, so the equilibrium is shifted to the left — not much hydroxide is present at equilibrium.

Determining pH

The amount of acidity in a solution is related to the concentration of the hydronium ion in the solution. The more acidic the solution is, the larger the concentration of the hydronium ion. In other words, a solution in which the $[H_3O^+]$ equals 1.0×10^{-2} is more acidic than a solution in which the $[H_3O^+]$ equals 1.0×10^{-7}. The *pH scale,* a scale based on the $[H_3O^+]$, was developed to more easily tell at a glance the relative acidity of a solution. *pH* is defined as the negative logarithm (abbreviated as *log*) of the $[H_3O^+]$ or in our short-hand notation $[H^+]$. Mathematically, it looks like this:

$$pH = -\log [H_3O^+]$$

Substances commonly found in everyday life cover a wide range of pH values. Table 9-2 lists some common substances and their pH values.

Table 9-2 Average pH Values of Some Common Substances

Substance	***PH***
Oven cleaner	13.8
Hair remover	12.8
Household ammonia	11.0

Substance	PH
Milk of magnesia	10.5
Chlorine bleach	9.5
Seawater	8.0
Human blood	7.3
Pure water	7.0
Milk	6.5
Black coffee	5.5
Soft drinks	3.5
Aspirin	2.9
Vinegar	2.8
Lemon juice	2.3
Auto battery acid	0.8

The following sections take a closer look at pH and explain the different ways you can calculate pH.

Calculating the pH of a solution

You can use the mathematical relationship to determine a pH of a solution. In pure water, the $[H_3O^+]$ equals 1.0×10^{-7}. Using this mathematical relationship of pH, you can calculate the pH of pure water:

$$pH = -\log [H_3O^+]$$

$$pH = -\log [1.0 \times 10^{-7}]$$

$$pH = -[-7]$$

$$pH = 7$$

The pH of pure water is 7. Chemists call this point on the pH scale *neutral.* A solution is called *acidic* if it has a larger $[H_3O^+]$ than water and a smaller pH value than 7. A *basic* solution has a smaller $[H_3O^+]$ than water and a larger pH value than 7.

The pH scale really has no end. You can have a solution of pH that registers less than 0. (A 10 M HCl solution, for example, has a pH of –1.) However, the 0 to 14 range is a convenient range to use for weak acids and bases and for dilute solutions of strong acids and bases.

The $[H_3O^+]$ of a 2.0 M acetic acid solution is 6.0×10^{-3}. Because this solution has a larger $[H_3O^+]$ than water, the solution is acidic. Calculate the pH of this solution:

$$pH = -\log [H_3O^+]$$

$$pH = -\log [6.0 \times 10^{-3}]$$

$$pH = -[-2.22]$$

$$pH = 2.22$$

Using the pOH to calculate the pH

Another expression, called the *pOH,* can be useful in calculating a solution's pH. The pOH is the negative logarithm of the $[OH^-]$. You can calculate the pOH of a solution just like the pH by taking the negative log of the hydroxide ion concentration. There is a relationship between the pH and the pOH: pH + pOH = 14. This equation makes going from pOH to pH quite easy.

When working with logarithms, numbers in front of the decimal point (the characteristic) aren't significant, while numbers after the decimal point (the mantissa) are significant.

Determining the antilog relationship

Just as you can you convert from $[H_3O^+]$ to pH, you can also go from pH to $[H_3O^+]$. To do this, you use what's called the *antilog relationship*, which is

$$[H_3O^+] = 10^{-pH}$$

Human blood, for example, has a pH of 7.3. Here's how you calculate the $[H_3O^+]$ from the pH of blood:

$$[H_3O^+] = 10^{-pH}$$

$$[H_3O^+] = 10^{-7.3}$$

$$[H_3O^+] = 5.01 \times 10^{-8}$$

You can use the same procedure calculate the $[OH^-]$ from the pOH.

Accepting Water's Autoionization

Water can act as either an acid or a base, depending on what it's being combined with. When an acid reacts with water, water acts as a base, or a proton acceptor. But in reactions with a base (like ammonia), water acts as an acid, or a proton donor. Substances that can act as either an acid or a base are called *amphoteric.*

But water can also react with itself. When two water molecules react with each other, one donates a proton and the other accepts it:

$$H_2O(l) + H_2O(l) \rightleftarrows H_3O^+(aq) + OH^-(aq)$$

You can also show this in short-hand form:

$$H_2O(l) \rightleftarrows H^+(aq) + OH^-(aq)$$

This reaction is an equilibrium. A modified equilibrium constant, called the K_w (which stands for *water dissociation constant*) is associated with this reaction. When you write the equilibrium expression, remember that not only is water a liquid, but it is also the solvent. Both of these are reasons to leave the concentration of water out of the equilibrium constant expression.

The K_w has a value of 1.0×10^{-14} at room temperature and has the following form:

$$1.0 \times 10^{-14} = K_w = [H_3O^+][OH^-]$$

or in short-hand form

$$1.0 \times 10^{-14} = K_w = [H^+][OH^-]$$

In pure water, the $[H_3O^+]$ equals the $[OH^-]$ from the balanced equation, so $[H_3O^+] = [OH^-] = 1.0 \times 10^{-7}$. This K_w value allows you to convert from $[H^+]$ to $[OH^-]$, and vice versa, in *any* aqueous solution, not just pure water. In aqueous solutions, the hydronium ion and hydroxide ion concentrations are rarely going to be equal. But if you know one of them, the K_w allows you to figure out the other one.

Look at a the 2.0 M acetic acid solution problem in the section "Calculating the pH of a solution" earlier in this chapter. I knew that the $[H_3O^+]$ was 6.0×10^{-3}. Now you have a way to calculate the $[OH^-]$ in the solution by using the K_w relationship:

$$K_w = 1.0 \times 10^{-14} = [H_3O^+][OH^-]$$

$$1.0 \times 10^{-14} = [6.0 \times 10^{-3}][OH^-]$$

$$1.0 \times 10^{-14} \div 6.0 \times 10^{-3} = [OH^-]$$

$$1.7 \times 10^{-12} = [OH^-]$$

Tackling More Equilibrium Problems

Working weak acid or weak base equilibrium problems are just like the homogeneous equilibrium problems I present in Chapter 8. In fact, all equilibrium problems are basically the same. I use ICE tables (see Chapter 8) and make appropriate approximations to simplify the calculations. In addition these relationships can really help you:

- Always write a balanced chemical equation for the reaction.
- The K_w relationship of $K_w = 1.0 \times 10^{-14} = [H+][OH^-]$ is always valid and can be used to convert from hydrogen ion to hydroxide ion concentration and vice versa.
- A relationship exists between the K_a of a weak acid and the K_b of its conjugate base: $K_w = K_aK_b$.
- A relationship exists between the pH and pOH: pH + pOH = 14.00

With these relationships in mind, you can work the following problems involving weak acids.

K_a Problems

One of the major goals in Chem II is to calculate the pH of a solution, especially one that is a solution of a weak acid or a weak base. Let's look at the weak acid situation first. Knowing the concentration of a weak acid solution and the weak acid's K_a allows us to calculate the solution's pH.

Calculate the pH of a 0.300 M acetic acid solution. Acetic acid's $K_a = 1.8 \times 10^{-5}$

To solve this problem, stick to these steps:

1. **Write the reaction.**

   $$CH_3COOH(aq) \rightleftarrows H^+(aq) + CH_3COO^-(aq)$$

   Use the short-hand notation where $H^+(aq)$ is the same as $H_3O^+(aq)$.

2. **Write the equilibrium constant expression for this reaction.**

   $$K_a = \frac{[H^+][CH_3COO^-]}{[CH_3COOH]} = 1.8 \times 10^{-5}$$

3. **Create an ICE table like Chapter 8 shows you.**

	CH_3COOH	H^+	CH_3COO^-
Initial	0.300	0	0
Change	$-x$	$+x$	$+x$
Equilibrium	$0.300-x$	$+x$	$+x$

4. **Enter the values from the equilibrium line into the equilibrium constant expression.**

$$K_a = \frac{[H^+][CH_3COO^-]}{[CH_3COOH]} = \frac{[x][x]}{[0.300-x]} = 1.8\times10^{-5}$$

5. **Introduce an approximation.**

You can take two approaches here in solving this equation: use the quadratic equation and solve for the $[H^+]$ and then find the pH, or you can introduce an approximation:

$0.300 - x$ » 0.300 M

6. **Make an assumption.**

$$\frac{[x][x]}{[0.300]} = 1.8\times10^{-5}$$

The hydrogen ion concentration is $x = [H^+] = 2.3 \times 10^{-3}$ M.

7. **Check to see if the assumption is reasonable.**

To do so, go back to the assumption:

$$(0.300 - x) \approx 0.300$$

See how the neglected x would have changed the 0.300 M.

$0.300 - 0.0023 = 0.298$

In this book, I say that the assumption is acceptable if no more than the last digit of the given concentration is affected, as it is in this case. Follow the rules that your instructor establishes.

Any time you make an assumption to neglect x, you must check the validity of your assumption. If the assumption proves not to be acceptable, you need to go back and use the quadratic formula.

The assumption usually works if the value of the K_a expression is at least three orders of magnitude smaller then the concentration of the weak acid or base.

$pH = -\log [H^+] = -\log [2.3 \times 10^{-3}]$

pH = 2.63 (an acidic solution)

The pH of any acid solution *must* be below 7 and the $[H^+]$ must be greater than 10^{-7} M. The pH of any base solution *must* be above 7 and the $[OH^-]$ must be greater the 10^{-7} M ($[H^+] < 10^{-7}$ M). If your results disagree with this guideline, you have made an error.

Now you can do a problem in determining the K_a of a weak acid:

> Find the K_a of nitrous acid given that a 0.0200 M solution of the acid has a hydrogen ion concentration of 2.8×10^{-3} M.

To solve, follow these steps:

1. **Write the reaction.**

$$HNO_2(aq) \rightleftarrows H^+(aq) + NO_2^-(aq)$$

2. **Write the K_a expression.**

$$K_a = \frac{[H^+][NO_2^-]}{[HNO_2]}$$

3. **Write the ICE table.**

	HNO_2	*H^+*	*NO_2^-*
Initial	0.0200	0	0
Change	$-x$	$+x$	$+x$
Equilibrium	$0.0200 - x$	$+x$	$+x$

In this case you know the $[H^+]$, 2.8×10^{-3} M.

4. **Because the nitrite ion concentration is also *x*, you can now add this information to the ICE table.**

	HNO_2	*H^+*	*NO_2^-*
Initial	0.0200	0	0
Change	$-x$	$+x$	$+x$
Equilibrium	$0.0200 - 2.8 \times 10^{-3}$	2.8×10^{-3}	2.8×10^{-3}

5. **Enter these values into the equilibrium expression and calculate the K_a value.**

$$K_a = \frac{[2.8 \times 10^{-3}][2.8 \times 10^{-3}]}{[0.0200 - (2.8 \times 10^{-3})]} = 4.6 \times 10^{-4}$$

Now I show you some weak base equilibrium problems.

K_b Problems

Now that you are familiar with determining the pH of a weak acid solution, let me show you how to handle calculating the pH of a weak base solution. These types of problems are also very common in Chem II.

A 0.500 M solution of ammonia has a pH of 11.48. What is the K_b of ammonia?

To solve this problem, follow these steps:

1. **Write the reaction.**

$$NH_3(aq)+H_2O(l)\rightleftarrows NH_4^+(aq)+OH^-(aq)$$

2. **Write the K_b expression.**

$$K_b=\frac{[NH_4^+][OH^-]}{[NH_3]}$$

The concentration of pure liquids or solvents (like water in this example) doesn't appear in the equilibrium constant expression.

3. **Create an ICE table.**

	NH_3	OH^-	NH_4^+
Initial	0.500	0	0
Change	$-x$	$+x$	$+x$
Equilibrium	$0.500-x$	$+x$	$+x$

4. **From the pH, determine the hydrogen ion concentration and then the hydroxide ion concentration.**

$pH = 11.48$

$[H^+] = 10^{-11.48}$

$[H^+] = 3.3 \times 10^{-12}$

5. **Use the hydrogen ion concentration and K_w expression and determine the hydroxide ion concentration in the solution.**

$K_w = [H^+][OH^-] = 1.00 \times 10^{-14}$

$$[OH^-]=\frac{K_w}{[H^+]}=\frac{1.00\times10^{-14}}{3.3\times10^{-12}}=3.0\times10^{-3}$$

From the ICE table you can see that the $[NH_4^+] = [OH^-] = 3.0 \times 10^{-3}$ and $[NH_4^+] = 0.500 - (3.0 \times 10^{-3})$

6. **Substitute those values into the equilibrium constant expression and solve for the K_b.**

$$K_b = \frac{[NH_4^+][OH^-]}{[NH_3]} = \frac{[3.0\times10^{-3}][3.0\times10^{-3}]}{[0.500-(3.0\times10^{-3})]} = 1.8\times10^{-5}$$

Well done. Allow me to walk you through one more weak base problem:

Find the pH of a 0.200 M calcium acetate solution.

Calcium acetate is a salt. Because it is a soluble salt, it is a strong electrolyte and it will dissociate as follows:

$Ca(CH_3COO)_2(aq) \rightarrow Ca^{2+}(aq) + 2\ CH_3COO^-(aq)$

The resulting solution has 0.200 M Ca^{2+} and 2 (0.200 M) = 0.400 M CH_3COO^-.

The Ca^{2+} is a spectator ion in this reaction. Ions such as CH_3COO^-, from a weak acid or a base, undergo *hydrolysis*, a reaction with water. The acetate ion is the conjugate *base* of acetic acid ($K_a = 1.74 \times 10^{-5}$). Because acetate is a weak base, this is a K_b problem.

To solve this problem, stick to these steps:

1. **Write the equilibrium.**

$$CH_3COO^-(aq) + H_2O(l) \rightleftarrows OH^-(aq) + CH_3COOH(aq)$$

2. **Write the equilibrium constant expression.**

$$K_b = \frac{[OH^-][CH_3COOH]}{[CH_3COO^-]}$$

3. **Create an ICE table.**

	CH_3COO^-	OH^-	CH_3COOH
Initial	0.400	0	0
Change	$-x$	$+x$	$+x$
Equilibrium	$0.400 - x$	$+x$	$+x$

4. **Find the value of the K_b for the acetate ion in order to complete the problem set-up.**

To do so, calculate it from the K_a of acetic acid which is 1.74×10^{-5} using the relationship:

$$K_w = K_a K_b$$

$$K_b = \frac{K_w}{K_a} = \frac{1.00\times10^{-14}}{1.74\times10^{-5}} = 5.7\times10^{-10}$$

5. **Enter the values from the ICE table and the value of the K_b into the equilibrium constant expression making the assumption that x is going to be very small so that 0.400 – x » 0.40.**

   $$5.7\times10^{-10} = \frac{[x][x]}{[0.400]}$$

   $x = [OH^-] = 1.5 \times 10^{-5}$

6. **Check your assumption.**

   Is $0.400 - 1.5 \times 10^{-5}$ » 0.400? Yes!

7. **Determine the pH from the hydroxide ion concentration.**

   You have a couple ways to do so. One method uses pOH while the other method uses the K_w.

   a. With the pOH method, determine the pOH of the solution:

   $pOH = -\log [OH^-] = -\log (1.5 \times 10^{-5}) = 4.82$

   Then use the relationship $pK_w = pH + pOH = 14.00$

   $pH = pK_w - pOH = 14.00 - 4.82 = 9.18$

   b. With the K_w approach, solve for the hydrogen ion concentration in the K_w expression:

   $K_w = [H^+] [OH^-] = 1.0 \times 10^{-14}$

   $$[H^+] = \frac{K_w}{[OH^-]} = \frac{1.00\times10^{-14}}{1.5\times10^{-5}} = 6.6\times10^{-10}$$

   $pH = -\log [H^+] = -\log (6.6 \times 10^{-10}) = 9.18$

   Both methods give the same answer and indicate that the solution is basic, which is expected from a base like the acetate ion.

Believing in Buffers

Buffers are solutions that resist a change in pH when an acid or base is added. A buffer contains both a weak acid (HA) and its conjugate base (A^-). The acid part neutralizes any base added, and the base part of the buffer neutralizes any acid added to the solution. This section examines buffers more closely with some practice problems.

Working a basic buffer problem

Buffer problems are another type that is an integral part of Chem II. Buffer calculations are also crucial in any course in biochemistry because living organism are very sensitive to changes in pH. Sharpen up your pencil and start to work buffer problems,

If a solution has a cyanic acid (pK_a = 3.46) concentration of 0.500 M and a sodium cyanate concentration of 0.750 M, what is the pH?

To solve this problem, follow these steps:

1. **Identify the chemical species involved and write the reactions involved.**

   Two processes are taking place: the dissolving of the sodium cyanate and the weak acid dissociation:

   Dissolving of sodium cyanate:

   $$NaOCN(aq) \rightarrow Na^+(aq) + OCN^-(aq)$$

   0.750 M 0.750 M + 0.750 M

   Weak acid dissociation:

   $$HOCN(aq) \rightleftarrows H^+(aq) + OCN^-(aq)$$

   $$0.500\text{ M} - x \quad +x \quad +x$$

   The ionization of cyanic acid is an equilibrium and has an associated K_a. Notice that the cyanate ion, OCN^-, appears in both equations. The presence of the weak acid (cyanic acid) indicates that this a K_a problem.

2. **Find the K_a value.**

   You can get it from the pK_a.

   K_a = = $10^{-3.46}$ = 3.5×10^{-4}

3. **Write the K_a expression.**

   $$K_a = 3.5 \times 10^{-4} = \frac{[H^+][OCN^-]}{[HOCN]}$$

4. **Show both sources of the cyanate ion in the equilibrium constant expression.**

   This problem has two sources of the cyanate ion. You get 0.750 M from the dissociation of the sodium cyanate and x amount from the ionization of the cyanic acid.

   $$K_a = 3.5 \times 10^{-4} = \frac{[H^+][OCN^-]}{[HOCN]} = \frac{[x][0.750 + x]}{[0.500 - x]}$$

5. **Assume that you can neglect the $+x$ and $-x$ and simplify the calculation (thank goodness!).**

   It's quite likely that x will be insignificant when compared to the 0.750 and 0.500, which allows you to make an assumption.

   $$K_a = 3.5\times10^{-4} = \frac{[H^+][OCN^-]}{[HOCN]} = \frac{[x][0.750]}{[0.500]}$$

   $$x = [H^+] = 2.3 \times 10^{-4}\ M$$

6. **Check the approximations.**

   The x is so small that ignoring it in the $[0.750 + x]$ and in the $[0.500 - x]$ will not have any appreciable effect on your answer.

7. **Get the pH.**

   $$pH = -\log [H^+] = -\log (2.3 \times 10^{-4}) = 3.63$$

Applying this general procedure is possible when using a K_b expression to represent a buffer solution. However, if you were doing a series of repetitive calculations of this type, having a simpler method of working this type of problem may be helpful. The Henderson-Hasselbalch equation provides you with this simpler method.

Introducing the Henderson-Hasselbalch equation

The *Henderson-Hasselbalch equation* is an equation used to calculate the pH or pOH of a buffer. The two forms of the Henderson-Hasselbalch equation are

$$pH = pK_a + \log\frac{[CB]}{[CA]}$$

$$pOH = pK_b + \log\frac{[CA]}{[CB]}$$

[CA] is the molar concentration of the conjugate acid and [CB] is the molar concentration of the conjugate base.

You can use the first of these two forms to solve the cyanic acid/sodium cyanate problem you completed in the previous section:

$$pH = pK_a + \log\frac{[CB]}{[CA]} = 3.46 + \log\frac{[0.750]}{[0.500]} = 3.63$$

The problem is simplified to a one-step problem after identifying the conjugate acid (HOCN) and the conjugate base (OCN^-).

These two methods show you how to find the pH of a solution containing a conjugate acid-base pair of a weak acid. However, simply determining the pH of the solution doesn't prove you have a buffer, a solution that resists a change in pH. In the next example, I prove to you that this solution is a buffer. This problem has two different parts beginning with the calculation of the pH of a simple buffer and then seeing if it really resists changes in pH.

(a) What is the pH of a solution containing 2.00 moles of ammonia and 3.00 moles of ammonium chloride in a volume of 1.00 L? K_b for ammonia = 1.81×10^{-5}

(b) What is the pH of this solution after the addition of 0.500 moles of hydrogen chloride (HCl)?

You can treat part (a) of the problem as a K_b problem involving the following equilibrium:

$$NH_3(aq) + H_2O(l) \rightleftarrows NH_4^+(aq) + OH^-(aq)$$

TIP

You can treat part (a) of this problem as a weak acid problem using the K_a for the NH_4^+ and you still get the same final answer.

The initial molarity of ammonia is:

$$[NH_3] = \frac{2.00 \text{ moles } NH_3}{1.00 \text{ liters}} = 2.00 \text{ M}$$

and the initial molarity of the ammonium ion is:

$$[NH_4^+] = \frac{3.00 \text{ moles } NH_4Cl}{1.00 \text{ liters}} \times \frac{1 \text{ mole } NH_4^+}{1 \text{ mole } NH_4Cl} = 3.00 \text{ M}$$

Ammonium chloride is a strong electrolyte and completely dissociates to ammonium ions and chloride ions.

Follow these steps to solve this problem.

1. **Put your concentration values in the equilibrium reaction.**

   $$NH_3(aq) + H_2O(l) \rightleftarrows NH_4^+(aq) + OH^-(aq)$$

   2.00 M 3.00 M

2. **Make sure some hydroxide ion (*x*) are formed in order to establish an equilibrium.**

   $$NH_3(aq) + H_2O(l) \rightleftarrows NH_4^+(aq) + OH^-(aq)$$

   2.00 M – *x* 3.00 M + *x* + *x*

   This is an alternate way of keeping track of the changes in equilibrium amounts instead of using an ICE table. Some people like to use an ICE table; some like the method above. I use both, depending on my mood.

3. **Write the equilibrium constant expression and enter the concentrations.**

   $$K_b = \frac{[NH_4^+][OH^-]}{[NH_3]} = 1.81 \times 10^{-5} = \frac{[3.00 + x][x]}{[2.00 - x]}$$

   The K_b is most probably small enough to justify ignoring the $+x$ and the $-x$ and avoiding to have to solve the problem using the quadratic equation.

   $$K_b = 1.81 \times 10^{-5} = \frac{[3.00][x]}{[2.00]}$$

4. **Solve for *x*.**

   $x = [OH^-] = 1.21 \times 10^{-5}$ M

   If you check you will see that the approximation is justified.

5. **Determine the pOH and then the pH.**

   pOH = $-\log [OH^-] = -\log (1.21 \times 10^{-5}) = 4.92$

   pH = 14.00 – pOH = 14.00 – 4.92 = 9.08

   You could also have found the pH of the solution by using the Henderson-Hasselbalch equation.

   In the second part of this problem, 0.500 moles of hydrogen chloride (HCl), a strong acid, is added to the buffer solution. If this solution were not a buffer, this amount of a strong acid would lower the pH to 0.300. (Check it out for yourself if you want.) If the solution, as expected, is a buffer the pH should not be significantly different from the answer in the first part of this problem (9.08).

To solve this second part of the problem, stick to these steps:

1. **Consider the reaction of the strong acid with the base that is present, ammonia (NH_3).**

   The hydrochloric acid reacts with the ammonia.

   $NH_3(aq) + HCl(aq) \rightarrow NH_4Cl(aq)$

   or

   $NH_3(aq) + H^+(aq) \rightarrow NH_4^+(aq)$

2. **Place the moles of each substance beneath its chemical formula.**

$NH_3(aq)$	+ $HCl(aq)$	$\rightarrow$ $NH_4Cl(aq)$
2.00 mol	0.500 mol	3.00 mol

   The hydrochloric acid is the limiting reagent in the reaction. Therefore, the reaction will proceed to the right until all of the acid has completely reacted. The reaction will also consume some of the ammonia and produce additional ammonium chloride.

3. **Because all species are reacting in a one-to-one ratio, simply subtract the moles from the reactant side and add moles to the product side.**

$NH_3(aq)$	+ $HCl(aq)$	$\rightarrow$ $NH_4Cl(aq)$
2.00 mol	0.500 mol	3.00 mol
–0.50	0 –0.500	+0.500
1.50 mol	0.000 mol	3.50 mol

4. **Use these values either in the K_b relationship or in the Henderson-Hasselbalch equation.**

   I show you how to use the Henderson-Hasselbalch equation here.

   $$pOH = pK_b + \log\frac{[CA]}{[CB]}$$

   $$pOH = 4.7423 + \log\frac{[3.50]}{[1.50]} = 5.110$$

   $pH = 14.000 - pOH = 14.000 - 5.110 = 8.890$

   The pH of the solution is lower than initially (9.08), but it's nowhere near the change to the 0.300 value that the strong acid would have given if this weren't a buffer. Recall that the definition of a buffer stated that the solution would resist changes in pH, not completely stop pH changes.

You may be wondering why I used the moles of the acid and base components instead of the molarity of these two species. I could have converted the moles to molarity, but this conversion is unnecessary because the two species are in the same volume of solution. Because of this, you can use either moles or molarity in the Henderson-Hasselbalch equation.

Grasping buffer capacity

A solution remains a buffer unless you add enough acid or base to react completely with one or the other of the conjugate acid-base pair. When the stoichiometry leads to a reduction of one of the buffer components to zero, you have exceeded the buffer capacity. The *buffer capacity* in the amount of acid or base that may be added to a buffer before the solution ceases to be a buffer.

The higher the concentration of buffer components is, the greater the buffer capacity. The removal of one of the components makes the solution either a weak base or a weak acid solution unless there is an excess of a strong acid or base added. If an excess of a strong acid or a strong base exists, this excess will control the pH. If the only component remaining is a weak acid, then you have a K_a problem, and if the only component remaining is a weak base, you have a K_b problem. Refer to the earlier "Tackling More Equilibrium Problems" section for examples of how to solve these two types.

Tackling Titration Curves and Indicators

An acid-base *titration* is a laboratory procedure that is used to determine the concentration of an unknown solution. You can add a base solution of known concentration to an acid solution of unknown concentration (or vice versa) until an acid-base *indicator* visually signals that the *endpoint* of the titration has been reached. The *equivalence point* is the point at which we have added a stoichiometric amount of the base to the acid.

If the acid being titrated is a weak acid, then equilibriums will be established and must be accounted for in the calculations. Typically, a plot of pH of the weak acid solution being titrated versus the volume of the strong base added starts at a low pH and gradually rises until close to the equivalence point in which the curve rises dramatically. After the equivalence point region, the curve returns to a gradual increase.

Four distinct regions are in the titration curve of a weak acid or a weak base with a strong base or a strong acid in which I would like to calculate the pH:

- The initial pH of the solution
- The pH at any point before the equivalence point
- The pH at the equivalence point
- The pH at any point after the equivalence point

In this section, I illustrate these types of calculations by working a problem.

A 100.0 mL sample of 0.150 M nitrous acid (pK_a = 3.35) was titrated with 0.300 M sodium hydroxide, NaOH. Determine the pH of the solution: a. initially; b. after adding 25.00 mL of the strong base; c. after adding 50.00 mL of the strong base; and d. after adding 55.00 mL of the strong base.

Follow these steps to figure this problem.

1. **Deal with the initial pH of the weak acid solution.**

   Because no base has been added, this is the initial pH section of the titration curve. The only substance present is nitrous acid, HNO_2, a weak acid, so it is just a K_a problem.

   1. **Set up the simple equilibrium problem as a K_a problem.**

      $$HNO_2(aq) \rightleftarrows H^+(aq) + NO_2^-(aq)$$

      $$0.150 - x \qquad x \qquad x$$

   2. **Find the K_a from the pK_a.**

      $K_a = 10^{-3.35} = 4.5 \text{ x } 10^{-4}$

   3. **Put these values into the K_a expression.**

      $$K_a = \frac{[H^+][NO_2^-]}{[HNO_2]} = 4.5\times10^{-4} = \frac{[x][x]}{[0.150 - x]}$$

   4. **Assume that 0.150–x ≈ 0.150.**

      $x = [H^+] = 8.2 \times 10^{-3}$ M

   5. **Determine the pH.**

      pH = $-\log [H^+] = -\log (8.2 \times 10^{-3}) = 2.09$

   The initial pH of this weak acid is, as expected, acidic.

2. **Focus on the second region of the titration curve, some point before the equivalence point.**

   After adding 25.00 mL of NaOH, the reaction is

   $HNO_2(aq) + NaOH(aq) \rightarrow Na^+(aq) + NO_2^-(aq) + H_2O(l)$

   1. **Calculate the moles of each reactant from the concentration and the volume of each solution in order to deal with the stoichiometry of this reaction.**

$$\text{moles } HNO_2 = \frac{0.150 \text{ moles } HNO_2}{L} \times \frac{1\text{ L}}{1000\text{ mL}} \times \frac{100.00\text{ mL}}{1} = 0.0150 \text{ moles}$$

$$\text{moles } NaOH = \frac{0.300 \text{ moles NaOH}}{L} \times \frac{1\text{ L}}{1000\text{ mL}} \times \frac{25.00\text{ mL}}{1} = 0.00750 \text{ moles}$$

   2. **Compare the moles to find the limiting reactant.**

      The coefficients in the reaction are all ones, so you can simply compare the moles to find the limiting reactant. In this case, the sodium hydroxide, with the smaller number of moles is limiting. Add the mole information to the balanced chemical equation.

	$HNO_2(aq)$ +	$NaOH(aq)$ →	$Na^+(aq)$ +	$NO_2^-(aq)$ +	$H_2O(l)$
Initial moles:	0.0150	0.00750	0	0	

   3. **Add the information by subtracting the moles of limiting reactant from each reactant and add the moles to each product.**

      During the reaction the moles of reactants present will decrease and the moles of products will increase.

	$HNO_2(aq)$ +	$NaOH(aq)$ →	$Na^+(aq)$ +	$NO_2^-(aq)$ +	$H_2O(l)$
Initial moles:	0.0150	0.00750		0	
Reacted moles:	–0.00750	–0.00750		+0.00750	

   4. **Add each column to get the post-reaction amounts.**

	$HNO_2(aq)$ +	$NaOH(aq)$ →	$Na^+(aq)$ +	$NO_2^-(aq)$ +	$H_2O(l)$
Initial moles:	0.0150	0.00750		0	
Reacted moles:	–0.00750	–0.00750		+0.00750	
After reaction:	0.00750	0.00000		0.00750	

After the reaction, the moles of sodium hydroxide, the limiting reagent, are zero; therefore, sodium hydroxide no longer affects the pH. The solution only contains unreacted nitrous acid, sodium ions, nitrite ions, and water. The nitrous acid and its conjugate base, the nitrite ion, NO_2^-, will influence the pH. Nitrous acid and the nitrite ion are a conjugate acid-base pair of a weak acid, so that you now have a buffer. Because you have a buffer, you have two ways to finish the problem.

You could use the moles of the nitrous acid and nitrite ion and the total volume of 125.00 mL (100.00 mL of acid solution + 25.00 mL of added base) and use the weak acid equilibrium constant expression to sole for the hydrogen ion concentration and finally the base.

Or you could use the Henderson-Hasselbalch equation using the moles of the weak acid and conjugate bases. Either method can give you the same answer, but the Henderson-Hasselbalch equation is faster.

5. **Enter the appropriate values into the Henderson-Hasselbalch equation.**

$$pH = pK_a + \log\frac{[CB]}{[CA]}$$

The pK_a appears in the problem (3.35), HA is nitrous acid, and A^- is the nitrite ion.

$$pH = 3.35 + \log\frac{[0.00750]}{[0.00750]} = 3.35$$

When the titration is halfway to the equivalence point (where the concentration of the conjugate acid equals the concentration of the conjugate base), the pH will equal the pK_a, and the pOH will equal the pK_b.

This is the pH for this point in the region before the equivalence point. The same procedure works for any point before the equivalence point. At any point between the initial pH and the equivalence point you will have a buffer system formed.

3. **Focus on the third region of the titration point, the equivalence point, after adding 50.00 mL of NaOH.**

   1. **Write the reaction.**

   $HNO_2(aq) + NaOH(aq) \rightarrow Na^+(aq) + NO_2^-(aq) + H_2O(l)$

2. Find the number of moles of hydroxide ion added as before.

You need to know the moles of each of the reactants in order to deal with the reaction stoichiometry. You already found the initial moles of nitrous acid (0.0150 moles) so you don't need to calculate it again. You can just focus on finding the number of moles of hydroxide ion added as before:

$$\text{moles NaOH} = \frac{0.300 \text{ moles NaOH}}{\text{L}} \times \frac{1 \text{ L}}{1000 \text{ mL}} \times \frac{50.00 \text{ mL}}{1} = 0.0150 \text{ moles}$$

Both the nitrous acid and the sodium hydroxide are present in equal moles.

3. Create a new reaction table.

	$HNO_2(aq)$ +	$NaOH(aq)$ →	$Na^+(aq)$ +	$NO_2^-(aq)$ +	$H_2O(l)$
Initial moles:	0.0150	0.0150		0	
Reacted moles:	–0.0150	–0.0150		+0.0150	
After reaction:	0.0000	0.0000		0.0150	

When both reactants are limiting (zero moles remaining), you're at the equivalence point.

The only substance remaining in the solution that can influence the pH is the nitrite ion, the conjugate base of a weak acid. Because a base is present, the pH will be above 7. This is a K_b problem.

4. Find the concentration of the nitrite ion.

$$[NO_2^-] = \frac{0.0150 \text{ moles } NO_2^-}{(100.00 + 50.00)\text{mL}} \times \frac{1000 \text{ mL}}{1 \text{ L}} = 0.100 \text{ M}$$

5. Deal with the equilibrium portion of the problem by finding the K_b from the relationship.

$K_w = K_a K_b = 1.00 \times 10^{-14}$

$pK_w = pK_a + pK_b = 14.00$

$pK_b = pK_w - pK_a = 14.00 - 3.35 = 10.65$

6. Take the antilog.

$K_b = 2.2 \times 10^{-11}$

The K_b equilibrium reaction is:

$$NO_2^-(aq) + H_2O(l) \rightleftarrows OH^-(aq) + HNO_2(aq)$$

$$K_b = \frac{[OH^-][HNO_2]}{[NO_2^-]} = 2.2 \times 10^{-11}$$

7. **Create the ICE table for this equilibrium.**

	$NO_2^-(aq)$	$OH^-(aq)$	$HNO_2(aq)$
Initial	0.100	0	0
Change	$-x$	$+x$	$+x$
Equilibrium	$0.100 - x$	x	x

Because the K_b is so very small, you can safely assume in assuming $0.100 - x \approx 0.100$

8. **Enter this information into the K_b expression and solve for x.**

$$K_b = \frac{[OH^-][HNO_2]}{[NO_2^-]} = 2.2\times10^{-11} = \frac{[x][x]}{[0.100]}$$

$x = 1.48 \times 10^{-6}$ M OH^-

9. **Use the hydroxide ion with the K_w to find the hydrogen ion concentration, and then find the pH, or you can find the pOH and use the pK_w to find the pH.**

I suggest you go with the latter method this time.

$pOH = -\log [OH^-] = -\log (1.48 \times 10^{-6}) = 5.828$ (unrounded)

$pK_w = pH + pOH = 14.00$

$pH = pK_w - pOH = 14.00 - 5.828 = 8.172$

4. **Focus on the fourth region of the titration curve, a poing after the equivalence point after adding 55.00 mL of NaOH.**

You know this is true because the preceding part was at the equivalence point. After the equivalence point, the titrant added will be in excess, and the other substance is limiting. In this case, the sodium hydroxide will be in excess and the nitrous acid will be limiting.

1. **Write the reaction.**

$$HNO_2(aq) + NaOH(aq) \rightarrow Na^+(aq) + NO_2^-(aq) + H_2O(l)$$

2. **Deal with the stoichiometry of this reaction.**

Good news. You need to know the moles of each of the reactants. You already found the initial moles of nitrous acid (0.0150 moles) so you don't need to find them again.

$$\text{moles NaOH} = \frac{0.300 \text{ moles NaOH}}{\text{L}} \times \frac{1 \text{ L}}{1000 \text{ mL}} \times \frac{55.00 \text{ mL}}{1} = 0.0165 \text{ moles}$$

3. **Create a new reaction table.**

$$HNO_2(aq) + NaOH(aq) \rightarrow Na^+(aq) + NO_2^-(aq) + H_2O(l)$$

	HNO_2	*NaOH*	*NO_{2-}*
Initial moles:	0.0150	0.0165	0
Reacted moles:	–0.0150	–0.0150	+0.0150
After reaction:	0.0000	0.0015	0.0150

4. **Find the sodium hydroxide ion concentration after the reaction.**

   Two bases are present after the reaction. Both could influence the pH. However, because the sodium hydroxide is a strong base, it is more important than the weaker base, the nitrite ion.

   $$[\text{NaOH}] = \frac{0.00150 \text{ moles NaOH}}{(100.00+55.00)\text{mL}} \times \frac{1000 \text{ mL}}{1 \text{ L}} = 9.68 \times 10^{-3} \text{ M}$$

   Sodium hydroxide is a strong base so [NaOH] = $[OH^-]$.

5. **Use the hydroxide ion concentration to determine the pH.**

   pOH = –log $[OH^-]$ = –log (9.68×10^{-3}) = 2.01

   pK_w = pH + pOH = 14.00

   pH = pK_w – pOH = 14.00 – 2.01 = 11.99

   The pH is basic as expected.

Take a deep breath and pat yourself on the back. You solved a complex problem. If you can do one like this, you know acid base equilibriums.

You can also determine the pH of a solution using an acid/base indicator. An *acid/base indicator* is a weak acid or base that changes color when it loses/gains a hydrogen ion. The pH where this occurs depends upon the pK_a of the indicator. When the indicator changes color, you know a change has occurred in the pH of the solution.

One common use of an acid/base indicator is in the titration procedure. If you add an indicator to the substance being titrated, it produces a color indicative of the pH of the solution. Then you can begin the titration. The continued addition of titrant eventually reduces the original acid or base to zero, and there will be excess titrant. The indicator changes color to reflect the pH change caused by the added excess titrant. This color change signals the endpoint of the titration, the experimental end of the titration. If you choose the correct indicator, the color change will indicate the point where the stoichiometric amount of reactant has been added, that is the equivalence point.

In choosing an indicator, attempt to have an indicator change color as close as possible to the equivalence point, the point at which you have added a stoichiometric amount of titrant. You can easily make this choice by comparing the pK_a of the indicator to the pH at the equivalence point. The closer these two values, the closer will be agreement between the equivalence point and the endpoint of the titration.

Chapter 10

Taking On Solubility and Complex Ion Equilibrium

In This Chapter

- Calculating solubility equilibrium
- Getting the lowdown on complex ions

The chapter introduces you to the final two types of equilibriums I discuss in this book — solubility equilibriums and complex ion equilibriums. (Chapters 8 and 9 have information on the other types of equilibriums.) I walk you through several different example problems so you're ready to do battle with equilibriums when your chemistry professor or teacher throws them your way.

Solving Solubility Equilibriums

Solubility equilibriums come about because no chemical compound is really totally insoluble. A more appropriate term might be *sparingly soluble*. In spite of the solubility rules you studied in your Chem I class, some material will always dissolve. The solubility rules are useful guidelines as to which salts are soluble and which are only sparingly soluble. You may want to grab a copy of my book *Chemistry For Dummies* 2nd edition (John Wiley & Sons, Inc.) or your textbook for a quick review of the solubility rules.

Solubility equilibriums are heterogeneous equilibriums. The most important property of solubility equilibriums, as *heterogeneous* equilibriums, is that a solid is always present. Working solubility equilibrium problems is relatively straightforward. Equilibrium is equilibrium. Remember that when working heterogeneous equilibriums problems the solids are never included in the equilibrium constant expressions. The problems are all the same, with just a few variations, which I show in the following sections.

Understanding the solubility product constant

In Chapters 8 and 9, I introduce to you the concept of using an equilibrium constant to express the mathematical relationship between the reactants and products when the system was at equilibrium. When I introduce homogeneous equilibrium, I introduce the equilibrium constant expression (K_c); with acids and bases, the weak acid dissociation constant (K_a), and the weak base dissociation constant (K_b). I show you that even water has a constant associated with it, the water dissociation constant (K_w). Now I consider the equilibrium constant associated with a sparingly soluble *(insoluble)* salt.

If you use the standard solubility rules, you would predict that a compound such as zinc phosphate ($Zn_3(PO_4)_2$) wasn't soluble in water. You could apparently prove it with experimentation. If you add some solid zinc phosphate to water, it settles to the bottom as unchanged. If you mix a solution of zinc nitrate with a potassium phosphate solution, a precipitate of zinc phosphate forms. However, if you apply the techniques and instrumentation of an environmental scientist, you would detect a minute quantity of dissolved zinc phosphate.

The small quantity of dissolved zinc phosphate results from the following equilibrium:

$$Zn_3(PO_4)_2(s) \rightleftarrows 3\ Zn^{2+}(aq) + 2\ PO_4^{3-}(aq)$$

The equilibrium constant expression for this equilibrium is:

$$K_{sp} = [Zn^{2+}]^3\ [PO_4^{3-}]^2$$

Because the reactant is a solid, it doesn't appear in the equilibrium expression. The subscript *sp* stands for *solubility product.* You can only write this type of reaction one way. The reaction must involve only the solid on the left of the reaction arrow, and the separated ions only on the right.

Calculating the concentration of dissolved ions

Using the solubility product constant I can determine the concentration of the dissolved ions in solution.

Suppose I want to determine the concentration of the ions resulting from the addition of zinc phosphate to water. If so, I need the K_{sp} value for this compound. The tabulated value for the K_{sp} of zinc phosphate is 9.1×10^{-33}

(a *very* small number). If I define x as the number of moles per liter of the solid zinc phosphate that dissolves, then I can now add the changes (x values) below the equation (or I could use an ICE table):

$$Zn_3(PO_4)_2(s) \rightleftarrows 3\,Zn^{2+}(aq) + 2\,PO_4^{3-}(aq)$$

$$+3x \qquad +2x$$

$$K_{sp} = [Zn^{2+}]^3[PO_4^{3-}]^2 = (3x)^3(2x)^2 = 9.1 \times 10^{-33}$$

No x value is necessary for the solid because the solid doesn't appear in the equilibrium constant expression. To solve this problem, stick to these steps:

1. **Substitute the concentration values into the K_{sp} expression.**

   The x value is really the moles per liter (*molarity*) of zinc phosphate that dissolves. According to the balanced equation, for every mole per liter of zinc phosphate that dissolves, three moles per liter of zinc ion are formed ($3x$) and two moles per liter of phosphate ions are formed ($2x$).

2. **Raise each concentration to the power equal to the coefficient in the balanced chemical equation.**

   Thus, you cube the $3x$ and square the $2x$ to get:

   $K_{sp} = [Zn^{2+}]^3[PO_4^{3-}]^2 = (27x^3)(4x^2) = 108\,x^5 = 9.1 \times 10^{-33}$

3. **Solve for x.**

   $x = 1.5 \times 10^{-7}$ (moles/liter of $Zn_3(PO_4)_2$ that dissolves so that:

   $[Zn^{2+}] = 3x = 4.5 \times 10^{-7}$ moles/L

   $[PO_4^{3-}] = 2x = 3.0 \times 10^{-7}$ moles/L

Calculating solubility — molar and otherwise

Another use for these K_{sp} calculations is in the determination of the solubility of a sparingly soluble salt.

Suppose I want to determine the grams of magnesium fluoride that will dissolve in 0.250 L of water. I need the K_{sp} value for magnesium fluoride, which is $K_{sp} = 8 \times 10^{-8}$.

To solve this problem, follow these steps:

1. **Assume that x = moles/liter of magnesium fluoride that dissolves and write the dissociation reaction and represent the concentration of dissolved ions.**

$$MgF_2(s) \rightleftarrows Mg^{2+}(aq) + 2\,F^-(aq)$$

$$+x \qquad\qquad +2x$$

2. **Write the K_{sp} expression:**

   $K_{sp} = [Mg^{2+}]\,[F^-]^2 = 8 \times 10^{-8}$

3. **Substitute the representations for the ion concentrations:**

   $K_{sp} = (x)\,(2x)^2 = 4x^3 = 8 \times 10^{-8}$

4. **Solve for x.**

   x = moles/liter of magnesium fluoride that dissolves = 3×10^{-3}

5. **Find the number of grams of magnesium fluoride per 250 mL solution:**

   $$\text{g } MgF_2 = \frac{3\times10^{-3}\text{moles } MgF_2}{\text{liter}} \times \frac{0.250 \text{ L}}{1} \times \frac{62.3 \text{ g } MgF_2}{\text{mole}} = 0.05 \text{ g } MgF_2$$

   The x value (3×10^{-3}) is the *molar solubility* of magnesium fluoride. The molar solubility is the concentration of a substance expressed as moles / liter (as molarity).

   In some cases, you want the *solubility* instead of the molar solubility. The solubility is the concentration of a substance expressed as g / liter. The solubility of magnesium fluoride is:

   $$MgF_2 \text{ solubility} = \frac{3\times10^{-3}\text{moles } MgF_2}{\text{liter}} \times \frac{62.3 \text{ g } MgF_2}{\text{mole}} = 0.2 \text{ g/L}$$

Predicting precipitation

You may encounter problems in your chemistry class that ask you to predict the amount of precipitation. Suppose I have a solution of barium chloride (a strong electrolyte) and I mix this solution with a solution of sodium sulfate (another strong electrolyte). There are two possible products, NaCl (soluble) and $BaSO_4$ (sparingly soluble). The question is whether the concentration of ions is large enough to cause *precipitation* (the formation of a solid salt resulting from the mixing of solutions).

To answer this question I need to calculate the reaction quotient, Q. (For a quick review of the reaction quotient, see Chapter 8.) The reaction quotient has exactly the same form as the relevant equilibrium constant expression (in this case the K_{sp}). As I explain in Chapter 8, the equilibrium constant value is the maximum value of Q. So in order to determine whether or not precipitation occurs, I can calculate the Q value and compare it to the K_{sp} of the possible precipitate. If the Q value is greater than the K_{sp}, precipitation will occur; if Q is less than the K_{sp}, precipitation won't occur.

If 10.0 mL of a 0.100 M $BaCl_2$ solution is added to 40.0 mL of a 0.0250 M Na_2SO_4 solution, will a precipitate form? The K_{sp} for $BaSO_4$ is 1.1×10^{-10}.

Barium chloride is a strong electrolyte, which produces one barium cation per barium chloride formula unit. Thus, the molarity of the barium ion is the same as that of the barium chloride (0.100 M). Similarly, the sodium sulfate is a strong electrolyte, which produces one sulfate ion per sodium sulfate formula unit. Thus, the molarity of sulfate ion is the same as the molarity of the sodium sulfate (0.0250 M). However, the molarity of these ions changes when these two solutions are mixed. The increase in volume results in a decrease of the concentrations because of dilution.

To solve this problem, follow these steps:

1. **Begin with a balanced chemical equation:**

   $$BaCl_2(aq) + Na_2SO_4(aq) \rightarrow BaSO_4(s) + 2\ NaCl(aq)$$

   You can also express it as the net ionic equation:

   $$Ba^{2+}(aq) + SO_4^{2-}(aq) \rightarrow BaSO_4(s)$$

   Because you're looking for the concentrations of the barium and sulfate ions, use the net ionic equation because it's slightly more convenient.

2. **Focus on the dilution by using the VC = VC approach.**

   The following formula is very useful in calculating the new concentration when a solution is diluted:

   $$V_{initial} \times C_{initial} = V_{final} \times C_{final}$$

3. **Calculate the barium ion concentration.**

   $$(10.00\ mL)(0.100\ M) = (50.00\ mL)(C_{final})$$

   The final volume is 50.00 mL because 10.00 mL of the barium chloride solution was added to 40.00 mL of the sodium sulfate solution.

   $$C_{final} = 0.0200\ M = [Ba^{2+}]$$

4. **Calculate the sulfate ion concentration.**

   $$(40.00\ mL)(0.0250\ M) = (50.00\ mL)(C_{final})$$

   $$C_{final} = 0.0200\ M = [SO_4^{2-}]$$

5. **Calculate the reaction quotient, Q.**

   $$Q = [Ba^{2+}]\ [SO_4^{2-}] = [0.0200][0.0200] = 4.00 \times 10^{-4}$$

6. **Compare it to the K_{sp} of $BaSO_4$.**

   $$4.00 \times 10^{-4} > 1.1 \times 10^{-10}$$

   $Q > K_{sp}$, therefore precipitation will occur.

In working precipitation problems, check to make sure that $Q > K_{sp}$. If Q isn't greater than the K_{sp} value, no precipitation exists. If the concentrations are relatively high and the K_{sp} value low, you may be able to get away with a simple visual inspection. When the concentrations are very low or the K_{sp} relatively large, a calculation is necessary.

Eyeing the common-ion effect

Solubility product (K_{sp}) equilibriums, like all equilibriums, are subject to the LeChatelier Principle, commonly called the *common-ion effect*, which involves an ion that is common to the equilibrium. See Chapter 8 for an extensive discussion of LeChatelier Principle. The common ion may be either the cation or the anion.

For example, if the magnesium fluoride ($K_{sp} = [Mg^{2+}] [F^-]^2$) in the earlier solubility example were added to a 0.100 M sodium fluoride solution, you simply need to add together the fluoride ion that came from the dissolving of the magnesium fluoride ($2x$) and the fluoride ion concentration of the sodium fluoride solution (1.00 M) and the equilibrium constant expression $K_{sp} = (x)(0.100 + 2x)^2$.

Calculate the molar solubility of Ag_2CrO_4 in (a) water (b) 1.00 M Na_2CrO_4. The K_{sp} for silver chromate is 1.9×10^{-12}.

To solve this problem, follow these steps:

1. **Write the reaction and the K_{sp} expression.**

   The reaction is as follows:

   $$Ag_2CrO_4(s) \rightleftarrows 2\,Ag^+(aq) + CrO_4^{2-}(aq)$$

   The K_{sp} expression is as follows:

   $$K_{sp} = [Ag^+]^2[CrO_4^{2-}] = 1.9 \times 10^{-12}$$

2. **Represent the concentration of the ions.**

   This part is a straightforward K_{sp} problem. Let x = moles/liter of silver chromate that dissolves, then the reaction stoichiometry gives $+\,2x$ for the silver cation and $+\,x$ for the chromate anion.

   $$Ag_2CrO_4(s) \rightleftarrows 2\,Ag^+(aq) + CrO_4^{2-}(aq)$$

   $$+\,2x \qquad\qquad +\,x$$

   $$K_{sp} = \left[Ag^+\right]^2\left[CrO_4^{2-}\right] = (2x)^2(x) = 4x^3 = 1.9 \times 10^{-12}$$

3. **Solve for x.**

   $x = 7.8 \times 10^{-5}$ moles/L = molar solubility of Ag_2CrO_4

   This is the molar solubility in water.

4. **Calculate the molar solubility in the sodium chromate solution.**

   The calculation of the molar solubility of the silver chromate in a sodium chromate solution differs from the calculation of the molar solubility in water because the sodium chromate solution includes the chromate ion, which is the common ion in this case. The chromate ion comes from the complete dissociation of the strong electrolyte sodium chromate.

5. **Determine the concentration of the common ion before adding the silver chromate.**

   $Na_2CrO_4\ (aq) \rightarrow 2\ Na^+(aq) + CrO_4^{2-}(aq)$

   1.00 M   2 (1.00 M) 1.00M

6. **Add the concentration of the common ion to the equilibrium.**

   The sodium ions, even though they are present, aren't part of the equilibrium, so you neglect them.

   $$Ag_2CrO_4(s) \rightleftarrows 2\ Ag^+(aq) + CrO_4^{2-}(aq)$$

   $+2x$ $1.00 + x$

   $$K_{sp} = [Ag^+]^2[CrO_4^{2-}] = (2x)^2(1.00 + x) = 1.9 \times 10^{-12}$$

7. **Simplify the calculation.**

   Because the K_{sp} is very small, you can assume $1.00 + x \approx 1.00$, which simplifies the calculation:

   $$K_{sp} = [Ag^+]^2[CrO_4^{2-}] = (2x)^2(1.00) = 1.9 \times 10^{-12}$$

8. **Solve for x.**

   $x = 6.9 \times 10^{-7}$ moles/L

   This is the molar solubility of silver chromate in 1.00 M sodium chromate.

According to the LeChatelier Principle, you can expect the equilibrium to be shifted to the left, reducing the solubility, which is exactly what you found in this calculation. It feels good to be right!

Following the Formation of Complex Ions

Complex ion equilibriums are usually homogeneous equilibriums. These equilibriums involve the reaction of some metal ion (a Lewis acid) with one or more other groups (called *ligands*), which act as Lewis bases. The formation of a complex ion is essentially a Lewis acid-base reaction (see Chapter 9) with the ligand furnishing an electron pair and the metal ion accepting it.

An example of a reaction forming a complex ion is:

$$Ni^{2+}(aq) + 6\ NH_3(aq) \rightleftarrows [Ni(NH_3)_6]^{2+}(aq)$$

The nickel(II) ions, electron deficient (as indicated by the positive charge), are able to gain one or more electron pairs. The ability of the nickel(II) ions to accept electron pairs makes these ions a Lewis acid. The ammonia molecules have a lone pair of electrons on the nitrogen atom. The ammonia is able to donate this lone pair; thus, ammonia is a Lewis base. The species resulting from the Lewis acid (Ni^{2+}) reacting with the Lewis base (the ligand NH_3) is the complex ion. Most complex ions are formed from the reaction of a metal ion with either four or six (as in this case) ligands. In general, the formula of the complex ion is written between brackets [ ].

You probably will encounter these types of equilibrium problems in your chemistry class, so having a good understanding of them is important. These sections focus on solving a couple types of these problems to help you.

Calculating the formation constant

You can calculate the overall formation *constant* (the product of the individual ones) and use the overall constant in your calculations in order to simplify these multiple equilibrium situations. The overall equilibrium process has an equilibrium constant utilizing the following expression:

$$K_f = \frac{\left[Ni(NH_3)_6^{2+}\right]}{\left[Ni^{2+}\right]\left[NH_3\right]^6}$$

The subscript *f* indicates a *formation constant*. The formation constant (K_f) is the equilibrium constant for the formation of a complex ion. You can find values for formation constants, such as K_a, in tables in textbooks. ***Note:*** Some books also refer to it as a *stability constant, K_{stab}.*

A formation constant always has the metal and ligands (Lewis acid and bases) on the left of the equilibrium arrow and the complex, and only the complex, on the right. This setup means that the equilibrium expression has the complex in the numerator, and the exponent on the numerator must be 1. The denominator can have only the metal ion and the ligands. Only the ligand may have an exponent that isn't a 1. Notice that when writing the equilibrium expression, the charge on the complex ion moves inside the square brackets so it's not mistaken for an exponent.

To help you understand quick complex ion equilibriums, you need to work practice problems to become better familiar with them. Here is one:

Calculate the concentration of chromium ions in a solution made by mixing 100.00 mL of a 1.00×10^{-4} M solution of chromium(III) chloride with 50.00 mL of a 1.000 M solution of potassium hydroxide. Assume the only important equilibrium is the complex ion formation equilibrium for $[Cr(OH)_4]^-$. The K_f for $[Cr(OH)_4]^-$ is 8×10^{29}.

1. **Write the reaction.**

   The very large K value indicates that the equilibrium lies very far to the right.

   $Cr^{3+}(aq) + 4\ OH^-(aq) \rightarrow [Cr(OH)_4]^-(aq)$

   Don't write this reaction as an equilibrium. At this point this problem is only simple stoichiometry. You need to do Step 2 to see how much complex could initially be formed.

2. **Find the number of moles of each of the chemical species involved in the complex formation and divide each by its coefficient in the balanced equation to determine the limiting reagent.**

   $$\text{moles } Cr^{3+} = \frac{1.00\times10^{-4}\text{ moles } Cr^{3+}}{L}\times\frac{0.100\text{ L}}{1} = 1.00\times10^{-5}\text{ mol } Cr^{3+}$$

   1.0×10^{-5} mol Cr^{3+} ÷ 1 (coefficient) = 1.00×10^{-5} mol/coef. for Cr^{3+}

   $$\text{moles } OH^- = \frac{1.00\text{ moles } OH^-}{L}\times\frac{0.0500\text{ L}}{1} = 5.00\times10^{-2}\text{ mol } OH^-$$

   5.00×10^{-2} mol OH^- ÷ 4 (coefficient) = 1.25×10^{-2} mol/coef. for OH^-

   Based upon the mole-to-coefficient ratios, the chromium(III) ions are the limiting reagent (it has the smallest mole/coefficient ratio).

3. **See what remains after the reaction of the limiting reagent.**

	$Cr^{3+}(aq)$ +	$4\ OH^-(aq)$ →	$[Cr(OH)_4]^-(aq)$
Initial:	1.00×10^{-5} mol	5.000×10^{-2} mol	0.000 mol
Reaction:	-1.00×10^{-5} mol	$-4(1.00 \times 10^{-5})$ mol	$+1.00 \times 10^{-5}$ mol
Post-Reaction:	0.000 mol	0.04996 mol	1.00×10^{-5} mol

   The ions are now present in (100.00 + 50.00) mL of solution, which gives a final volume of 0.15000 L

4. **Divide the moles after the reaction by the total volume to get the molarity of each of the ions according to the stoichiometry.**

   $[Cr^{3+}]$ = 0.000 M

   $[OH^-]$ = (0.04996 mol / 0.15000 L) = 0.3331 M

   $[Cr(OH)_4^-]$ = (1.00×10^{-5} mol / 0.15000 L) = 6.67×10^{-5} M

5. **Use the calculated molarities in the equilibrium portion of the problem.**

   Indicate the changes by x values. (The $+x$ must be for the Cr^{3+} ions because it's not possible to reduce 0 M any further.)

   $$Cr^{3+}(aq) + 4\ OH^-(aq) \rightleftarrows [Cr(OH)_4]^-(aq)$$

   $+x$ $\quad 0.3331 + 4x \quad 6.67 \times 10^{-5} - x$

6. **Enter these results into the equilibrium constant (K_f) expression.**

$$K_f = \frac{[Cr(OH)_4^-]}{[Cr^{3+}][OH^-]^4}$$

$$8 \times 10^{29} = \frac{[6.67 \times 10^{-5} - x]}{[x][0.3331 + 4x]^4}$$

7. **Make appropriate approximations and put them into the K_f expression.**

   The very large value of K_f tells you that there will only be a very small change in concentration. Therefore:

   $0.3331 + 4x \approx 0.3331$ M and $6.67 \times 10^{-5} - x \approx 6.67 \times 10^{-5}$ M

$$8 \times 10^{29} = \frac{[6.67 \times 10^{-5}]}{[x][0.3331]^4}$$

8. **Solve for x.**

   $x = 7 \times 10^{-33}$ M $= [Cr^{3+}]$

   The very small value of x justifies the approximation of dropping the $+ 4x$ and $-x$.

Solving them from the dissociation

Another way of representing complexes is to examine the dissociation of the complex instead of its formation. *Dissociation* is the breaking up of a compound (normally a salt or a complex) into its components. The dissociation equilibrium reaction for the complex I have been using as an example is:

$$[Ni(NH_3)_6]^{2+}(aq) \rightleftarrows Ni^{2+}(aq) + 6\ NH_3(aq)$$

This dissociation equilibrium has an equilibrium constant known as a *dissociation constant*, K_d. The dissociation constant is the equilibrium constant for the breaking down of a complex into its components:

$$K_d = \frac{[Ni^{2+}][NH_3]^6}{[Ni(NH_3)_6^{2+}]}$$

The dissociation equilibrium reaction is the reverse of the formation equilibrium reaction like that I solved in the previous section. The K_d expression is the inverse of the K_f expression. Some textbooks use K_{instab} (instab = instability) in place of K_d.

Part III

A Plethora of Chemistry II Concepts

In this part . . .

In this part I start off in Chapter 11 by examining thermodynamics, building on that little taste of thermochemistry you studied in Chem I. I talk about enthalpy and entropy and Gibbs Free Energy, as well as the three laws of thermodynamics, with the goal of being able to predict whether a reaction will occur spontaneously. Chapter 12 focuses on electrochemistry, beginning with a discussion of redox reactions. I show you how to balance redox reactions, which can be a bane of Chem II students, and then show how redox reactions are related to electrochemical cells — batteries and electroplating. I discuss calculation of cell voltage and electrolysis, the process that can decompose water.

The last three chapters of this part give you a rest from calculations. I start off by giving you a glimpse into the world of organic chemistry, the chemistry of carbon, in Chapter 13. I discuss hydrocarbons in a little detail and give you a brief introduction to other functional groups, such as the alcohols. This discussion gives you just a hint as to what your next chemistry course (should you be luck enough to need another one) may be like. In Chapter 14, I show you an application of organic chemistry — polymers. I discuss some of the different types of polymers (plastics) in terms of their structure and usage. In Chapter 15, I introduce you to the world of biochemistry, the chemistry of living things. I show you amino acids, proteins, and a host of other biologically important compounds.

Chapter 11

Getting Hot with Thermodynamics

In This Chapter

- Examining the ins and outs of thermochemistry
- Figuring out what enthalpy is
- Eyeing entropy
- Taking on thermodynamics
- Comprehending Gibbs Free Energy
- Noting what nonstandard conditions are
- Touching on the Haber process

In your first Chem I course you probably studied a topic called *thermochemistry* and examined some heat changes associated with reactions. Thermochemistry is a part of a larger topic called thermodynamics. *Thermodynamics* is the study of energy and its interconversions from one type of energy to another.

With thermodynamics, you can predict whether a chemical process is either spontaneous or nonspontaneous. Why are these two distinctions relevant?

- A *spontaneous process* proceeds without any outside help. Open a bottle of cologne and you quickly notice the fragrance. You did nothing else to cause the diffusion of the gas particles. They spontaneously mixed with the other gas molecules in the air. A spontaneous process happens; however, chemists don't know how long it will take to occur. The time factor involved in the reaction is related to the area of kinetics (refer to Chapter 7 for more information). A spontaneous process may take place rapidly or very slowly.
- A *nonspontaneous process* only proceeds with outside help. Suppose that you are cooking breakfast. You crack an egg into your frying pan and wait patiently for it to cook. Nothing happens until you apply heat and keep applying heat until it's fully cooked. This process is nonspontaneous. A nonspontaneous process never occurs without outside help.

This chapter examines what thermodynamics is and the different concepts you'll probably cover in your Chemistry II class.

Determining the Change in Energy

When scientists investigate the transfer of energy, they look at energy either entering or leaving a *system* (the part of the universe they are studying) from its *surroundings* (the rest of the universe). The quantity of energy entering or leaving the system depends upon the initial and final states of the system and not upon what happens in between these extremes. The *state of the system* is an exact description of the properties of the system. This exact description depends upon a series of variables known as state functions. *State functions* depend only upon the initial and final values, not their history. The temperature, *T,* of the system is an example of a state function.

To determine the change in any state function, you subtract the initial value from the final value (final – initial). A capital Greek letter delta, Δ, indicates this process. In the case of a temperature change, ΔT, the value is $T_{final} - T_{initial}$. Nearly all thermodynamic functions that I discuss in this chapter (and the ones that you will encounter in your Chem II course) are Δ functions. The presence of a degree symbol, such as in ΔH°, serves to indicate that the value is in the thermodynamic standard state.

If the function isn't a state function (and most of the time you will be told whether a particular thermodynamic is or isn't a state function), you need to know more about the process occurring during the change. According to the First Law of Thermodynamics (the Law of Conservation of Energy; refer to the section, "Remaining constant: The first law" later in this chapter for more on this law), energy can't be created or destroyed. In your Chem I class, you probably saw that you could find the change in the internal energy by utilizing the following relationship:

$$\Delta E = q + w$$

In this relationship, ΔE is the change in the internal energy, q is the heat, and work is w. Heat lost by the system is negative, and heat gained by the system is positive. Work done by the system is negative, and work done on the system is positive. The determination of the change in the internal energy is usually under conditions of constant pressure or constant volume. Under constant volume conditions, the work is zero ($w = 0$). Thus, $\Delta E = q$. If the volume isn't constant, $w = -P\Delta V$. Under conditions of constant pressure, the heat is the enthalpy change, ΔH. The *enthalpy change (ΔH)* is the heat lost or gained by the system during a chemical reaction while under constant pressure conditions (check out the next section for more on enthalpy). Neither heat nor work is a state function; however, the internal energy is a state function.

The two units that you use in thermochemistry calculations are the *joule* and the *calorie*.

- **Joule (J)** is the SI unit of energy. It has the units of kg m^2/s^2. Remember that 1 J = 1 kg m^2/s^2.
- **Calorie (cal)** is the amount of energy needed to raise the temperature of 1 gram of water 1 degree Celsius. By definition a calorie is exactly 4.184 J.

Discussing Enthalpy

A large majority of the reactions that chemists study are reactions at constant pressure. Many chemists like to study these kinds of reactions because the pressure can be held constant at atmospheric pressure by simply having the beaker or flask or any other type of reaction vessel open to the atmosphere. Because these reactions are so common, chemists give this energy change a special name, *enthalpy*. The following conventions apply to the sign associated with the ΔH:

If $\Delta H < 0$, then the reaction is *exothermic*.

If $\Delta H > 0$, then the reaction is *endothermic*.

For example, consider the reaction of solid carbon and oxygen gas to form a mole of carbon dioxide under constant volume conditions.

$$C(s) + O_2(g) \rightarrow CO_2(g) \quad \Delta H = -393.5 \text{ kJ}$$

The negative sign on the enthalpy tells you that this reaction is exothermic, energy is released. Often this energy change is referred to as the enthalpy (heat) of reaction, $\Delta H_{reaction}$ or ΔH_{rxn}.

This type of equation is an example of a thermochemical equation. A *thermochemical equation* is a balanced chemical equation that not only shows the mole relationship between reactants and products, but also the enthalpy change associated with that specific reaction. In a thermochemical equation having fractional coefficients (unlike in ordinary chemical reactions) is perfectly okay; the *coefficients* always refer to moles of reactants or products, not individual molecules.

Be sure to use the correct state of matter for the reactants and products. The two following conventions also apply to thermochemical equations:

- **If you use a multiplier on the equation, you use that same multiplier on the *ΔH*.** For example, suppose you want to write the thermochemical equation for the production of 2 moles of carbon dioxide. You have to multiply the preceding entire equation, including the *ΔH*, by 2.

  $$2 \times [C(s) + O_2(g) \rightarrow CO_2(g)] \qquad \Delta H = -393.5 \text{ kJ}$$

  $$2\ C(s) + 2\ O_2(g) \rightarrow 2CO_2(g) \qquad \Delta H = -787.0 \text{ kJ}$$

- **If you reverse a thermochemical equation, you must reverse the sign on the enthalpy change.** For example, suppose we wanted to write the thermochemical equation for the decomposition of a mole of carbon dioxide. We could simply reverse the initial equation for the formation of a mole of carbon dioxide and change the sign of the ΔH.

  $$CO_2(g) \rightarrow C(s) + O_2(g) \qquad \Delta H = +393.5 \text{ kJ}$$

You can measure enthalpies of reaction using a calorimeter. However, you can also calculate the values. *Hess's Law* states that if you express a reaction in a series of steps, then the enthalpy change for the overall reaction is simply the sum of the enthalpy changes of the individual steps. If, in adding the equations of the steps together, reversing one of the given reaction is necessary, then you need to reverse the sign of the *ΔH*. In addition, you need to pay particular attention if you must adjust the reaction stoichiometry.

It really doesn't matter whether or not the steps that I use are the actual ones in the mechanism (pathway) of the reaction because $\Delta H_{reaction}$ is a state function. The pathway is unimportant; all that matters is the initial and final states.

Using the following information:

$$C(s) + O_2(g) \rightarrow CO_2(g) \quad \Delta H = -393.5 \text{ kJ}$$

$$2\ H_2(g) + O_2(g) \rightarrow 2\ H_2O(l) \qquad \Delta H = -483.6 \text{ kJ}$$

$$C_3H_8(g) + 5O_2(g) \rightarrow 3\ CO_2(g) + 4\ H_2O(l) \quad \Delta H = -2043.0 \text{ kJ}$$

Find the enthalpy change for $3\ C(s) + 4\ H_2(g) \rightarrow C_3H_8(g)$.

To solve this problem, follow these steps:

1. **Multiply the first equation by 3 because you need 3 carbon atoms.**

   $$3\ (C(s) + O_2(g) \rightarrow CO_2(g))\ \Delta H = 3(-393.5 \text{ kJ})$$

2. **Reverse the third equation and change the sign of ΔH because C_3H_8 appears on the product side.**

   $$3\ CO_2(g) + 4\ H_2O(l) \rightarrow C_3H_8(g) + 5\ O_2(g)\ \Delta H = -(-2043.0 \text{ kJ})$$

3. **Multiply the second equation by 2 because you need 4 hydrogen molecules.**

$$2\,(2\,H_2(g) + O_2(g) \rightarrow 2\,H_2O(l)) \qquad \Delta H = 2(-483.6\text{ kJ})$$

4. **Add the reactions and ΔH's.**

$$3\,C(s) + 4\,H_2(g) \rightarrow C_3H_8(g) \qquad \Delta H = -105.2\text{ kJ}$$

Investigating Entropy

A negative value of the enthalpy of reaction means that the reaction is exothermic and should be spontaneous. However, a significant number of endothermic reactions are also spontaneous. In addition to considering enthalpy, you also need to consider entropy, which I discuss in the following sections.

Defining entropy

Entropy is a measure of the degree that energy disperses from a localized state to a more widely spread-out state. You may also think of entropy (S) as a measure of the disorder of a system. The Second Law of Thermodynamics states that all processes that occur spontaneously move in the direction of an increase in entropy of the universe (system + surroundings) and that the entropy of the universe is continually increasing. (Refer to the later section, "Checking for spontaneity: The second law" for more details about his law.) For a reversible process, a system at equilibrium, $\Delta S_{universe} = 0$. Mathematically, you can state this as:

$$\Delta S_{universe} = \Delta S_{system} + \Delta S_{surroundings} > 0 \text{ for a spontaneous process}$$

Entropy, S, is a state function. Like all state functions, only the final and initial states are important. The greater the number of arrangements of the particles (microstates) present, the greater the entropy. A *microstate* is one single combination of positions and energies that the particles might occupy. As the number of possible microstates increases, so does the entropy.

Gases have a great number of freely moving particles. Solids, on the other hand, severely restrict the motion of the particles present to maintain a lattice. The greater freedom that gas particles have leads to gases having more microstates and thus more entropy than solids. Liquids tend to have intermediate entropy values, which are nearer to those of solids. Large complicated molecules tend to have greater entropy than simpler molecules in the same phase. Solids have some entropy because the particles present tend to vibrate in place. This motion, though small, leads to some finite amount of entropy. If you could force the molecules to remain motionless, they would have zero entropy.

The qualitative entropy change (increase or decrease of entropy) for a system can sometimes be determined using a few simple rules:

- Entropy increases when the number of molecules increases during a reaction.
- Entropy increases with an increase in temperature.
- Entropy increases when a gas forms from either a liquid or solid.
- Entropy increases when a liquid forms from a solid.

You tabulate the standard molar entropies ($S°$) of elements and compounds the same way as you tabulate the $\Delta H°$ values (see the previous section). This entropy is associated with 1 mole of a substance in its standard state. Unlike the enthalpies, the entropies of elements aren't zero.

Determining spontaneity

In many processes, knowing the change in entropy, ΔS, is important. When the total entropy change is positive, the process has a greater likelihood of being spontaneous. One way to determine this value is by subtracting the total entropy of the reactants from the total entropy of the products. Doing so is similar to the method used to determine the enthalpy change for a reaction (refer to the previous section, "Discussing Enthalpy").

For a reaction, you can calculate the standard entropy change in the same fashion as the enthalpies of reaction:

$$\Delta S° = \Sigma\Delta S°_{\text{products}} - \Sigma\Delta S°_{\text{reactants}}$$

The following example problem shows how to determine the entropy change in a reaction.

Calculate $\Delta S°$ for the following reaction:

$C_3H_8(g) + 5O_2(g) \rightarrow 3\ CO_2(g) + 4\ H_2O(l)$

To solve this problem, stick to these steps:

1. **Find the standard entropy values for each of the substances in the reaction.**

These values come from the appropriate tables in your chemistry textbook. (I place these standard entropy values below the appropriate substances in the equations.)

$C_3H_8(g) + 5O_2(g) \rightarrow 3\ CO_2(g) + 4\ H_2O(l)$

[270.3 + 5(205.2) 3(213.8) + 4(70.0)] J/mol × K

Unlike enthalpy values, all standard entropy values are non-zero and positive.

2. **Add together the entropy values of the products and then the reactants.**

   Products: [3(213.8) + 4(70.0)] = 931.4 J/K

   Reactants: [270.3 + 5(205.2)] = 1,296.3 J/K

3. **Subtract the total entropy value of the reactants from the total entropy value of the products.**

   [931.4J/K – 1,296.3 J/K)] = –364.9 J/K

Tackling the Laws of Thermodynamics

Three laws define and restrict the field of thermodynamics. You may have studied the first law during your Chemistry I course. No matter though, I want to make sure you have a good grasp of them because you'll need to refer to them as you solve thermodynamics problems during your Chemistry II class.

Remaining constant: The first law

The *First Law of Thermodynamics* states that the total energy of the universe is constant. In other words, this law is simply the Law of Conservation of Energy. You can mathematically state this law as

$$\Delta E_{universe} = \Delta E_{system} + \Delta E_{surroundings} = 0$$

The thermochemistry problems you worked in Chem I were an application of the First Law.

Checking for spontaneity: The second law

The *Second Law of Thermodynamics* states that a spontaneous process must be accompanied by an increase in entropy. According to this law, the entropy of the universe is continually increasing. I've already worked some Second Law problems in this chapter, but in this section, I delve into the implications of the Second Law in a little more detail.

Application of the second law provides a means of determining if a process is spontaneous or nonspontaneous. To determine if a process is spontaneous, you must determine the total entropy change. Thus, you need to know the entropy changes in the system, ΔS_{system}, and the surroundings, $\Delta S_{surroundings}$. The total entropy change is $\Delta S_{universe}$. For a spontaneous process, the following relationship must be true:

$$\Delta S_{universe} = \Delta S_{system} + \Delta S_{surroundings} > 0$$

The ΔS_{system} or $\Delta S_{surroundings}$ may be negative (nonspontaneous), and the process may still be spontaneous as long as the total change is positive. The entropy change of the system depends upon the process taking place within the system. The entropy change of the surroundings depends upon the heat exchange between the system and the surroundings. Heat energy leaving the system causes an increase in the entropy of the surroundings, and heat energy entering the system results in a decrease in the entropy of the surroundings. You can determine the entropy change of the surroundings by the following relationship:

$$\Delta S_{surroundings} = -\Delta H/T$$

If you consider the freezing of a liquid, you can see how this process may be classified as spontaneous or nonspontaneous. When a liquid freezes, it changes from the liquid state to the solid state. The entropy of the liquid state is higher than the entropy of the solid. Therefore, the change of a substance from the liquid state to the solid state involves a decrease (negative) in entropy. Thus, if you only consider the entropy change of the system, the process should always be nonspontaneous.

However, when a liquid freezes, you must remove a quantity of energy from the liquid. This energy is the enthalpy of crystallization, which is the reverse (opposite sign) of the enthalpy of fusion. You can represent the energy lost by the system as $-\Delta H$. Substituting this into the entropy of the surroundings gives you:

$$\Delta S_{surroundings} = -(-\Delta H)/T > 0$$

The entropy change of the surroundings depends upon both the enthalpy change and the temperature. Because the enthalpy of crystallization is always exothermic, the entropy change of the surrounding is always positive.

Now consider the overall entropy change:

$$\Delta S_{universe} = \Delta S_{system} + \Delta S_{surroundings}$$

The ΔS_{system} is negative, while the $\Delta S_{surroundings}$ is positive. The value of the second term varies with the temperature. As the temperature lowers, you divide the enthalpy change by a smaller number, thus the entropy change of the surroundings increases as the temperature decreases.

When the temperature becomes sufficiently low, the entropy change of the surroundings cancels the entropy change of the system. That is:

$$\Delta S_{universe} = \Delta S_{system} + \Delta S_{surroundings} = 0$$

At an even lower temperature:

$$\Delta S_{universe} = \Delta S_{system} + \Delta S_{surroundings} > 0$$

The name for the temperature at which the overall entropy change first reaches zero is the *freezing point*. Therefore, you can see that a substance spontaneously freezes when the temperature is at or below the freezing point. The process is nonspontaneous above the freezing point.

The repeating structure of a crystalline solid, the *lattice,* restricts the movement of the particles in a solid, which leads to solids having lower entropy than either liquids or gases. If you lower the temperature of a solid, you lower the kinetic energy of the particles present. As the kinetic energy of the particles decreases so does their movement. If the temperature is sufficiently low (absolute zero), all motion of the particles ceases. If the particles aren't moving, their entropy is zero, which leads to the next section.

Zeroing in: The third law

The *Third Law of Thermodynamics* states that for a pure crystalline substance at 0 K, the entropy is zero. The implication of the Third Law of Thermodynamics is that all substances have zero entropy at absolute zero. At any temperature above absolute zero, the substance will have entropy greater than zero.

Because of the zero point, you can determine absolute entropy values. You couldn't do so with enthalpies of formation, because no actual zero point exists. All heats of formation are relative values. The relative values have a Δ, while the absolute values don't, which is why you see ΔH_f° values tabulated versus S° values.

Predicting Spontaneity for Enthalpy and Entropy Changes

If you at least knew the signs of the enthalpy change and the magnitude of the entropy change, you can begin to make predictions about the spontaneity of a process. Doing so is important because it allows you to judge the practicality of conducting a specific reaction. It's also important because your instructor will want you to be able to do these determinations. Four possible combinations of the signs exist for the enthalpy and entropy changes:

Enthalpy (ΔH)	*Entropy (ΔS) (system)*	*Implication*
–	+	Always spontaneous
+	–	Never spontaneous
–	–	Spontaneous at low temperature
+	+	Spontaneous at high temperature

As this table shows you, if the enthalpy is negative and the entropy is positive, both factors favor a spontaneous process and the process is spontaneous under all conditions. If the reverse signs are present where enthalpy is positive and entropy is negative, both factors favor a nonspontaneous process so that the process isn't spontaneous under all conditions.

If the signs of the enthalpy change and the entropy change are both either positive or negative, the two processes are working against each other. When both are negative, the enthalpy change favors a spontaneous process and the entropy change favors a nonspontaneous process. Thus, the process can only be spontaneous if the entropy contribution is less than the enthalpy contribution. The entropy contribution increases with increasing temperature as I show you earlier in this chapter. For this reason, decreasing the entropy contribution by lowering the temperature is possible. When the temperature is sufficiently low, the enthalpy contribution predominates, and the process becomes spontaneous.

When both the enthalpy and entropy contributions are positive, the entropy change favors a spontaneous process, while the enthalpy change favors a nonspontaneous process. If you wish to make the process spontaneous, you must increase the entropy contribution. To increase the entropy contribution, you need to increase the temperature. As soon as the temperature becomes sufficiently high, the entropy contribution becomes greater than the enthalpy contribution, and the process changes from nonspontaneous to spontaneous.

Grasping Gibb's Free Energy

Instead of treating enthalpy and entropy separately when predicting spontaneity, combining them into one term is more convenient. To do so, you use the Gibb's Free Energy (ΔG), which is $G = H - TS$. The *Gibb's Free Energy* is the maximum amount of work that is available from a spontaneous process or the minimum amount of work required to force a nonspontaneous process to occur. The standard Gibb's Free Energy for an element in its standard state, like the standard enthalpy of formation, is zero by definition.

In general, the change in the Gibb's Free Energy is needed in order to predict spontaneity. Two general relationships allow you to determine the change in the Gibb's Free Energy, ΔG. These relationships are

$$\Delta G = \Delta H - T\Delta S$$

$$\Delta G = \Delta G_{final} - \Delta G_{initial}$$

The Gibb's Free Energy indicates the spontaneity of a process. A spontaneous process has $\Delta G < 0$, while a nonspontaneous process has $\Delta G > 0$. When $\Delta G = 0$, the process is at equilibrium.

If you want to calculate the standard Gibb's Free Energy change for a specific reaction, you can separately calculate the enthalpy change and the entropy change. You can then enter these values into

$$\Delta G^{\circ}_{reaction} = \Delta H^{\circ}_{reaction} - T\Delta S^{\circ}_{reaction}$$

The degree symbol, °, indicates standard (state) conditions. Standard (state) conditions are 1 atm (or 1 bar) for gases and 1 M for solutions at the temperature of interest, which you can assume to be 25°C.

You can also find the change in the Gibb's Free Energy by using tables of thermodynamic data commonly found in your textbook or the CRC Handbook. These tables list standard values for the Gibb's Free Energy of a variety of substances. After you have the individual values, you use this relationship:

$$\Delta G^{\circ}_{reaction} = \Sigma\Delta G^{\circ}_{products} - \Sigma\Delta G^{\circ}_{reactants}$$

Calculate the Gibb's Free Energy change, ΔG°, for the following reaction and predict whether the reaction will be spontaneous or nonspontaneous:

$$2\ NH_4Cl(s) + CaO(s) \rightarrow CaCl_2(s) + H_2O(l) + 2\ NH_3(g)$$

To solve this problem, follow these steps:

1. **Find the appropriate standard Gibb's Free Energy of formation below each substance.**

   You can find them in your textbook or in the CRC Handbook.

   $$2\ NH_4Cl(s) + CaO(s) \rightarrow CaCl_2(s) + H_2O(l) + 2\ NH_3(g)$$

   [–203.9 –604.2 –750.2 –237.2 –16.6] kJ/mole

2. Use the following relationship and solve.

$$\Delta G^\circ_{reaction} = \Sigma\Delta G^\circ_{products} - \Sigma\Delta G^\circ_{reactants}$$

Doing so gives you:

$$\Delta G^\circ_{reaction} = [-750.2 + (-237.2) + 2(-16.6)] - [2(-203.9) + (-604.2)]$$

$$= -8.6 \text{ kJ/mole}$$

Because the Gibb's Free Energy is less than zero, the reaction is spontaneous.

Checking Out Nonstandard Conditions

Under nonstandard conditions, the thermodynamic values differ from the standard values that I discuss in the previous section. In many cases, the deviation in nonstandard values is small, but you can make corrections for nonstandard conditions.

You can determine a nonstandard value for the Gibb's Free Energy by adjusting the standard value. The following equation allows you to adjust a standard free energy change, ΔG°, to a nonstandard free energy change, ΔG:

$$\Delta G = \Delta G^\circ + RT \ln Q$$

ΔG is the free energy under nonstandard conditions, ΔG° is the standard free energy, R is the gas constant (8.314 J/mol K), T is the temperature in Kelvin, and Q is the reaction quotient.

The *reaction quotient, Q,* is a fraction with information concerning the products in the numerator and information concerning the reactants in the denominator. In the numerator, the product's concentrations are multiplied together with each one being raised to the power of the coefficient in the balanced chemical equation. The denominator consists of the product of the reactant concentrations, each one of them raised to the power of the coefficient in the balanced chemical equation. The reaction quotient depends upon the process under consideration.

Calculate ΔG for the following reaction at 500. K:

$$2\ NO(g) + O_2(g) \rightarrow 2\ NO_2(g)$$

The concentrations of the gases are: 2.00 M NO, 0.500 M O_2, and 1.00 M NO_2. Assume that you can neglect the variation of the Gibb's Free Energy with temperature.

To solve this problem, stick to these steps:

1. **Substitute the concentrations of the gas into the formula to get the following.**

   $2\ NO(g) + O_2(g) \rightarrow 2\ NO_2(g)$

   2.00 M 0.500 M 1.00 M T = 500. K

2. **Add the Gibb's Free Energy information from an appropriate table of thermodynamics data from either your textbook or the CRC Handbook to get this.**

   $2\ NO(g) + O_2(g) \rightarrow 2\ NO_2(g)$

   2.00 M 0.500 M 1.00 M T = 500. K

   [86.71 0.000 51.84] kJ/mole

3. **Calculate the Gibb's Free Energy change using the relationship.**

   $\Delta G^\circ_{reaction} = \Sigma \Delta G^\circ_{products} - \Sigma \Delta G^\circ_{reactants}$

   $\Delta G^\circ_{reaction} = 2(51.84) - [2(86.71) + 0.000] = -69.74$ kJ/mole

   For this reaction, the reaction quotient is

   $$Q = \frac{[NO_2]^2}{[NO]^2[O_2]}$$

4. **Adjust the Gibb's Free Energy for nonstandard conditions using.**

   $$\Delta G = \Delta G^\circ + RT \ln Q$$

   $$\Delta G = \Delta G^\circ + RT\ \ln \frac{[NO_2]^2}{[NO]^2[O_2]}$$

5. **Enter the values to get the following.**

   $$\Delta G = \left(-69.74 \frac{kJ}{mole}\right) \times \left(\frac{1000\ J}{kJ}\right) + \left(8.314 \frac{J}{mole} \times K\right) \times (500.\ K) \times \ln \frac{[1.00]^2}{[2.00]^2[0.500]}$$

   $\Delta G = -7.262 \times 10^4$ J/mole

Revisiting the Haber-Bosch Process

The *Haber-Bosch Process,* sometimes just called the Haber Process, is the industrial chemical process by which atmospheric nitrogen gas combines with hydrogen gas to produce ammonia. The following equation shows this process:

$$N_2(g) + 3\ H_2(g) \rightleftarrows 2\ NH_3(g)$$

Understanding this process is essential because ammonia is useful in the production of many important substances, including fertilizers and explosives. An examination of the thermodynamics of this process gives some insight to the nature of the chemical system.

You can calculate the standard enthalpy change for the Haber process by the following:

$$\Delta H^\circ_{reaction} = \Sigma\Delta H^\circ_{products} - \Sigma\Delta H^\circ_{reactants}$$

$$N_2(g) \quad +3\,H_2(g) \quad \rightleftarrows 2\,NH_3(g)$$

$$[1\text{ mol}(0) + 3\text{ mol }(0)\,]\text{kJ/mol} \quad [\,2\text{ mol }(-46.11]\text{ kJ/mol}$$

$$[0]\text{ kJ} \quad [-92.22]\text{ kJ}$$

$$\Delta H^\circ = -92.22\text{ kJ}$$

The negative value of the enthalpy indicates a spontaneous reaction (depending on the temperature).

You can calculate the entropy change in a similar fashion:

$$N_2(g) \quad +3\,H_2(g) \quad \rightleftarrows \quad 2\,NH_3(g)$$

$$[1\text{ mol }(191.5) + 3\text{ mol }(130.6)\,]\text{J/mol K} \quad [2\text{ mol}(192.3]\text{ J/mol K}$$

$$[583.3]\text{ J/K} \quad [384.6]\text{ J/K}$$

$$\Delta S^\circ = [384.6\text{ J/K}] - [583.3\text{ J/K}] = -198.7\text{ J/K}$$

The negative entropy value indicates a nonspontaneous reaction. However, the negative enthalpy implies the opposite. These two opposing factors lead to the reaction being spontaneous at low temperatures. The question is how low must the temperature be for the reaction to be spontaneous? Calculating this temperature using the Gibb's Free Energy relationship is possible with this equation:

$$\Delta G = \Delta H - T\Delta S$$

To do so, follow these simple steps:

1. **Set $\Delta G = 0$ and put in the values for the enthalpy and entropy changes and the appropriate conversions.**

   $$\Delta G = 0 = \Delta H - T\Delta S$$

2. **Rearrange this equation to**

   $$T = \frac{\Delta H}{\Delta S}$$

3. **Enter the enthalpy and entropy values and calculate the temperature.**

$$T = \frac{(-92.22\ kJ)}{-198.7\ J/K} \times \frac{1000\ J}{kJ} = 464.1\ K\ (188.1°C)$$

This is the highest temperature at which this reaction will be spontaneous.

Industrially, the reaction conditions for the Haber process are a pressure of 200 atmospheres and at a temperature between 380 and 450°C. The high pressure is an application of Le Chatelier's principle to shift the equilibrium toward the product (see Chapter 8 for a discussion of chemical equilibrium and Le Chatelier's principle). Although contrary to Le Chatelier's principle, engineers run the reaction at a moderately high temperature in order to increase the speed of the reaction. Engineers also use a catalyst to speed up the reaction even further. With these conditions and careful engineering, the Haber process can operate at greater than 99 percent efficiency.

Chapter 12

Causing Electrons to Flow: Electrochemistry

In This Chapter

- Grasping redox reactions
- Explaining cells and cell potentials
- Taking down the Nernst equation
- Figuring out what electrolysis is
- Putting together pieces of power: Electrochemical cells
- Comprehending electroplating
- Burning fuels and food

Many of the things people deal with in real life are related either directly or indirectly to electrochemical reactions. Think of all the things around you that contain batteries — flashlights, watches, automobiles, calculators, PDAs, pacemakers, cell phones, toys, garage door openers, and so on.

Do you drink from an aluminum can? The aluminum was extracted by an electrochemical reaction. Do you have a car with a chrome bumper? That chrome is electroplated onto the bumper, just like the silver on Grandmother Grace's tea service or the gold on that five-dollar gold chain. Do you watch television, use electric lights or an electric blender, or have a desktop computer? More than likely the electricity you use for these things is generated from the combustion of some fossil fuel. Combustion is a redox reaction. So are respiration, photosynthesis, and many other biochemical processes that you and I depend upon for life. Electrochemical and redox reactions surround everyone.

In this chapter, I explain redox reactions, go through the balancing of this type of equation, and then show you some applications of redox reactions in an area of chemistry called electrochemistry.

Following Those Pesky Electrons: Redox Reactions

Redox reactions — reactions in which there are a simultaneous transfer of electrons from one chemical species to another — are really composed of two different reactions: *oxidation* (a loss of electrons) and *reduction* (a gain of electrons). These reactions are coupled, as the electrons that are lost in the oxidation reaction are the same electrons that are gained in the reduction reaction. In fact, these two reactions (reduction and oxidation) are commonly called *half-reactions,* because it takes these two halves to make a whole reaction, and the overall reaction is called a *redox* (*red*uction/*ox*idation) reaction. These sections take a closer look at redox reactions, including the individual processes and how they combine into an overall reaction and how to balance these redox reactions.

Losing electrons: Oxidation

One part of a redox reaction is the oxidation half. Oxidation has three definitions, which can help you understand what goes on during this process:

- The loss of electrons
- The gain of oxygen
- The loss of hydrogen

Because I typically deal with electrochemical cells, I normally use the definition that describes the loss of electrons. The other definitions are useful in processes such as combustion and photosynthesis.

Loss of electrons

One way to define oxidation is where a chemical substance loses electrons going from the reactant to the product during a reaction. For example, when sodium metal reacts with chlorine gas to form sodium chloride (NaCl), the sodium metal loses an electron, which the chlorine gains. The following equation shows sodium losing the electron:

$$Na(s) \rightarrow Na^+ + e^-$$

When sodium loses the electron, chemists say that the sodium metal has been oxidized to the sodium cation. (A *cation* is an ion with a positive charge due to the loss of electrons; see Chapter 3.)

Reactions of this type are quite common in *electrochemical reactions,* reactions that produce or use electricity. (For more info about electrochemical reactions, flip to the section "Power on the Go: Common Electrochemical Cells," later in this chapter.)

Gain of oxygen

Sometimes, in certain oxidation reactions, oxygen has been gained in going from reactant to product. Reactions where the gain of oxygen is more obvious than the gain of electrons include combustion reactions (*burning*) and the *rusting* of iron. Here are two examples:

$$C(s) + O_2(g) \rightarrow CO_2(g) \quad \text{(burning of coal)}$$

$$2\,Fe(s) + 3\,O_2(g) \rightarrow 2\,Fe_2O_3(s) \quad \text{(rusting of iron)}$$

In these cases, chemists say that the carbon and the iron metal have been oxidized (gained oxygen).

Loss of hydrogen

In other reactions, oxidation can best be seen as the loss of hydrogen. Methyl alcohol (wood alcohol) can be oxidized to formaldehyde as the following equation shows:

$$CH_3OH(l) \rightarrow CH_2O(l) + H_2(g)$$

In going from methanol to formaldehyde, the compound went from having four hydrogen atoms to having two hydrogen atoms.

Finding electrons: Reduction

The other half of the redox reaction includes reduction. Like oxidation, reduction has three definitions you can use to describe the process:

- The gain of electrons
- The loss of oxygen
- The gain of hydrogen

Gain of electrons

Reduction is often seen as the gain of electrons. In the process of electroplating silver onto a teapot (see the section on electroplating, later in this chapter), for example, the silver cation is reduced to silver metal by the gain of an electron. The following equation shows the silver cation gaining the electron:

$$Ag^+ + e^- \rightarrow Ag$$

When it gains the electron, chemists say that the silver cation has been reduced to silver metal.

Loss of oxygen

In other reactions, during reduction the loss of oxygen happens in going from reactant to product. For example, iron ore (primarily rust, Fe_2O_3) is reduced to iron metal in a blast furnace by a reaction with carbon monoxide:

$$Fe_2O_3(s) + 3\ CO(g) \rightarrow 2\ Fe(s) + 3\ CO_2(g)$$

The iron has lost oxygen, so chemists say that the iron ion has been reduced to iron metal.

Gain of hydrogen

In certain cases, you can also describe a reduction as the gain of hydrogen atoms in going from reactant to product. For example, carbon monoxide and hydrogen gas can be reduced to methyl alcohol:

$$CO(g) + 2\ H_2(g) \rightarrow CH_3OH(l)$$

In this reduction process, the CO has gained the hydrogen atoms.

One's loss is the other's gain

Neither oxidation nor reduction can take place without the other. When those electrons are lost, something has to gain them. So when the substance that undergoes oxidation loses an electron, the substance that undergoes reduction gains it.

To help yourself remember what happens to electrons, memorize the phrase "LEO goes GER" (*L*ose *E*lectrons *O*xidation; *G*ain *E*lectrons *R*eduction).

Consider, for example, the *net-ionic equation* (the equation showing just the chemical substances that are changed during a reaction) for a reaction with zinc metal and an aqueous copper(II) sulfate solution:

$$Zn(s) + Cu^{2+} \rightarrow Zn^{2+} + Cu(s)$$

This overall reaction is really composed of two half-reactions:

$Zn(s) \rightarrow Zn^{2+} + 2e^-$ (oxidation half-reaction involving the loss of electrons)

$Cu^{2+} + 2e^- \rightarrow Cu(s)$ (reduction half-reaction involving the gain of electrons)

Zinc loses two electrons; the copper(II) cation gains those same two electrons. Zn is being oxidized. Without Cu^{2+} present, nothing will happen. The following two are important for the redox equation:

- **Oxidizing agent:** The *oxidizing agent* is the substance that's being reduced. The copper cation is the oxidizing agent, which is necessary for the oxidation process to proceed. The oxidizing agent accepts the electrons from the chemical species that is being oxidized.
- **Reducing agent:** The *reducing agent* is the substance that's being oxidized; this substance furnishes the electrons. Cu^{2+} is reduced as it gains electrons. In this case, the reducing agent is zinc metal.

Both the oxidizing and reducing agents are on the left (reactant) side of the redox equation.

Playing the numbers: Oxidation numbers

Oxidation numbers are book keeping numbers. They allow chemists to do things such as balance redox equations. Oxidation numbers are positive or negative numbers, but don't confuse them with charges on ions or valences. These rules help you keep track of the oxidation numbers:

- **Rule 1:** The oxidation number of an element in its free (uncombined) state is zero (for example, Al(*s*) or Zn(*s*)). This rule is also true for elements found in nature as *diatomic* (two-atom) elements (H_2, O_2, N_2, F_2, Cl_2, Br_2, or I_2) and for sulfur, found as S_8.
- **Rule 2:** The oxidation number of a *monatomic* (one-atom) ion is the same as the charge on the ion (for example, $Na^+ = +1$, $S^{2-} = -2$).
- **Rule 3:** The sum of all oxidation numbers in a neutral compound is zero. The sum of all oxidation numbers in a *polyatomic* (many-atom) ion is equal to the charge on the ion. This rule often allows chemists to calculate the oxidation number of an atom that may have multiple oxidation states, if the other atoms in the ion have known oxidation numbers.
- **Rule 4:** The oxidation number of an alkali metal (IA family) in a compound is +1; the oxidation number of an alkaline earth metal (IIA family) in a compound is +2.
- **Rule 5:** The oxidation number of oxygen in a compound is usually –2. If, however, the oxygen is in a class of compounds called *peroxides* (for example, hydrogen peroxide, or H_2O_2), then the oxygen has an oxidation number of –1. If the oxygen is bonded to fluorine, the number is +1.
- **Rule 6:** The oxidation state of hydrogen in a compound is usually +1. If the hydrogen is part of a *binary metal hydride* (compound of hydrogen and some metal), then the oxidation state of hydrogen is –1.

- **Rule 7:** The oxidation number of fluorine is always –1. Chlorine, bromine, and iodine usually have an oxidation number of –1, unless they're in combination with an oxygen or fluorine. (For example, in ClO^-, the oxidation number of oxygen is –2 and the oxidation number of chlorine is +1; remember that the sum of all the oxidation numbers in ClO^- have to equal –1.)

These rules give you another way to define oxidation and reduction — in terms of oxidation numbers. For example, consider this reaction, which shows oxidation by the loss of electrons:

$$Zn(s) \rightarrow Zn^{2+} + 2e^-$$

Notice that the reactant (zinc metal) has an oxidation number of zero (rule 1), and the product (zinc cation) has an oxidation number of +2 (rule 2). In general, you can say that a substance is oxidized when its oxidation number *increases.*

Reduction works the same way. Consider this reaction:

$$Cu^{2+} + 2e^- \rightarrow Cu(s)$$

The copper is going from an oxidation number of +2 to zero. A substance is reduced if its oxidation number *decreases.*

Balancing redox equations

Redox equations are often so complex that the *inspection method* (the fiddling with coefficients method) of balancing chemical equations doesn't work well with them. So chemists have developed two different methods of balancing redox equations:

- **Oxidation number method:** One method is called the *oxidation number method.* It's based on the changes in oxidation numbers that take place during the reaction. Personally, I don't think this method works nearly as well as the second method because it's sometimes difficult to determine the exact change in the numerical value of the oxidation numbers.
- **Ion-electron method:** This method is also called the *half-reaction* method. This method is easier to figure. With this method, you convert the unbalanced redox equation to the ionic equation and then break it down into two half-reactions — oxidation and reduction. You separately balance and then combine each of these half-reactions to give the balanced ionic equation. Finally, you put the spectator ions into the balanced ionic equation, converting the reaction back to the molecular formula. (For a discussion of molecular, ionic, and net-ionic equations, see my book, *Chemistry For Dummies,* 2nd edition [John Wiley & Sons, Inc.].)

Balance this redox equation with the ion-electron method:

$$Cu(s) + HNO_3(aq) \rightarrow Cu(NO_3)_2(aq) + NO(g) + H_2O(l)$$

Follow these steps to solve. Following these steps precisely and in the order listed is important. Otherwise, you may not be successful in balancing redox equations.

1. **Convert the unbalanced redox reaction to the ionic form.**

   In this reaction, you show the nitric acid in the ionic form, because it's a strong acid (for a discussion of strong acids, see Chapter 9). Copper(II) nitrate is soluble (indicated by *(aq)*), so it's shown in its ionic form. Because NO(g) and water are molecular compounds, they remain shown in the molecular form:

   $Cu(s) + H^+ + NO_3^- \rightarrow Cu^{2+} + 2\ NO_3^- + NO(g) + H_2O(l)$

2. **If necessary, assign oxidation numbers and then write two half-reactions (oxidation and reduction) showing the chemical species that have had their oxidation numbers changed.**

   In some cases, being able to tell what has been oxidized and reduced is easy; but in other cases, it isn't as easy. Start by going through the example reaction and assigning oxidation numbers. You can then use the chemical species that have had their oxidation numbers changed to write your unbalanced half-reactions:

   $Cu(s) + H^+ + NO_3^- \rightarrow Cu^{2+} + 2\ NO_3^- + NO(g) + H_2O(l)$

   0 +1 +5(−2)3 +2 +5(−2)3 +2 -2 (+1)2 -2

   Look closely. Copper changed its oxidation number (from 0 to 2) and so has nitrogen (from +5 to +2). Your unbalanced half-reactions are

   $Cu(s) \rightarrow Cu^{2+}$

   $NO^{3-} \rightarrow NO$

3. **Balance all atoms, with the exception of oxygen and hydrogen.**

   Waiting until the end to balance hydrogen and oxygen atoms is a good idea, so always balance the other atoms first. You can balance them by inspection — fiddling with the coefficients. (You can't change subscripts; you can only add coefficients.) However, in this particular case, both the copper and nitrogen atoms already balance, with one each on both sides:

   $Cu(s) \rightarrow Cu^{2+}$

   $NO_3^- \rightarrow NO$

4. **Balance the oxygen atoms.**

   - In acid solutions, take the number of oxygen atoms needed and add that same number of water molecules to the side that needs oxygen.
   - In basic solutions, add 2 OH^- to the side that needs oxygen for every oxygen atom that is needed. Then, to the other side of the equation, add half as many water molecules as OH^- anions used.

An acidic solution will have some acid or H^+ shown; a basic solution will have an OH^- present. The example equation is in acidic conditions (nitric acid, HNO_3, which in ionic form is $H^+ + NO_3^-$). You don't have to do anything on the half-reaction involving the copper, because no oxygen atoms are present. But you do need to balance the oxygen atoms in the second half-reaction:

$Cu(s) \rightarrow Cu^{2+}$

$NO_3^- \rightarrow NO + 2\ H_2O$

5. **Balance the hydrogen atoms.**
   - In acid solutions, take the number of hydrogen atoms needed and add that same number of H^+ cations to the side that needs hydrogen.
   - In basic solutions, add one water molecule to the side that needs hydrogen for every hydrogen atom that's needed. Then, to the other side of the equation, add as many OH^- anions as water molecules used.

   The example equation is in acidic conditions. You need to balance the hydrogen atoms in the second half-reaction:

   $Cu(s) \rightarrow Cu^{2+}$

   $4\ H^+ + NO_{3-} \rightarrow NO + 2\ H_2O$

6. **Balance the ionic charge on each half-reaction by adding electrons.**

   The electrons should end up on opposite sides of the equation in the two half-reactions. Remember that you're using ionic charge, not oxidation numbers.

   $Cu(s) \rightarrow Cu^{2+} + 2\ e^-$ (oxidation)

   $3\ e^- + 4\ H^+ + NO_{3-} \rightarrow NO + 2\ H_2O$ (reduction)

7. **Balance the electron loss with the electron gain between the two half-reactions.**

   The electrons that are lost in the oxidation half-reaction are the same electrons that are gained in the reduction half-reaction. The number of electrons lost and gained must be the same. But Step 6 shows a loss of 2 electrons and a gain of 3. So you must adjust the numbers using appropriate multipliers for both half-reactions. In this case, you have to find the lowest common denominator between 2 and 3, which is 6, so multiply the first half-reaction by 3 and the second half-reaction by 2.

   $3 \times [Cu(s) \rightarrow Cu^{2+} + 2\ e^-] = 3\ Cu(s) \rightarrow 3\ Cu^{2+} + 6\ e^-$

   $2 \times [3\ e^- + 4\ H^+ + NO_3^- \rightarrow NO + 2\ H_2O] = 6\ e^- + 8\ H^+ + 2\ NO_3^- \rightarrow 2\ NO + 4\ H_2O$

8. **Add the two half-reactions together and cancel anything common to both sides.**

   The electrons should always cancel.

   $3\ Cu + \mathit{6}\ e^- + 8\ H^+ + 2\ NO_3^- \rightarrow 3\ Cu^{2+} + \mathit{6\ e^-} + 2\ NO + 4\ H_2O$

9. **Convert the equation back to the molecular form by adding the spectator ions (if you started out with the equation in the molecular form).**

   If adding spectator ions to one side of the equation is necessary, add the same number to the other side of the equation. For example, the left side of the equation has 8 H^+. In the original equation, the H^+ was in the molecular form of HNO_3 so you need to add the NO_3^- spectator ions back to it. You have 2 already there on the left, so you simply add 6 more. But to keep everything balanced, you add 6 NO_3^- to the right hand side. Those six nitrate anions will be the spectator ions that you'll need for the Cu^{2+} cation to convert it back to the molecular form that you want.

   $$3\ Cu(s) + 8\ HNO_3(aq) \rightarrow 3\ Cu(NO_3)_2(aq) + 2\ NO(g) + 4\ H_2O(l)$$

10. **Check to make sure that all the atoms are balanced and all the charges are balanced (if working with an ionic equation at the beginning).**

    3 **Cu** = 3

    8 **N** [(3 × 2) + 2] = 8

    24 **O** [(3 × 6) + 2 + 4] = 24

    8 **H** (4 × 2) = 8

    Zero **charge** zero

Reactions that take place in base are just as easy, as long as you follow the rules. Take a quick look at a half-reaction that is occurring in basic solution:

$$MnO_4^- \rightarrow MnO_2(s) \text{ (basic solution)}$$

To figure out, follow these steps:

1. **Add 2 OH^- to the side that needs oxygen for every oxygen atom that is needed.**

   All atoms are balanced with the exception of oxygen. In this case the equation needs two oxygens on the right, so we add 4 OH^- to the right hand side:

   $$MnO_4^- \rightarrow MnO_2(s) + 4\ OH^-$$

2. **To the other side of the equation, add half as many water molecules as OH^- anions used.**

   You added 4 OH^- so now you add 2 H_2O to the left side, ending up with:

   $$2\ H_2O + MnO_4^- \rightarrow MnO_2(s) + 4\ OH^-$$

3. **Balance the charge by adding 3 electrons to the left side.**

   $$3\ e^- + 2\ H_2O + MnO_4^- \rightarrow MnO_2(s) + 4\ OH^-$$

The half-reaction is now balanced by charge and by atoms.

Going Indirect: Clarifying Cells and Cell Potentials

A *indirect* electron transfer is one in which the electrons are forced to flow through a wire to get from the oxidation half-reaction to the reduction half-reaction. For example, in the section "One's loss is the other's gain," I discuss a *direct* electron reaction in which I put a piece of zinc metal into a copper(II) sulfate solution. The copper metal begins spontaneously plating out on the surface of the zinc. The equation for the reaction is

$$Zn(s) + Cu^{2+} \rightarrow Zn^{2+} + Cu$$

In this example, zinc gives up two electrons (becomes oxidized) to the Cu^{2+} ion that accepts the electrons (reducing it to copper metal). Nothing happens if you place a piece of copper metal into a solution containing Zn^{2+}, because zinc gives up electrons more easily than copper.

If you were able to separate those two half-reactions so that when the zinc is oxidized, the electrons it releases are forced to travel through a wire to get to the Cu^{2+} (an indirect electron transfer), you'd have something useful. You'd have a *galvanic* or *voltaic cell,* a redox reaction that produces electricity. In this section, I show you how that Zn/Cu^{2+} reaction may be separated out so you have an *indirect* electron transfer and can produce some useable electricity.

Galvanic cells are commonly called batteries, but sometimes this name is somewhat incorrect. A battery is composed of two or more cells connected together. You put a *battery* in your car, but you put a *cell* into your flashlight.

Looking at the Daniell cell

Here we take a look at a galvanic cell that utilizes this indirect transfer to produce electricity (potential or voltage). Granted, this cell would be inconvenient to put in your mp3 player, but it makes for a nice example.

Figure 12-1 shows a Daniell cell that uses the Zn/Cu^{2+} reaction to produce electricity. (British chemist John Frederic Daniell invented it in 1836.)

The Daniell cell utilizes one container with a piece of zinc metal immersed in a solution of zinc sulfate and a second container that uses a piece of copper metal placed in a solution of copper(II) sulfate. These strips of zinc and copper are called the cell's *electrodes.* They act as a *terminal,* a holding place, for electrons and a physical surface onto which you can attach a wire that connects the electrodes. However, nothing happens until you put a salt bridge between the two containers. The *salt bridge,* normally a U-shaped hollow tube filled with a concentrated salt solution, provides a path for ions to move from one container to the other. This is necessary in order to keep the solutions electrically neutral. It's like connecting only one jumper cable to a battery; the jumper won't work unless you connect a second wire to complete the circuit.

After the salt bridge in connected, electrons start to flow. It's the same basic direct electron exchange as the one I show you at the beginning of this section. Zinc is being oxidized, releasing electrons. These electrons flow through the wire to the copper electrode, where they combine with the Cu^{2+} ions in solution to form copper metal. Copper ions from the copper(II) sulfate solution plate out on the copper electrode, while the zinc electrode is being consumed. The cations in the salt bridge migrate to the container containing the copper electrode to replace the copper ions being consumed, while the anions in the salt bridge migrate toward the zinc side, where they keep the solution containing the newly formed Zn^{2+} cations electrically neutral.

The copper electrode is called the *cathode,* the electrode at which reduction takes place, and is labeled with a "+" sign. The solution in which the cathode is immersed is called the *cathode compartment.* The zinc electrode is called the *anode,* the electrode at which oxidation takes place, and is labeled with a "–" sign. The solution in which the anode is immersed is called the *anode compartment.*

This cell will produce a little over one volt. You can get just a little more voltage if you make the solutions that the electrodes are in very concentrated. But what can you do if you want more voltage, for example, three volts? You have a couple of choices.

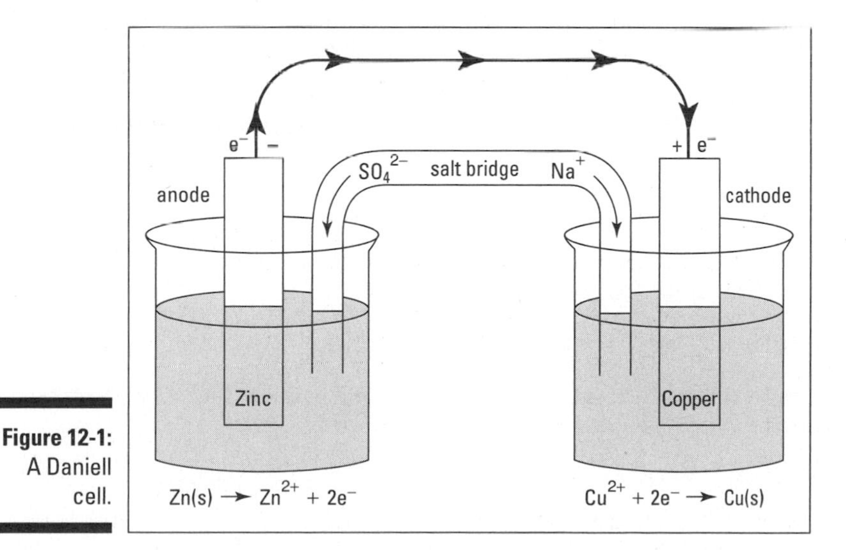

Figure 12-1: A Daniell cell.

- You can hook three of these cells up together and produce two volts.
- You can choose two different metals from the activity series chart that are farther apart than zinc and copper. The farther apart the metals are on the activity series, the more voltage the cell will produce.

Sometimes the half-reaction(s) involved in the cell do not have a solid conductive part to act as the electrode. Maybe there is a gas involved or an ion. In a case such as this an *inert* (inactive) electrode, a solid conducting electrode that does not take part in the redox reaction, is used. Graphite and platinum are common inert electrodes.

Writing cell notation

Cell notation is a shorthand notation of representing a galvanic cell. It is used to keep from drawing the cell as we did with the Daniell cell, replacing a picture with a line of notation. Pretty good trade off for us lazy chemists. To write the cell notation for the Daniell cell you follow these steps:

1. **Write the chemical formula of the anode.**

   $Zn(s)$

2. **Draw a single vertical line to represent the phase boundary between the anode and the anode compartment.**

   $Zn(s)|$

3. **Write the reactive part of the anode compartment with its initial concentration (if known) in parenthesis (assume 1 M in this case).**

   $Zn(s)|Zn^{2+}(1\,M)$

4. **Draw a double vertical line to represent the salt bridge connecting the two electrode compartments.**

   $Zn(s)|Zn^{2+}(1\,M)\|$

5. **Write the reactive part of the cathode compartment with its initial concentration (if known) as shown in parenthesis:.**

   $Zn(s)|Zn^{2+}(1\,M)\|Cu^{2+}(1\,M)$

6. **Draw a single vertical line representing the phase boundary between the cathode compartment and the cathode.**

   $Zn(s)|Zn^{2+}(1\,M)\|Cu^{2+}(1\,M)|$

7. **Write the chemical formula of the cathode.**

   $Zn(s)|Zn^{2+}(1\,M)\|Cu^{2+}(1\,M)|Cu(s)$

If an inert electrode is present, then show where the inert electrode is with its phase boundary. If the electrode components are in the same phase, then separate them by commas; if not, use a vertical phase boundary line. For example, consider the following redox reaction:

$$Ag^{+}(aq) + Fe^{2+}(aq) \rightarrow Fe^{3+}(aq) + Ag(s)$$

The oxidation of the ferrous ion to ferric doesn't involve a solid, so you need to use an inert electrode, such as platinum. Use the steps in this section to solve this problem. The cell notation for this example is:

$$Pt(s)|Fe^{2+}(aq), Fe^{3+}(aq)||Ag^{+}(aq)|Ag(s)$$

Getting a grip on standard reduction potentials

What if I want a cell that produces 1.5 volts or 2.3 volts and so on? There should be a way that helps me decide what half reactions I need to produce a certain voltage or what voltage I can expect if I connect two half-reactions. A table of *standard reduction potentials* helps me answer those questions.

For example, in the discussion of the Daniell cell earlier in this chapter, I indicate that this cell produces a voltage of 1.10 V. This voltage is the difference in potential between the two half-cells. The cell potential (really the half-cell potentials) is dependent upon the concentration and temperature, but initially I simply look at the half-cell potentials at the standard state of 298 K (25°C) and all components in their standard states (1 M concentration of all solutions, 1 atm pressure for any gases and pure solid electrodes).

Half-cell potentials appear in tables as the reduction potentials, that is, the potentials associated with the reduction reaction. The hydrogen half-reaction ($2 H^{+}(aq) + 2 e^{-} \rightarrow H_2(g)$) is the standard and has a value of exactly 0.00 V. You measure all the other half-reactions relative to it; some are positive and some are negative. See Table 12-1 for some selected reduction potentials.

When reviewing Table 12-1, keep the following points in mind:

- All reactions appear in terms of the reduction reaction relative to the standard hydrogen electrode (in bold in the middle of the table).
- The more positive the value of the voltage associated with the half-reaction (E°) the more readily the reaction occurs.
- The strength of the oxidizing agent increases as the value becomes more positive, and the strength of the reducing agent increases as the value becomes more negative.

Table 12-1 Selected Half-Cell Reduction Potentials

Reduction Reactions	***E° (volts)***
$F_2(g) + 2\ e^- \rightarrow 2\ F^-(aq)$	+2.87
$MnO_4^-(aq) + 8\ H^+(aq) + 5\ e^- \rightarrow Mn^{2+}(aq) + 4\ H_2O(l)$	+1.49
$Cl_2(g) + 2\ e^- \rightarrow 2\ Cl^-(aq)$	+1.36
$Cr_2O_7^{2-}(aq) + 14\ H^+(aq) + 6\ e^- \rightarrow 2\ Cr^{3+}(aq) + 7\ H_2O(l)$	+1.33
$O_2(g) + 4\ H+(aq) + 4\ e^- \rightarrow 2\ H_2O(l)$	+1.23
$Br_2(l) + 2\ e^- \rightarrow 2\ Br^-(aq)$	+1.09
$Hg^{2+}(aq) + 2\ e^- \rightarrow Hg(l)$	+0.85
$Ag^+(aq) + 1\ e^- \rightarrow Ag(s)$	+0.80
$Hg_2^{2+}(aq) + 2\ e^- \rightarrow 2\ Hg(l)$	+0.79
$Fe^{3+}(aq) + 1\ e^- \rightarrow Fe^{2+}(aq)$	+0.77
$I_2(s) + 2\ e^- \rightarrow 2\ I^-(aq)$	+0.54
$Cu^+(aq) + 1\ e^- \rightarrow Cu(s)$	+0.52
$Cu^{2+}(aq) + 2\ e^- \rightarrow Cu(s)$	+0.34
$Cu^{2+}(aq) + 1\ e^- \rightarrow Cu^+(aq)$	+0.15
$Sn^{4+}(aq) + 2\ e^- \rightarrow Sn^{2+}(aq)$	+0.15
$2\ H^+(aq) + 2\ e^- \rightarrow H_2(g)$	**+0.0000**
$Fe^{3+}(aq) + 3\ e^- \rightarrow Fe(s)$	–0.04
$Pb^{2+}(aq) + 2\ e^- \rightarrow Pb(s)$	– 0.13
$Sn^{2+}(aq) + 2\ e^- \rightarrow Sn(s)$	– 0.14
$PbSO_4(s) + 2\ e^- \rightarrow Pb(s) + SO_4^{2-}(aq)$	– 0.36
$Fe^{2+}(aq) + 2\ e^- \rightarrow Fe(s)$	– 0.44
$Cr^{3+}(aq) + 3\ e^- \rightarrow Cr(s)$	– 0.74
$Zn^{2+}(aq) + 2\ e^- \rightarrow Zn(s)$	– 0.76
$2\ H_2O(l) + 2\ e^- \rightarrow H_2(g) + 2\ OH^-(aq)$	– 0.83
$Cr^{2+}(aq) + 2\ e^- \rightarrow Cr(s)$	– 0.91
$Al^{3+}(aq) + 3\ e^- \rightarrow Al(s)$	– 1.66
$Mg^{2+}(aq) + 2\ e^- \rightarrow Mg(s)$	– 2.37
$Ca^{2+}(aq) + 2\ e^- \rightarrow Ca(s)$	– 2.87
$K^+(aq) + 1\ e^- \rightarrow K(s)$	– 2.92
$Li^+(aq) + 1\ e^- \rightarrow Li(s)$	– 3.05

I know that table looks imposing, but it's useful. You can use it to write the overall cell reaction and to calculate the standard cell potential, the potential (voltage) associated with the cell at standard conditions. When calculating the standard cell potential from Table 12-1, remember these additional pointers:

- Because the standard cell potential is for a galvanic cell, it must be a positive value, $E^\circ > 0$.
- Because one half-reaction must involve oxidation, you must reverse one of the half-reactions shown in Table 12-1 in order to indicate the oxidation. If you reverse the half-reaction, you must also reverse the sign of the standard reduction potential.
- Because oxidation occurs at the anode and reduction at the cathode, you can calculate the standard cell potential from the standard reduction potentials of the two half reactions involved in the overall reaction by using the equation:

  $$E^\circ_{cell} = E^\circ_{cathode} - E^\circ_{anode} > 0$$

  Both the $E^\circ_{cathode}$ and E°_{anode} values are shown as reduction potentials, used directly from the table without reversing.

After you calculate the standard cell potential, you can then write the reaction by reversing the half reaction associated with the anode (show it as oxidation) and adding the two half-reactions.

Don't forget that the number of electrons lost must equal the number of electrons gained. If they aren't equal, use appropriate multipliers to ensure that they are equal.

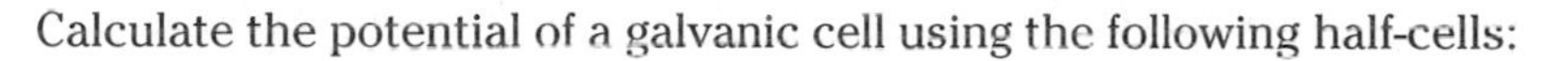

Calculate the potential of a galvanic cell using the following half-cells:

$Zn^{2+} + 2\,e^- \rightarrow Zn(s)$ $\quad E^\circ = -0.76\ V$

$Ag^+ + e^- \rightarrow Ag(s)$ $\quad E^\circ = 0.80\ V$

To solve this problem, stick to these steps:

1. **Calculate the cell potential using**

   $$E^\circ_{cell} = E^\circ_{cathode} - E^\circ_{anode} > 0$$

   Because the cell potential must be positive (a galvanic cell), only one arrangement of -0.25 and 0.80 V can result in a positive value:

   $$E^\circ_{cell} = 0.80\ V - (-0.76\ V) = 1.56\ V$$

   This means that the Zn electrode is the anode and is involved in oxidation.

2. **Reverse the reduction half-reaction involving Zn.**

   To do so, change the sign of the standard half-cell potential and add it to the silver half-reaction. You must multiply the silver half-reaction by two to equalize the electron loss and gain, but not the half-cell potential:

   $Zn(s) \rightarrow Zn^{2+}(aq) + 2\ e^-$ $\quad E^\circ = 0.76\ V$

   $2(Ag^+(aq) + e^- \rightarrow Ag(s))$ $\quad E^\circ = 0.80\ V$

   $Zn(s) + 2\ Ag^+(aq) \rightarrow Zn^{2+}(aq) + 2\ Ag(s)$ $\quad E^\circ_{cell} = 1.56\ V$

Tackling the Nernst Equation

Many times the cell you're studying isn't at standard conditions — commonly the concentrations aren't 1 M. In that case, you calculate the actual cell potential, E_{cell}, by using the *Nernst equation* (named after Walther Nernst, a German chemist, who received the Nobel Prize in 1920 for his work in thermodynamics). He also did pioneering work in electrochemistry and developed the equation that bears his name.

$$E_{cell} = E^\circ_{cell} - \left(\frac{RT}{nF}\right) \ln Q$$

In this equation, R is the ideal gas constant, T is the Kelvin temperature, n is the number of electrons transferred, F is Faraday's constant, and Q is the activity quotient.

The Nernst equation also has a more useful second form, which is as follows:

In the second version of the equation, Faraday's constant, the ideal gas constant values, and a temperature of 298 K are combined to give a new constant 0.0592. If you know the cell reaction, the concentrations of ions (used in calculating the Q value — see Chapters 9 and 10 for a review of the Q value), and the E°_{cell}, then you can calculate the actual cell potential.

The following sections outline different types of problems that you can calculate using the Nernst equation.

Solving a basic problem with this equation

When using the Nernst equation on a cell reaction in which only the half-reactions and concentrations are supplied but not the overall reaction, you have two methods to solve this problem (both methods should lead you to the correct answer):

- **Write the overall redox reaction based upon E° values and then apply the Nernst equation.** If the E_{cell} turns out to be negative, it indicates that the reaction isn't a spontaneous one (an electrolytic cell) or that the reaction is written backwards if it's supposed to be a galvanic cell. If it was supposed to be a galvanic cell, then all you need to do is reverse the overall reaction and change the sign on the E_{cell} to positive.
- **Use the Nernst equation with the individual half-reactions and then combine them depending on whether or not it is a galvanic cell.** The only disadvantage to the second method is that you must use the Nernst equation twice.

Calculate the potential of the following half-cell containing 0.10 M $K_2Cr_2O_7$, 0.20 M $Cr^{3+}(aq)$, and 1.0×10^{-4} M $H^+(aq)$:

$Cr_2O_7^{2-}(aq) + 14\,H^+(aq) + 6\,e^- \rightarrow 2\,Cr^{3+}(aq) + 7\,H_2O(l)$ E° = 1.33 V

To solve this problem, I suggest you use the first method. Follow these steps:

1. **Using the reaction, insert the Q value into the Nernst equation.**

$$E = E^\circ - \frac{0.0592}{6}\log\left(\frac{\left[Cr^{3+}\right]^2}{\left[Cr_2O_7^{2-}\right]\left[H^+\right]^{14}}\right)$$

   Remember that water doesn't appear in the reaction quotient or equilibrium constant.

2. **Put in the values of the concentration and the value for *n*, the number of electrons exchanged, and calculate the cell potential.**

$$E = 1.33\ V - \frac{0.0592}{6}\log\left(\frac{[0.20]^2}{[0.10]\left[1\times10^{-4}\right]^{14}}\right) = 0.78\ V$$

Calculating equilibrium constants

You can also use the Nernst equation to calculate equilibrium constants (Chapters 8 through 10 show the relationship between the concentrations of reactants and products when the reaction is at equilibrium.) The trick here is to realize that if a cell is at equilibrium, then the cell potential, $E_{cell} = 0$ and the Q becomes the K_{eq} so that:

$$E_{cell} = E^\circ_{cell} - \left(\frac{0.0592}{n}\right)\log Q$$

$$0 = E^\circ_{cell} - \left(\frac{0.0592}{n}\right)\log K_c$$

$$\log K_c = \frac{nE^\circ_{cell}}{0.0592}$$

Calculate the equilibrium constant for this reaction:

$Zn(s) + 2\ Ag^+(aq) \rightarrow Zn^{2+}(aq) + 2Ag(s) \quad E^\circ_{cell} = 1.56\ V$

To make this calculation, just stick to these steps:

1. **Write down the appropriate form of the Nernst equation involving the equilibrium constant.**

   The number of electrons transferred, *n,* will equal 2 in this case.

   $$\log K_c = \frac{nE^\circ_{cell}}{0.0592}$$

2. **Put in the values on *n* and the E°_{cell} and compute the value of log K_c.**

   $$\log K_c = \frac{2(1.56)}{0.0592} = 52.7$$

3. **Use the antilog function on your calculator to generate the K_c value.**

   $K_c = 5.01 \times 10^{52}$

   The large positive value of the equilibrium constant indicates that the equilibrium favors the products of the reaction.

Considering other uses for the equation

In addition to solving the actual cell potential, you can use the Nernst equation to calculate the following types of problems:

- **Concentration of one of the reactants from the cell potential measurements:** If you know the actual cell potential and the E°_{cell}, you can calculate Q, the activity quotient. Knowing Q and all but one of the concentrations allows you to calculate the unknown concentration.
- **Concentration cells:** A *concentration cell* is an electrochemical cell in which the same chemical species are used in both cell compartments, but differing in concentration. Because the half reactions are the same, the $E^\circ_{cell} = 0.00$ V. You simply substitute the appropriate concentrations into the Q value of the Nernst equation and calculate the cell potential.

Checking Out Electrolysis

The electricity in an electrolytic cell can decompose water. This process of producing chemical changes by passing an electric current through an electrolytic cell is called *electrolysis* (yes, just like the permanent removal of hair). The overall cell reaction is

$$2\ H_2O(l) \rightarrow 2\ H_2(g) + O_2(g)$$

Electrolytic cells use electricity from an external source to produce a desired redox reaction. Electroplating and the recharging of an automobile battery are examples of electrolytic cells. (Check out the nearby sidebar on electroplating.)

In the operation of both galvanic and electrolytic cells, a reaction occurs on the surface of each electrode. For example, the following reaction takes place at the cathode of a cell:

$$Cu^{2+}(aq) + 2\ e^- \rightarrow Cu(s)$$

The rules of stoichiometry also apply in this case. In electrochemical cells, you must consider not only the stoichiometry related to chemical formulas, but also the stoichiometry related to electric currents. The half-reaction under consideration not only involves 1 mol of each of the copper species, but also 2 mol of electrons. You can construct a mole ratio that includes the term *moles of electrons* or you can construct a mole ratio using *faradays*. A *faraday (F)* is a mole of electrons. Thus, you can use either of the following ratios for the copper half-reaction:

$$\frac{1\ \text{mole Cu}^{2+}}{2\ \text{mole}^-} = \frac{1\ \text{mole Cu}^{2+}}{2\ \text{faradays}}$$

The SI base unit for electric current is the *ampere (A)*. In addition to being an SI base unit, an ampere is a coulomb (C) per second, and a faraday is 96,485 C/mol of electrons. Therefore:

$$1\ \text{A} = \frac{1\ \text{coulomb}}{1\ \text{second}} \text{ and } 1\ \text{faraday}(\text{F}) = \frac{96{,}485\ \text{coulombs}}{1\ \text{mole electrons}}$$

If molten titanium(IV) chloride is electrolyzed by a current of 2.000 A for 4.00 h, how many grams of titanium will be produced?

To solve this problem, follow these steps:

1. **Write the half-reaction.**

   $$Ti^{4+}(l) + 4\ e^- \rightarrow Ti(s)$$

2. **Simply the solution.**

   To simplify the solution write amperes as its definition of coulomb/second.

$$\text{g Ti} = \frac{4.00\ \text{hr}}{1} \times \frac{3600\ \text{s}}{1\ \text{hr}} \times \frac{2.000\ \text{coul}}{\text{s}} \times \frac{1\ \text{mole e}^-}{96{,}485\ \text{coul}} \times \frac{1\ \text{mole Ti}}{4\ \text{moles e}^-} \times \frac{47.87\ \text{g Ti}}{1\ \text{mole Ti}} = 3.57\ \text{g Ti}$$

Making a silver teapot: Electroplating

Electrolytic cells are also used in a process called *electroplating.* In electroplating, a more-expensive metal is plated (deposited in a thin layer) onto the surface of a cheaper metal by electrolysis. Back before plastic auto bumpers became popular, chromium metal was electroplated onto steel bumpers. Those cheap silver teapots you can buy are really made of some cheap metal with an electroplated surface of silver. The following figure shows the electroplating of silver onto a teapot.

A battery is commonly used to furnish the electricity for the process. The teapot acts as the cathode and a bar of silver acts as the anode. The silver bar furnishes the silver ions that are reduced onto the teapot's surface. Many metals and even some alloys can be plated out in this fashion. Everybody loves those plated surfaces, especially without the high cost of the pure metal. (Reminds me of an Olympic athlete who was so proud of his gold metal that he had it bronzed!)

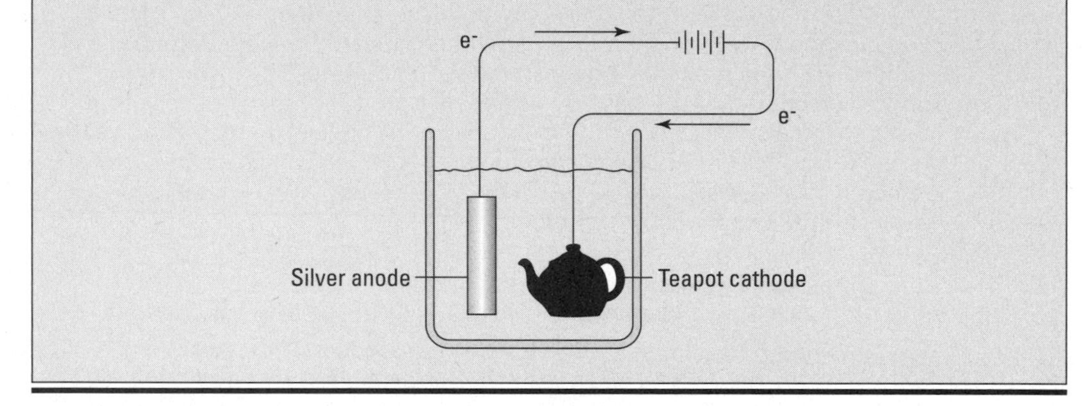

If you electrolyze this solution using a current of 2.000 amps for 4.00 hours, you can form 3.57 grams of titanium. Seems reasonable and the significant figures are correct.

Power on the Go: Common Electrochemical Cells

One of the primary uses of electrochemical reactions is the production of power and in particular portable power. Who would want to use a cellphone or a flashlight or a pacemaker that had to be connected to an extension cord? Consumers want portable power — batteries. The following two sections describe two popular types of cells, the flashlight battery and the automobile battery.

Let the light shine: Flashlight cells

The common flashlight cell (see Figure 12-2), a dry cell (it's not in a solution like a Daniell cell), is contained in a zinc housing that acts as the anode. The other electrode, the cathode, is a graphite rod in the middle of the cell. A layer of manganese oxide and carbon black (one of the many forms of carbon) surrounds the graphite rod, and a thick paste of ammonium chloride and zinc chloride serves as the electrolyte. The cell reactions are

$Zn(s) \rightarrow Zn^{2+} + 2\,e^-$ (anode reaction/oxidation)

$2\,MnO_2(s) + 2\,NH_4^+ + 2\,e^- \rightarrow Mn_2O_3(s) + 2\,NH_3(aq) + H_2O(l)$
(cathode reaction/reduction)

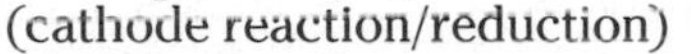

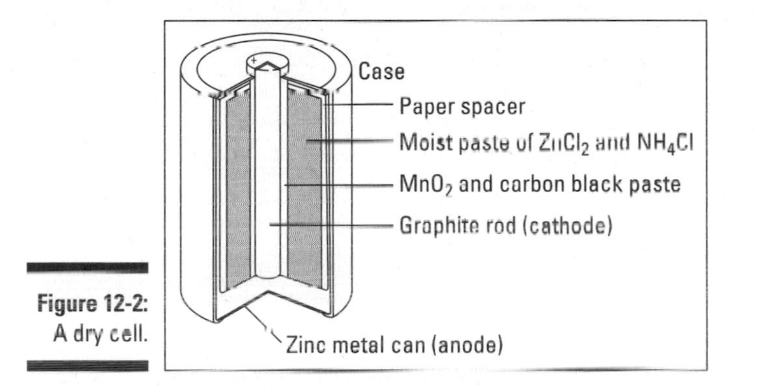

Figure 12-2: A dry cell.

Note that the battery's case is actually one of the electrodes; it's being used up in the reaction. If the case has a thin spot, a hole could form, and the cell could leak the corrosive contents. In addition, the ammonium chloride tends to corrode the metal case, again allowing for the possibility of leakage.

In the alkaline dry cell (alkaline battery), basic (alkaline) potassium hydroxide replaces the acidic ammonium chloride of the regular dry cell. Adding this chemical greatly reduces the corrosion of the zinc case.

Another common electrochemical cell is the small mercury battery commonly used in watches, pacemakers, and so on. In this battery, the anode is zinc and the cathode is steel. Mercury(II) oxide (HgO) and some alkaline paste form the electrolyte. You should dispose of this type of battery carefully, to keep the toxic mercury metal from being released into the environment.

There are cells (batteries) that can be recharged, meaning that the redox reaction can be reversed to regenerate the original reactants. Nickel-cadmium (Ni-Cad) and lithium batteries fall into this category. The most familiar type of rechargeable battery is probably the automobile battery.

Starting your engine: Automobile batteries

The ordinary automobile battery, or lead storage battery, consists of six cells connected in series (see Figure 12-3). The anode of each cell is lead, while the cathode is lead dioxide (PbO_2). The electrodes are immersed in a sulfuric acid (H_2SO_4) solution. When you start your car, the following cell reactions take place:

$$Pb(s) + H_2SO_4(aq) \rightarrow PbSO_4(s) + 2\ H^+ + 2\ e^-\ \text{(anode)}$$

$$2\ e^- + 2\ H^+ + PbO_2(s) + H_2SO_4(aq) \rightarrow PbSO_4(s) + 2\ H_2O(l)\ \text{(cathode)}$$

$$Pb(s) + PbO_2(s) + 2\ H_2SO_4(aq) \rightarrow 2\ PbSO_4 + 2\ H_2O(l)\ \text{(overall reaction)}$$

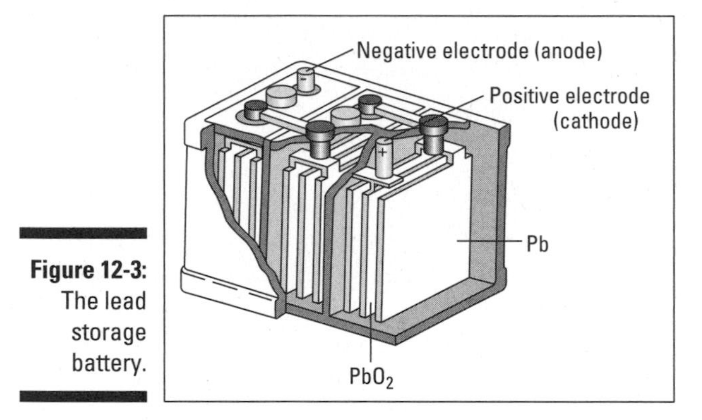

Figure 12-3: The lead storage battery.

When this reaction takes place, both electrodes become coated with solid lead (II) sulfate, and the sulfuric acid is used up.

As soon as the auto has been started and is moving down the highway, the alternator takes over the job of producing electricity for the electrical system and also recharges the battery. The alternator reverses the flow of electrons into the battery and reverses the original redox reactions, regenerating the lead and lead dioxide:

$$2\ PbSO_4(s) + 2\ H_2O(l) \rightarrow Pb(s) + PbO_2(s) + 2\ H_2SO_4(aq)$$

During charging, the automobile battery acts like a second type of electrochemical cell, an *electrolytic cell,* which uses electricity to produce a desired redox reaction. The lead storage battery can be discharged and recharged many times. But the shock of running over bumps in the road (or dead armadillos in Texas) or into the curb, flakes off a little of the lead (II) sulfate and eventually causes the battery to fail.

Burning the Bacon! Combustion of Fuels and Foods

Another extremely important type of redox reaction is combustion. *Combustion reactions* (burning) are reactions involving the combination with oxygen, producing heat. These types of redox reactions are absolutely essential for life and civilization because heat keeps yours and my homes warm and provides the majority of society's electricity. The combustion of gasoline, jet fuel, and diesel fuel powers the transportation systems. And the combustion of food powers our complex bodies.

In order to measure the heat output from a fuel or food, an instrument called a *bomb calorimeter* is used. Figure 12-4 shows the major components of a bomb calorimeter.

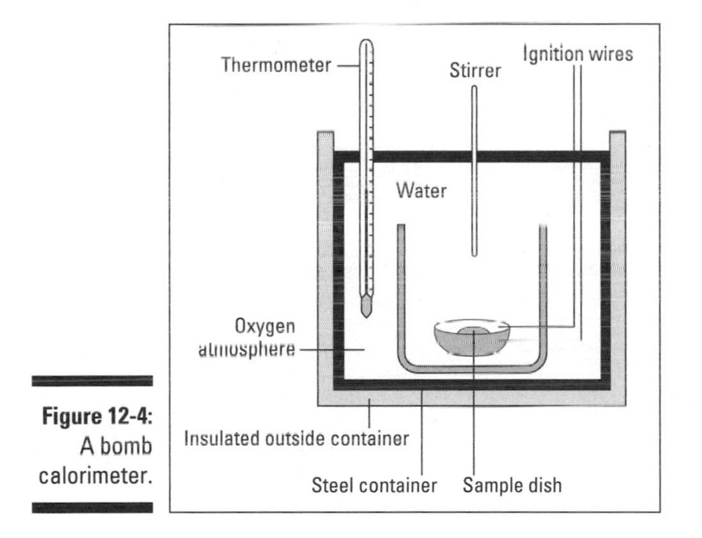

Figure 12-4: A bomb calorimeter.

A known mass of the material to be measured is placed into a sample cup and sealed. The air is removed from the sample cup and replaced with pure oxygen. The cup is then placed in the calorimeter with a known amount of water covering it. The initial temperature of the water is measured, and then the sample is ignited electrically. The rise in the temperature is measured, and the number of calories of energy that is released is calculated. One unit of heat output that is commonly used is the calorie. A *calorie* is the amount of energy needed to raise the temperature of 1 gram of water 1 degree Celsius. The complete combustion of a large kitchen match, for example, gives you about 1,000 calorie of heat.

If the caloric content of food is being determined nutritional Calories are normally reported. A nutritional Calorie is equal to a chemist's kilocalorie (1,000 calories). A 300 Calorie candy bar produces 300,000 calories of energy. Many times not all that energy is required immediately, so some is stored in chemical compounds in the body, normally as fats. I'm carrying around the result of *many* candy bars.

Chapter 13

Going the Carbon Route: Organic Chemistry

In This Chapter

- Introducing hydrocarbons
- Checking out the different functional groups

The largest and most systematic area of chemistry is *organic chemistry,* which is the chemistry of carbon. Of the 12 to 13 million chemical compounds known, about 90 percent are organic compounds. Organic compounds are everywhere. People burn organic compounds as fuel, eat organic compounds, wear organic compounds, and are made of organic compounds. The entire world consists of organic compounds.

In the early years of chemistry, scientists believed that living organisms could only produce organic compounds with a vital force involved. In 1828, the German scientist Friedrich Wohler changed the field of chemistry forever by making an organic compound, urea, by accident while trying to make an inorganic compound. This work was the beginning of today's modern field of organic synthesis.

Scientists have since discovered many compounds of carbon. So many exist because carbon contains four valence electrons and so can form four covalent bonds to other carbons or to other elements. (Organic chemistry students commonly make a mistake when drawing structures by not making sure that every carbon has four bonds attached to it.) The bonds that carbon forms are strong covalent bonds, and carbon has the ability to bond to itself in long chains and rings. It can form double and triple bonds to another carbon or to another element. These properties allow carbon to form the vast multitude of compounds needed to make an amoeba or a butterfly or a baby.

Organic chemistry is one reason why chemistry gets the reputation of being smelly. You have lab experiments, especially synthesis labs, where you can build complex molecules from simpler ones. The downside is the odor.

This chapter provides a basic introduction into the world of organic chemistry. I explain hydrocarbons, which are compounds of carbon and hydrogen, as well as some of the other different classes of organic compounds and their uses.

Hustling Hydrocarbons — the Simplest Organic Compounds

Hydrocarbons are the simplest organic compounds; they're composed of carbon and hydrogen. Economically, the hydrocarbons are extremely important to society, primarily as fuels. Gasoline is a mixture of hydrocarbons. You and I use methane (natural gas) and propane and butane, all hydrocarbons, for their ability to burn and release a large amount of energy.

Hydrocarbons may contain only single bonds (the alkanes) or double bonds (the alkenes) or triple bonds (the alkynes). And they may form rings containing single or double bonds (cycloalkanes, cycloalkenes, aromatics). The following sections examine these different types of bonded carbons in more depth.

From gas grills to gasoline: Alkanes

Alkanes are the simplest of the hydrocarbons. They're called *saturated* hydrocarbons, which means each carbon is bonded to four other atoms. No double or triple bonds exist in the alkanes. Straight-chained alkanes (no rings) have the general formula of C_nH_{2n+2}, where n is a whole number. (If a ring is present, the formula becomes C_nH_{2n}.) If n = 1, then the ring has four hydrogen atoms, and the result is CH_4, methane. Table 13-1 lists the names of the first ten normal (straight-chained) alkanes:

Table 13-1 The First Ten Normal Alkanes (C_nH_{2n+2})

n	*Formula*	*Name*
1	CH_4	methane
2	C_2H_6	ethane
3	C_3H_8	propane
4	C_4H_{10}	butane
5	C_5H_{12}	pentane
6	C_6H_{14}	hexane
7	C_7H_{16}	heptane

n	Formula	Name
8	C_8H_{18}	octane
9	C_9H_{20}	nonane
10	$C_{10}H_{22}$	decane

Figure 13-1 shows models of the first four of these alkanes.

The alkanes shown in Figure 13-1 are called *normal* or *straight-chained alkanes*. However, they really aren't straight. Carbon bonds in a tetrahedral (think of a tripod with a fourth leg sticking straight up) fashion with bond angles at 109.5 degrees. (Check out my book *Chemistry For Dummies,* 2nd edition [John Wiley & Sons, Inc.] for a discussion of molecular geometry.) Every carbon except the end ones is bonded to two other carbons. A lot of time when I draw the structures I show them in a straight line.

Focusing on the molecular and structural formulas

The molecular formula tells what kinds of atoms are present in the compound and the actual number of them. Table 13-1 shows the molecular formula of these alkanes. They're all normal or straight-chained hydrocarbons, but the structural formula can better illustrate this makeup. The structural formula tells the kinds of atoms present, the actual number of each, and the bonding pattern, the what is bonded to what.

The structural formula can be shown in a number of different ways.

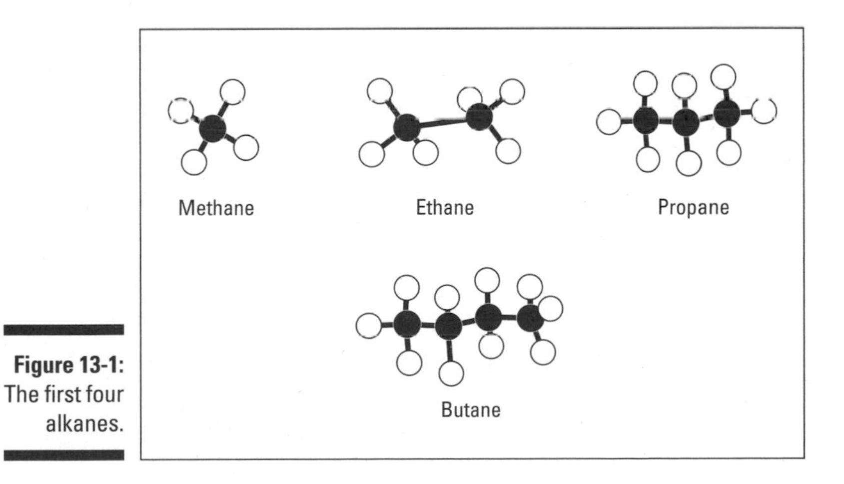

Figure 13-1: The first four alkanes.

- **Expanded structural formula:** This method basically shows each covalent bond as a line. With organic compounds like the hydrocarbons, if you want to show the way that the carbons are bonded, you can sometimes omit the hydrogen atoms on the expanded form and just indicate them by the covalent-bond line.
- **Condensed form:** This method groups parts of the molecule and still indicates the bonding pattern, which you can do in several ways. Figure 13-2 shows an expanded and a condensed form of the structural formula of propane, C_3H_8.

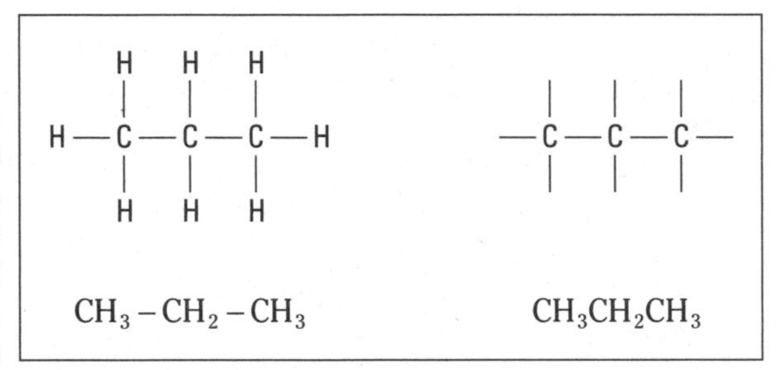

Figure 13-2: Structural formulas of propane.

Giving alkanes their names

An international group of chemists sets rules for naming organic compounds among other things. This group, called the International Union of Pure and Applied Chemistry (IUPAC), has an extremely systematic set of rules for the naming of compounds. On a regular basis, the group determines how to name new types of compounds as they're discovered in nature or made in the laboratory. Naming the compounds is complex, so for now I show you just the rules to name simple alkanes. (Even the naming of alkanes can get complicated, so remember to avoid overcomplicating it.)

To help you understand how simple alkanes are named, check out this abbreviated list of IUPAC rules:

- **Rule 1:** Find the longest continuous carbon chain in the alkane (*longest* refers to the greatest number of carbon atoms and *continuous* refers to starting at one end of the chain and connecting the carbons with your pencil without picking it up or backtracking). The straight-chained hydrocarbon that has that number of carbons is the parent or base name of the alkane. The parent name will end in the suffix *-ane*.
- **Rule 2:** The parent name is modified by adding the names of substituent groups that are attached as branches to the parent compound. Substituent *groups* are those groups that have been substituted for a hydrogen atom in the alkane parent. For alkane hydrocarbons, these substituent groups are alkane branches that attach to the parent. They are named by taking the alkane name, dropping the *-ane* and substituting *-yl,* so that methane becomes methyl, ethane becomes ethyl, and so on.

- **Rule 3:** The position of a particular substituent group on the parent carbon chain is indicated by location numbers. These are assigned by consecutively numbering the carbons of the parent carbon chain from one end to the other, so that the sum of the location numbers of all substituent groups will be as small as possible. The location number of the carbon to which the group is attached is placed in front of the substituent group name and separated from the name by a hyphen.
- **Rule 4:** The names of the substituent groups are placed in front of the parent name in alphabetical order. If there are a number of identical substituent groups then the numbers of all the carbons to which these groups are attached, separated by commas, are used and the common Greek prefixes such as *di-*, *tri-*, *tetra-*, *penta-* are used. These prefix names are not used to determine alphabetical order. The last substituent alkyl group is used as a prefix to the parent alkane name.

For example, *isomers* are compounds that have the same molecular formula but different structural formulas. An isomer of butane has the same molecular formula as our straight-chained compound, C_4H_{10}, but a different bonding pattern. This isomer is mostly called by the common name *isobutane* and is what I call a branched hydrocarbon. Check out Figure 13-3 to see it shown in a variety of ways.

So how do you differentiate which butane you're talking about when faced with the formula C_4H_{10}? Use a unique name that stands for that one compound and that compound only. For the straight-chained compound, you can say butane or normal-butane or better yet, n-butane. The **n-** makes it perfectly clear to a chemist that you're talking about the straight-chained isomer. What about the other isomer, isobutane? You could use the common name, but that isn't really accepted everywhere. That's where the IUPAC comes into play. All chemists around the world can agree upon a common name to avoid confusion and to help communication.

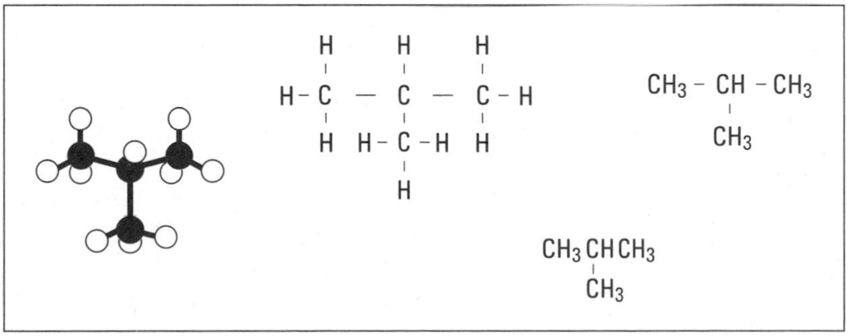

Figure 13-3: Isobutane.

Naming some compounds: A couple examples

Although you may be a bit overwhelmed by naming compounds, it really isn't that difficult. Here I provide you a couple examples. After you get into it, you may realize like many of my students do that naming organic compounds are one of the most fun things in organic chemistry.

Figure 13-4 shows an alkane that I'm going to name by following the rule numbers:

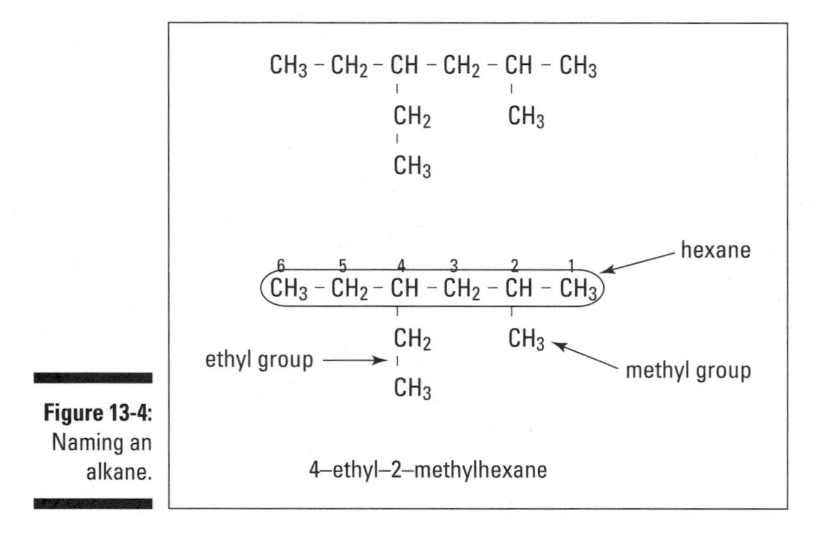

Figure 13-4: Naming an alkane.

- **Rule 1:** Using the condensed structural formula, the longest continuous carbon chain is composed of six carbons. Three different 6-carbon chains can actually be used (with the same name resulting eventually) but start with the horizontal one. The chain has six carbons, so the parent name would be hexane.
- **Rule 2:** You have two substituent groups, one composed of two carbons (ethyl) and another with one carbon (methyl).
- **Rule 3:** Number the parent chain from right to left, giving you alkyl groups at carbons # 2 & 4 (sum of 6). If numbered from left to right, the groups would have been on carbons 3 & 5 (sum of 8).
- **Rule 4:** So you have a 4-ethyl and a 2-methyl. Placing them alphabetically, the result is 4-ethyl-2-methylhexane.

Let me provide one more example to help you master this concept. Figure 13-5 shows another alkane.

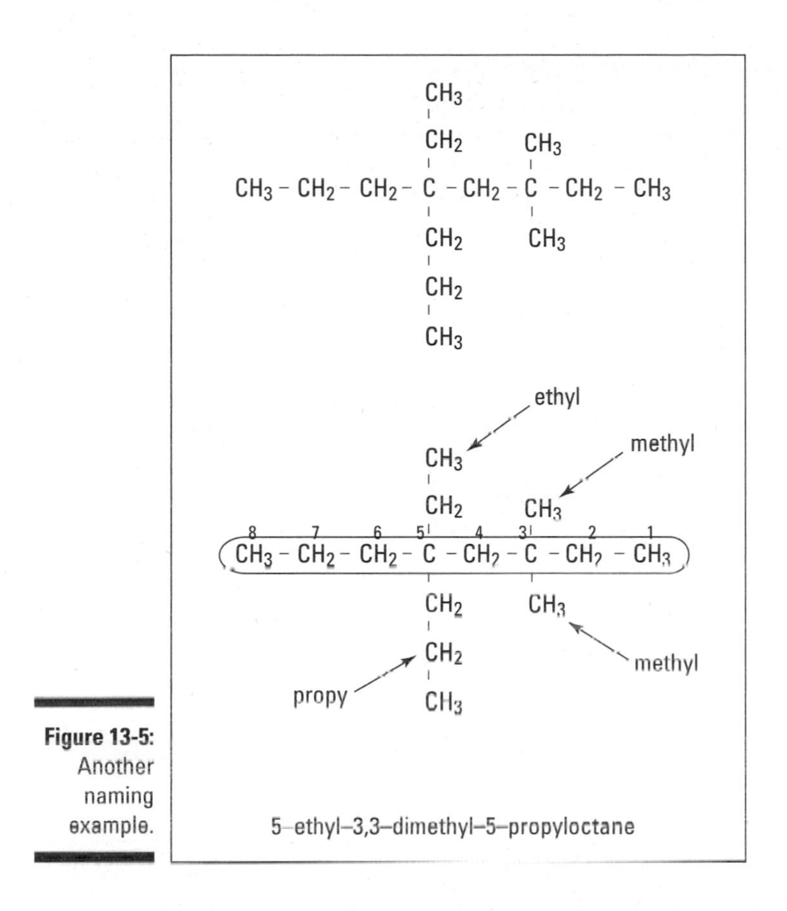

Figure 13-5: Another naming example.

- **Rule 1:** Using the condensed structural formula, the longest continuous carbon chain is composed of eight carbons, so you have an octane parent name.
- **Rule 2:** You have four substituent groups, two methyl groups (dimethyl), an ethyl group, and a propyl group.
- **Rule 3:** Number the parent chain right to left (3+3+5+5=16) instead of left to right (4+4+6+6=20), so that you have a 3,3-dimethyl, a 5-ethyl, and a 5-propyl group.
- **Rule 4:** Arrange alphabetically, remembering that the *di-* of dimethyl doesn't count: 5-ethyl-3,3-dimethyl-5-propyloctane.

As you may suppose, the more carbons you have, the more isomers that are possible. For an alkane with the formula $C_{20}H_{42}$ there are more than 300,000 possible isomers and for $C_{40}H_{82}$ there are about 62 *trillion* possible isomers!

Cycloalkanes – fellowship of the rings

Alkanes may also form ring systems to make compounds called *cycloalkanes*. The naming of these compounds is very similar to the branched alkanes except the cyclo- prefix is used on the parent name and the ring is numbered to give the lowest sum of location numbers. Many times in the condensed structural formula the ring is simply drawn as lines where the intersection of two straight lines indicates a carbon atom and the hydrogen atoms are not shown at all. Figure 13-6 shows the expanded, condensed, and very condensed structural formulas of cyclohexane.

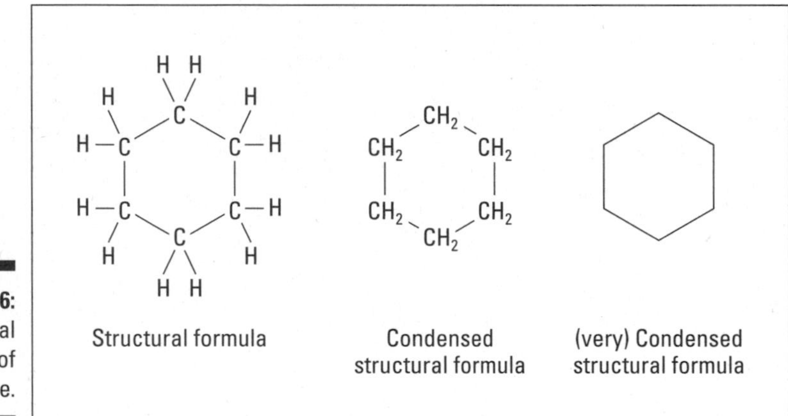

Figure 13-6: Structural formulas of cyclohexane.

Combustion is the primary reaction of alkanes. The alkanes, both straight-chained and some cycloalkanes, are used primarily as fuels. Methane is the primary component in natural gas and, like most hydrocarbons, is odorless. The gas companies add a stinky organic compound containing sulfur called a *mercaptan* to the natural gas to help alert you to gas leaks. Butane is used in lighters and propane in gas grills. Some of the heavier hydrocarbons are found in petroleum. Combustion is the primary reaction of alkanes.

Halogenated hydrocarbons

The *halogenated hydrocarbons* are hydrocarbons, including alkanes, in which one or more of the hydrogen atoms have been replaced by some halogen, normally chlorine or bromine. Halogen substituents are named as chloro-, bromo-, and so on. Members of this class of compound include chloroform, once used as an anesthetic; carbon tetrachloride, used at one time in dry cleaning solvent; and freons, (chlorofluorocarbons, CFCs) elements that have played a major role in the depletion of the ozone layer.

Unsaturated hydrocarbons: Alkenes

Alkenes are hydrocarbons that have at least one carbon-to-carbon double bond (C=C). These compounds are called *unsaturated hydrocarbons* because they don't have the maximum number of hydrogen atoms possible attached to the carbons. (I'm sure that you have heard the terms *saturated* and *unsaturated* in regards to fats and oils in nutrition discussions. They mean exactly the same thing there — saturated fats and oils contain no carbon-to-carbon double bonds, unsaturated ones do and polyunsaturated ones have more than one C=C per molecule.) Alkenes that have only one double bond have the general formula C_nH_{2n}. For every additional double bond, subtract two hydrogen atoms.

Naming alkenes

Alkenes have a parent name ending with the -ene suffix. You use a location number to indicate the first of the double-bonded carbons that you encounter. When selecting the longest carbon chain, you number so that the location number of the double bond is as low as possible.

The first two members of the alkene family are *ethene,* written as $H_2C\text{-}CH_2$ or $CH_2\text{=}CH_2$, and *propene,* $CH_3CH\text{=}CH_2$. These two alkenes are often called by their common names, ethylene and propylene, respectively. They are two of the most important chemicals produced by the chemical industry in the United States. Ethylene is used in the production of *polyethylene,* one of the most useful plastics produced, and in the production of *ethylene glycol,* the principal ingredient in most antifreeze. Propylene is used in the production of *isopropyl alcohol* (rubbing alcohol) and some plastics. Figure 13-7 shows a couple of ways of representing the structural formula of ethene (ethylene).

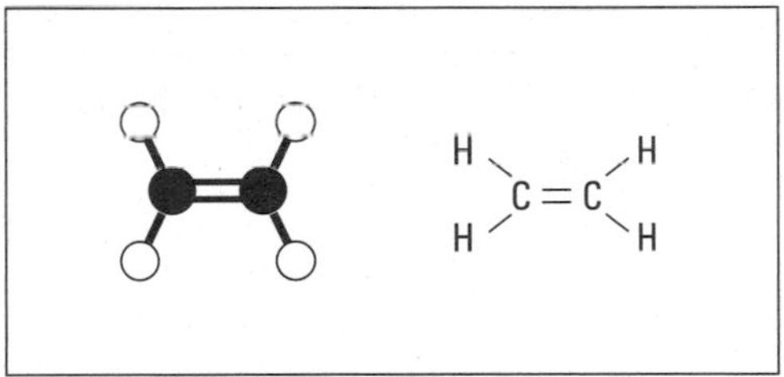

Figure 13-7: Ethene.

The reactions of alkene

The primary reaction of alkenes is *addition reactions.* A double bond is very reactive and the pi-bond can easily be broken (leaving the sigma bond intact). The two carbons which were originally double-bonded can now form new single bonds to other atoms. One of the most economically important is in the process of *hydrogenation,* in which hydrogen is added across the double bond. Here is the hydrogenation of butene:

$$CH_3\,CH_2CH{=}CH_2 + H_2 \rightarrow CH_3\,CH_2CH_2CH_3$$

This hydrogenation reaction is used in the food industry to convert unsaturated vegetable oils to solid fats (vegetable oil to margarine, for example) and requires the use of a nickel metal catalyst.

Another important addition reaction of alkenes is *hydration,* the addition of a water molecule across the double bonding, yielding an alcohol. Here is the hydration of butene that gives butyl alcohol (I bolded the water molecule so that you could tell where the -OH ended up):

$$CH_3\,CH_2CH{=}CH_2 + H\text{-}\mathbf{OH} \rightarrow CH_3CH_2CH\text{-}CH_2\,\text{-}\mathbf{OH}$$

Ethyl alcohol is produced from the hydrogenation of ethane and is identical to the ethyl alcohol produced by the fermentation process. However, by federal law, it can't be sold for human consumption in alcoholic beverages. The fermentation group must have a strong Washington lobby.

Undoubtedly, the most important reaction of the alkenes is *polymerization,* in which the double bond reacts to produce long chains of the once-alkenes bonded together. This is the process used to produce plastics and is discussed in Chapter 14.

A triple play: Alkynes and their triple bond

Hydrocarbons that have at least one carbon-to-carbon triple bond are called *alkynes.* These compounds are named with suffix -yne. Alkynes with only a single triple bond have the general formula of C_nH_{2n-2}. The simplest alkyne is ethyne, commonly called *acetylene.* Figure 13-8 shows the structure of acetylene.

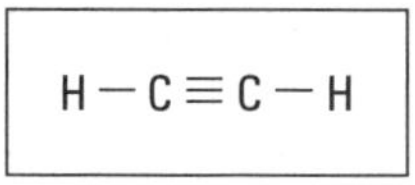

Figure 13-8: The structure of ethyne (acetylene).

$$H-C\equiv C-H$$

Chemists can produce acetylene by a number of different ways. One way is to react coal with calcium oxide to produce calcium carbide, CaC_2. Calcium carbide is then reacted with water to produce acetylene. Miners' lamps used to be powered by this reaction. Water was dripped on calcium carbide and the acetylene burned to produce light. Today, most of the acetylene produced is either used in oxyacetylene torches in cutting and welding and in the synthesis of a wide variety of polymers (plastics).

Benzene and other aromatic (smelly) compounds

Hydrocarbons that have alternating single and double bonds are called *aromatic hydrocarbons*. The simplest cyclic aromatic compound is *benzene,* C_6H_6. It contains a cyclohexene-type ring system with alternating single and double bonds. Benzene is far less reactive than chemists would expect having those three sets of double bonds. In the current model for benzene, six electrons, two from each of the three double bonds, are donated to an electron cloud associated with the entire benzene molecule. These electrons are *delocalized* over the entire ring instead of simply being located between two carbon atoms. This electron cloud is above and below the planar ring system. Figure 13-9 shows a couple of traditional ways of representing the benzene molecule and a couple of ways showing the delocalized structure.

One may attach a wide variety of groups to this benzene ring, making many new aromatic compounds. For example, an -OH may replace a hydrogen atom and the resulting compound is called *phenol.* Phenol is used as a disinfectant and in the manufacture of plastics, drugs, and dyes. Two benzene rings fused together make *naphthalene,* commonly called mothballs.

Benzene and its related compounds will burn, but they burn with a sooty flame. And it also has been shown that benzene and some of its related compounds are either known or suspected carcinogens.

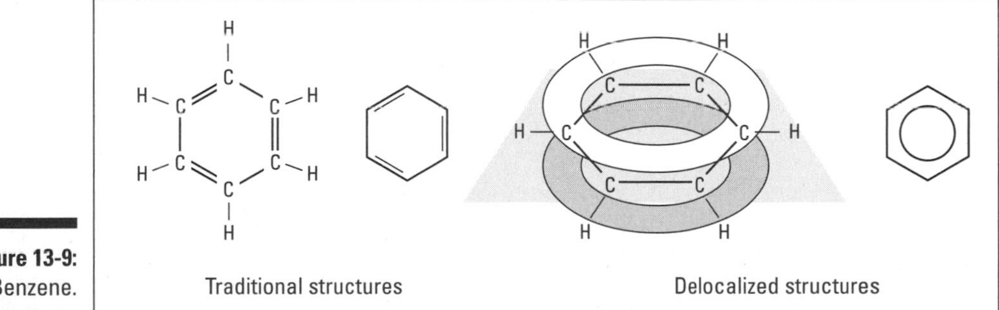

Figure 13-9: Benzene.

Following Functional Groups

The preceding section covers hydrocarbons, compounds of just carbon and hydrogen. Can you imagine how many new organic compounds can be generated if a nitrogen atom or a halogen or a sulfur atom or some other element(s) is thrown in?

Consider some well-known alcohols — ethyl alcohol (the drinking alcohol), methyl alcohol, wood alcohol, and isopropyl alcohol, rubbing alcohol. They're quite different, yet remarkably the same in terms of the kinds of chemical reactions that they undergo. The reactions all involve the -OH group on the molecule, that part of the molecule that really defines the identity of an alcohol, just as the double bond really defines the identity of an alkene.

In many cases, it doesn't really matter what the rest of the molecule turns out to be. In reactions, one alcohol is pretty much the same as another. That atom or group of atoms that defines the reactivity of the molecule is called the *functional group.* For alcohols it's the -OH, for alkenes it's the C=C, and so on. Classification by functional group makes studying and classifying the properties of organic compounds much easier. The following sections identify some general properties of the most common functional groups.

Alcohols (make mine ethanol): R-OH

Alcohols are organic compounds that contain the -OH functional group. In fact, many times alcohols are generalized as *R-OH,* where the *R* stands for the *R*est of the molecule (like that radio newscaster's "rest of the story"). Alcohols are named using the -ol suffix to replace the -ane of the corresponding alkane. Here are two important alcohols:

- **Methanol (methyl alcohol):** This alcohol is sometimes called *wood alcohol* because years ago its primary synthesis involved heating wood in the absence of air. The more current method of synthesis of methanol involves reacting carbon monoxide and hydrogen with a special catalyst at elevated temperatures:

  $$CO(g) + 2\,H_2(g) \rightarrow CH_3OH(l)$$

  About half of the methanol produced in the United States is used in the production of *formaldehyde,* which is then used as embalming fluid and in the plastics industry. It's also sometimes added to ethanol to make it unfit for human consumption, a process called *denaturing.* Methanol is also being considered as a replacement to gasoline, but there are still major problems to be overcome. But there is a process that uses methanol in the production of gasoline. New Zealand currently has such a plant that produces about a third of their gasoline.

- **Ethanol (ethyl alcohol or grain alcohol):** This alcohol is produced primarily in one of two ways. If the ethanol is to be used in alcoholic beverages, it's produced by the fermentation of carbohydrates and sugars by the enzymes in yeast:

  $$C_6H_{12}O_6(aq) \rightarrow 2\,CH_3CH_2OH(l) + CO_2(g)$$

The industrial preparation of ethanol by the hydration of ethylene using an acid catalyst is as follows:

$$H_2C{=}CH_2 + H_2O \rightarrow CH_3\text{-}CH_2\text{-}OH$$

In addition to being used in alcoholic beverages, ethanol is used as a solvent in perfumes and medicines, as well as an additive to gasoline making it *gasohol*.

Bad odor carboxylic acids: R-COOH

Chemists use the -COOH or -CO_2H to indicate this functional group. These compounds are named with an *-oic acid* suffix. Acetic acid, shown in Figure 13-10, would also be called *ethanoic acid*. This figure shows the structure of the carboxylic acid functional group.

```
    O                  O
    ||                 ||
R — C — OH      CH3 — C — OH
                   acetic acid
```

Figure 13-10: The carboxylic acid functional group and acetic acid.

Carboxylic acids are commonly prepared by the oxidation of an alcohol. For example, leave a bottle of wine in contact with the air or some other oxidizing agent and the ethanol is oxidized to acetic acid:

$$CH_3CH_2OH(l) + O_2(g) \rightarrow CH_3COOH(l) + H_2O(l)$$

Formic acid, also known as methanoic acid, can be isolated by the distillation of ants. The sting resulting from the bite of an ant is due to formic acid. That is why applying some base, like baking soda, helps to neutralize the acid and relieve the pain.

Many of these organic acids have a distinct odor associated with them. I'm sure you are familiar with the odor of vinegar (acetic acid), but other acids have distinct odors, such as those mentioned in Table 13-2.

Table 13-2 Nasty Smells and What They Are

$CH_3(CH_2)_2COOH$	Butyric acid	odor of rancid butter
$CH_3(CH_2)_3COOH$	Pentanoic acid	odor of manure
$CH_3(CH_2)_4COOH$	Hexanoic acid	odor of goats

Good odor esters: R-COOR'

The ester functional group is very similar to the carboxylic acid function group except that another -R group has replaced the hydrogen atom. Esters are made by reacting a carboxylic acid with an alcohol, producing an ester and water. Figure 13-11 shows the synthesis of an ester.

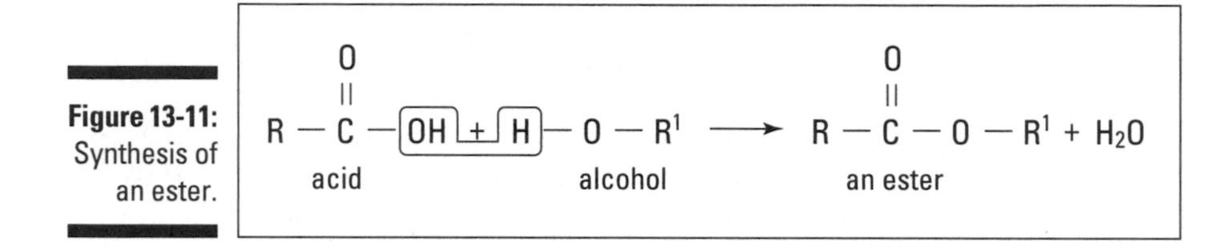

Figure 13-11: Synthesis of an ester.

Even though many of the carboxylic acids from which esters are made have foul odors, many esters have pleasant odors. Oil of wintergreen is an ester and other esters have the odor of bananas, apples, rum, roses, and pineapples. Esters often find a lot of use in the flavoring and perfume industry.

Aldehydes (R-COH) and ketones (R-CO- R'): Related to alcohols

Both aldehydes and ketones are produced by the oxidation of alcohols. These functional groups are shown in Figure 13-12.

aldehydes	ketones
R — C(=O) — H	R — C(=O) — R^1

Figure 13-12: Aldehyde and ketone functional groups.

Formaldehyde, HCHO, is an economically important aldehyde. It's used as a solvent and for preservation of biological specimens. Formaldehyde is also used in the synthesis of certain polymers, such as Bakelite and *Melmac* (used in melamine dishes). Other aldehydes, especially those with a benzene ring in their structure, have pleasing odors and, like esters, are used in the perfume and flavoring industry.

Acetone, CH_3-CO-CH_3, is the simplest ketone and finds many uses as a solvent, especially for paint. Many of us are familiar with acetone-based fingernail polish remover. And methyl ethyl ketone is the solvent in model airplane glue.

Sleepy ethers: R-O-R'

Ethers contain an oxygen atom bonded to two hydrocarbon groups, R-O-R. The R- groups may or may not be the same. Diethyl ether was once used as an anesthetic, but its high flammability has caused it to be largely replaced in operating rooms. Since ethers are fairly unreactive (except for combustion) they are commonly used as solvents in organic reactions. They will however slowly react with the oxygen in the atmosphere to form explosive compounds called *peroxides*.

I can synthesize ethers by the reaction of alcohols, with the loss of water (a *dehydration* reaction). Diethyl ether can be made by reacting ethyl alcohol in the presence of sulfuric acid:

$$2\ CH_3CH_2OH(l) \rightarrow CH_3CH_2\text{-}O\text{-}CH_2CH_3(l) + H_2O(l)$$

If I use two different alcohols I get what is called a mixed ether where the two R groups are not the same.

The organic bases: Amines ($R\text{-}NH_2$) and amides ($R\text{-}CO\text{-}NH_2$)

Amines and amides are derived from ammonia and contain nitrogen in their functional groups. Figure 13-13 shows the amine and amide functional groups.

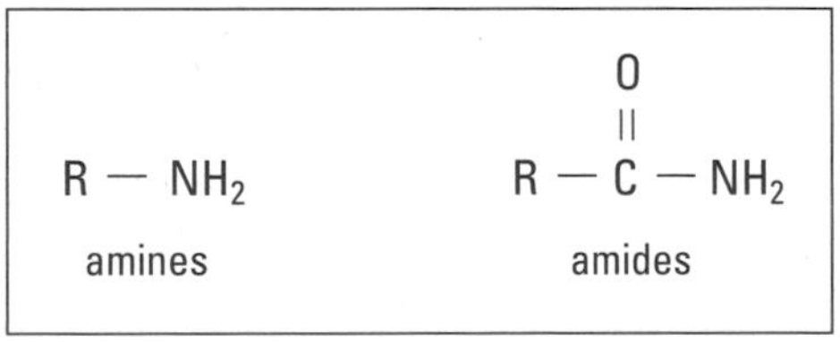

Figure 13-13: Amine and amide functional groups.

$$R - NH_2 \qquad\qquad R - \overset{\overset{\large O}{\|}}{C} - NH_2$$

amines amides

Look at Figure 13-13. Any of the hydrogen atoms attached to the nitrogen on both the amine and amide can be replaced by some other R group.

More than you wanted to know about carbon

Carbon is the seventeenth most abundant element in the Earth's crust. In the Earth, carbon is found primarily as carbonates; however, in the biosphere there are innumerable carbon compounds from very simple molecules, such as CO_2, to complicated molecules, such as DNA. Elemental carbon is also present as the minerals graphite and diamond. Carbon is present in all living organisms and is the basis for both organic chemistry and biochemistry.

Carbon forms two important oxides—carbon monoxide, CO, and carbon dioxide, CO_2. Carbon monoxide has many industrial uses, both as the starting material in numerous synthetic procedures and as a reducing agent. In a blast furnace, carbon monoxide serves to reduce iron oxides to iron metal. Carbon dioxide is the product of many oxidation processes such as burning and respiration and is one of the greenhouse gases.

Amines and amides, like ammonia, tend to be weak bases. Amines are used in the synthesis of disinfectants, insecticides, and dyes. They are in many drugs, both naturally occurring and synthetic. *Alkaloids* are naturally occurring amines found in plants. And most amphetamines are amines.

Chapter 14
Pondering Polymers

In This Chapter

- Eyeing natural monomers and polymers
- Grasping what synthetic monomers and polymers are
- Handling plastics in different ways

One way of classifying historic stages is by the composition of the tools men used. Humanity has gone through the Bronze Age, the Iron Age, and now humanity is in the Polymers Age. Polymers are often referred to as *plastics,* but no matter what you call them, they belong to a class of compounds called *macromolecules* — very large molecules. Nature has been building macromolecules forever. Proteins, cotton, wool, and cellulose are all macromolecules. Now man has learned to produce macromolecules in the lab, changing the face of society forever.

Polymers (plastics) are an integral part of today's world. I dress in polymers, I ride around in automobiles that are increasingly constructed of polymers, my home is filled with plastic bottles of all shapes, sizes, and hardness, and even some body parts are being replaced by plastic ones. I cook with pans that have nonstick surfaces, I use a nylon spatula, I watch a TV that has a plastic case, and I go to sleep on a foam mattress. Polymers have replaced metal in trains, planes, and automobiles. This world is truly becoming a plastic world.

In this chapter, I show you how the process of making polymers and how chemists go about designing polymers with certain desired characteristics. In addition, I discuss some ways for getting rid of those plastics before society buries itself in a mountain of plastic milk jugs and disposable diapers.

Examining Natural Monomers and Polymers

Nature has been building polymers for a long time. Cellulose (wood) and starch are prime examples of naturally occurring polymers. Take a look at Figure 14-1 at the structures of cellulose and starch.

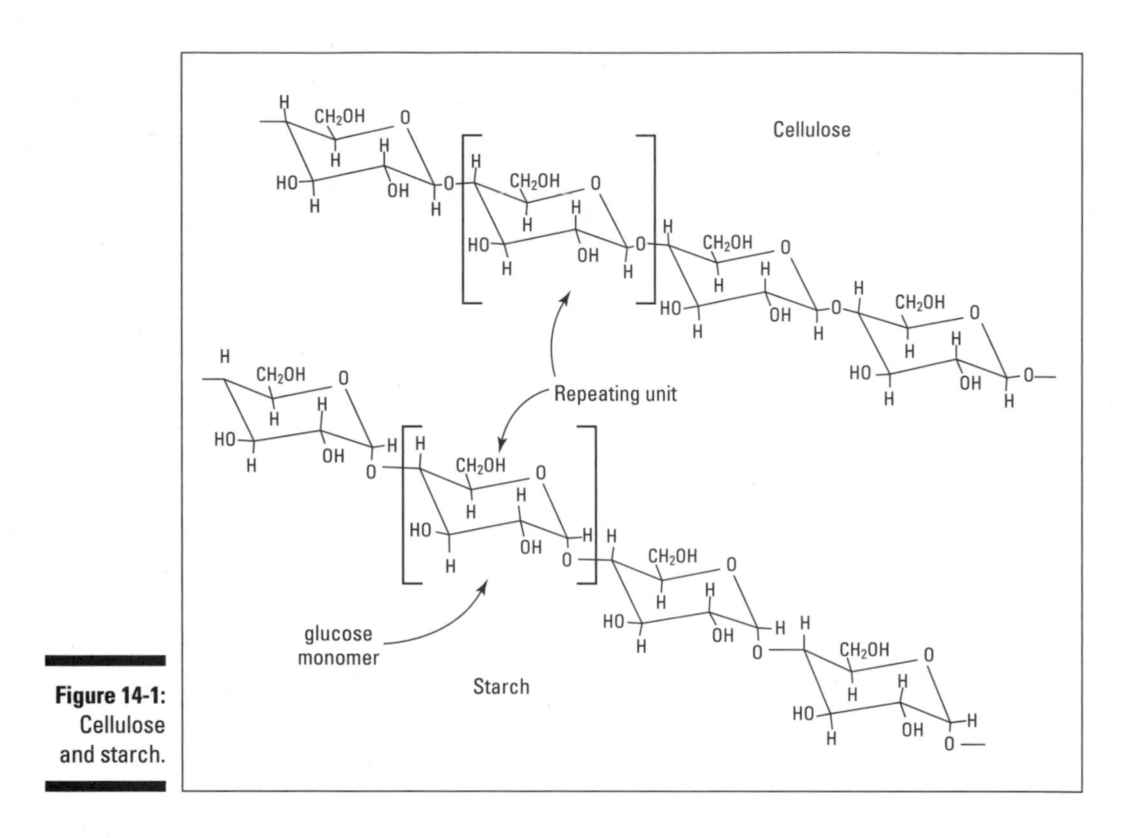

Figure 14-1: Cellulose and starch.

Notice anything similar about the two structures in Figure 14-1? They are both made up of repeating units. In fact, the repeating unit in both cases is a glucose unit. Both starch and cellulose are natural *macromolecules* (large molecules), but they're also examples of naturally occurring *polymers.* Polymers are macromolecules in which there is a repeating unit called a monomer. (Polymer should stand for "many mer." The "mer" in this case is the mono*mer.*) In the case of starch and cellulose, the monomer is a glucose unit. The structure of polymers is similar to taking a bunch of paper clips (monomers) and hooking them together to make a big long chain (polymer).

Notice another thing about cellulose and starch. The only way they differ is in how the glucose units are attached to each other. This minor change makes the difference between a potato and a tree. (Okay, it's not *quite* that simple.) Human beings can digest (metabolize) starches but not cellulose. A termite can digest cellulose just fine. In natural polymers, just like in synthetic ones, a minor change sometimes makes a big difference in the properties of the polymer.

Classifying Unnatural (Synthetic) Monomers and Polymers

Chemists took this idea of hooking together small units into very large ones from nature and developed a number of different ways of doing this in the laboratory. Now many different types of synthetic polymers exist. In this section, I introduce you to some synthetic polymers. I talk about their structures, properties, and uses.

Because chemists are big on classifying things together, they've put polymers into different classes. That works quite well. Grouping gives chemists something to do and makes it easier for normal folks to learn about the various kinds of polymers out there.

Providing a little structure

One way of classifying polymers is by the structure of their polymer chain. The following are three types of structures:

- **Linear:** They are composed of many long strands thrown together like pieces of rope.
- **Branched:** They have short branches coming off the main polymer strand. Imagine taking those long pieces of rope and tying short pieces of rope to them along the entire length.
- **Crosslinked:** These polymers have the individual polymer chains linked together by the side chains. It is as if someone had taken those pieces of rope and made them into a hammock.

Feel the heat

Another way of classifying polymers is by their behavior under heat. Polymers can be classified in two ways based on their heat:

- **Thermoplastic:** These polymers become soft when heated. This type of polymer is composed of long linear or branched strands of monomer units hooked together. Have you ever left a child's plastic toy on the dashboard of your car in middle of the summer? These plastics become really soft. Because they do soften and melt, they can be remolded time and time again. This makes thermoplastics much easier to recycle. A vast majority of the plastics that are produced in the United States are of the thermoplastic type.

- **Thermosetting:** These polymers don't soften when heated, nor can they be remolded. When this type of plastic is being produced, crosslinking (bridges between the polymer strands) is created in the plastic by heating it. Bakelite is a good example of a thermosetting plastic. It is a hard, strong nonconductor. These properties make it ideal as an insulator and as handles for frying pans and toasters.

Used and abused

A third way of classifying polymers is by their use by the consumer. The following are ways polymers are classified based on use:

- **Plastics:** A plastic refers to the ability to be molded. Whether they're thermoplastics or thermosetting, polymers are molded during the manufacture of the end product. These polymers are used to make our dishes, our toys, and so on.
- **Fibers:** They are linear strands held together by intermolecular forces such as hydrogen bonding between the polymer strands. These polymers are generally called *textiles.* They're used to make clothes and carpets.
- **Elastomers:** Sometimes called *rubber,* elastomers are thermoplastic materials that become slightly crosslinked during their formation. As a result, they will stretch and bounce. Natural rubber (latex) is classified as an elastomer along with its synthetic counterparts.

Chemical process

One the best ways of classifying synthetic polymers is by the chemical process that is used to create the polymer. These processes normally fall into one of two categories:

- Addition polymerization
- Condensation polymerization

These sections introduce these two processes and show you the important polymers in each category.

Hooking up: Addition polymerization

Many of the common polymers that you come into contact with every day are *addition* polymers — polymers that are formed in a reaction called *addition polymerization*. In this type of reaction, all the atoms that start out in the monomer are incorporated into the polymer chain. The monomers involved in this type of polymerization normally have a carbon-to-carbon double bond that is partially broken during polymerization. This broken bond forms a *radical reactive site,* a highly reactive atom that has an unpaired electron. The radical then gains an electron by joining up with another radical and a chain is started. A little perplexed? Following are some examples of addition polymerization that might help clear the fog a little.

Polyethylene: Sandwich wrap and milk jugs

Polyethylene is the simplest of the addition polymers. It's also one of the most economically important addition polymers. Ethane is heated at high temperature in the presence of a metal catalyst like palladium. Ethane loses two atoms of hydrogen (which make hydrogen gas) and forms a double bond:

$$CH_3\text{-}CH_3(g) + \text{heat and catalyst} \rightarrow CH_2\text{=}CH_2(g) + H_2(g)$$

The ethylene (ethene) that is produced is the monomer that is used in the production of polyethylene. The ethylene is then subjected to high heat with a catalyst in the absence of air. The high heat and catalytic action causes one of the carbon-to-carbon double bonds (C=C) to break, with one electron going to each carbon. Both carbons now have an unpaired electron, so they become radicals. Radicals are extremely reactive and will attempt to gain an electron. In terms of this polymerization reaction, the radicals can gain an electron by joining up with another radical to form a covalent bond. This happens at both ends of the molecule, and the chain begins to grow. Polyethylene molecules up to a molecular weight of 1 million grams/mol may be produced in this way (see Figure 14-2).

Different catalysts and pressures are used to control the structure of the final product. The polymerization of ethylene can yield three different types of polyethylene:

- **Low-density polyethylene (LDPE):** LDPE has some branches on the carbon chain, so it doesn't pack together as close and as tight as the linear polymer. It forms a tangled network of branched polymer strands. This type of polyethylene is soft and flexible. It's often used for food wrap, sandwich bags, grocery bags, and trash bags. And it, like all the forms of polyethylene, is resistant to chemicals.
- **High-density polyethylene (HDPE):** HDPE is composed of linear chains that are closely packed. This type of polymer is rigid, hard, and tough. Milk jugs, toys, and TV cabinets are made from HDPE. The Hula-Hoop was one of the first products ever made from this form of polyethylene.
- **Crosslinked polyethylene (CLPE):** CLPE has crosslinking between the linear strands of monomers that are bonded together, producing a polymer that is extremely tough. The lid on that HDPE milk carton is probably CLPE. Soft drink bottle caps are also CLPE. The soft drink bottles are made of another type of polymer that I discuss a little later in the chapter.

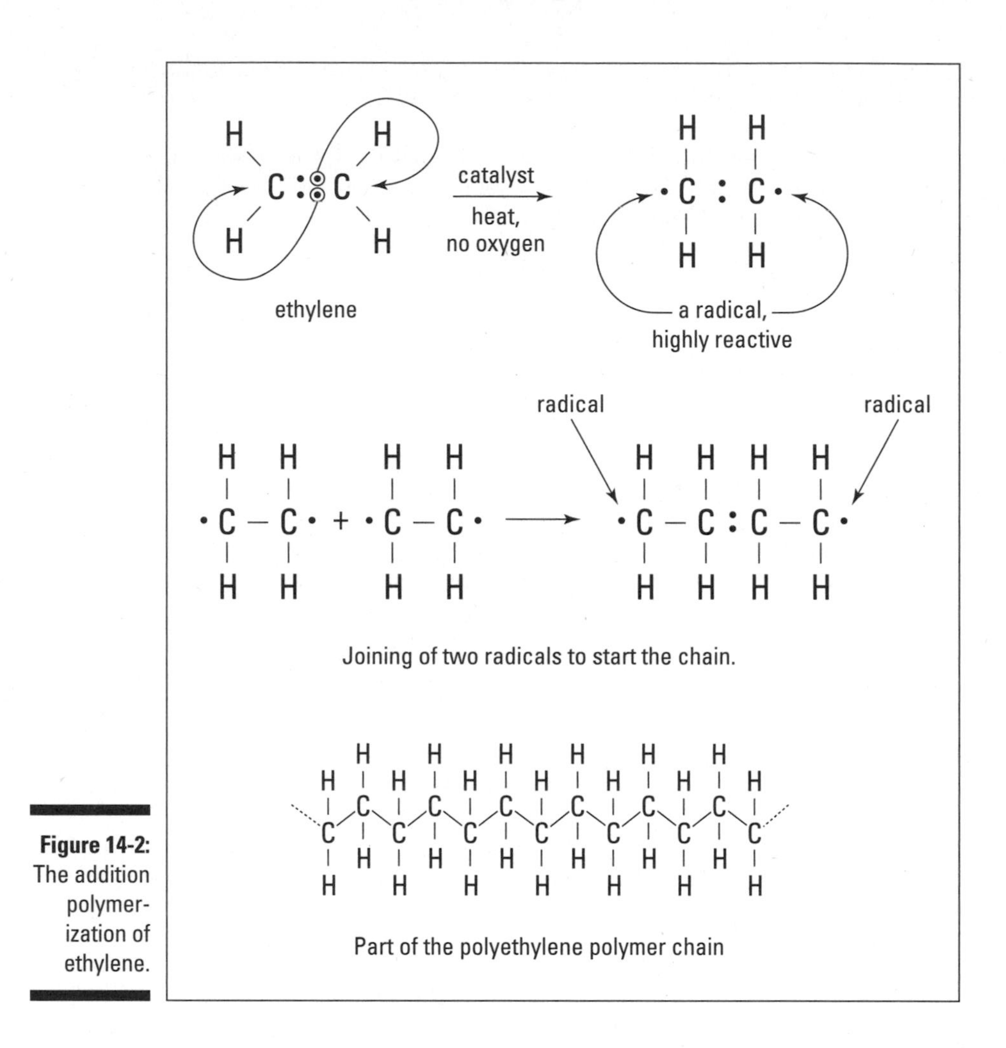

Figure 14-2: The addition polymerization of ethylene.

Polypropylene: Plastic ropes and indoor-outdoor carpeting

If you substitute a methyl group for a hydrogen atom, you get propylene. Propylene, just like ethylene, has the double bond, so it can undergo addition polymerization in the same way as ethylene. The result is polypropylene. See Figure 14-3.

The small *n* in Figure 14-3 indicates a number of the repeating units. Notice that this polymer has a methyl group side chain. Any time the molecule's structure is changed, the molecule's properties change. By carefully adjusting the reaction conditions, polymers that have the side chains on the same side of the molecule, on alternating sides of the molecule, or distributed randomly can be constructed. The addition of these side chains changes the properties of the polymer somewhat so that polypropylene can be used for a wide variety of purposes, such as indoor-outdoor carpeting, battery cases, ropes, bottles, and automotive trim.

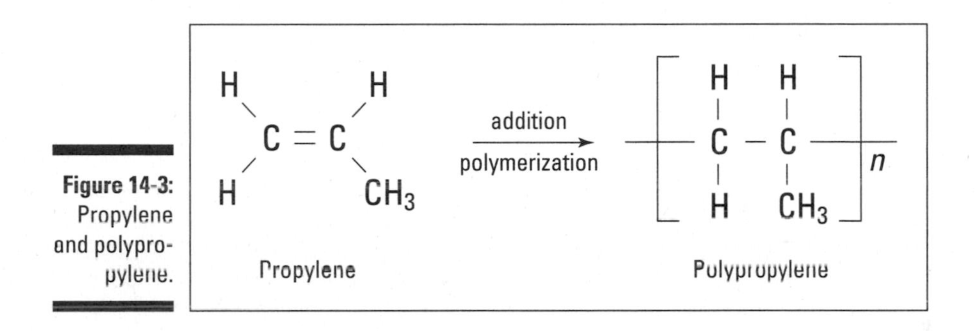

Figure 14-3: Propylene and polypropylene.

Polystyrene: Coffee cups

If you substitute a benzene ring for one of the hydrogen atoms on ethylene, you make styrene. Addition polymerization gives you polystyrene, as shown in Figure 14-4.

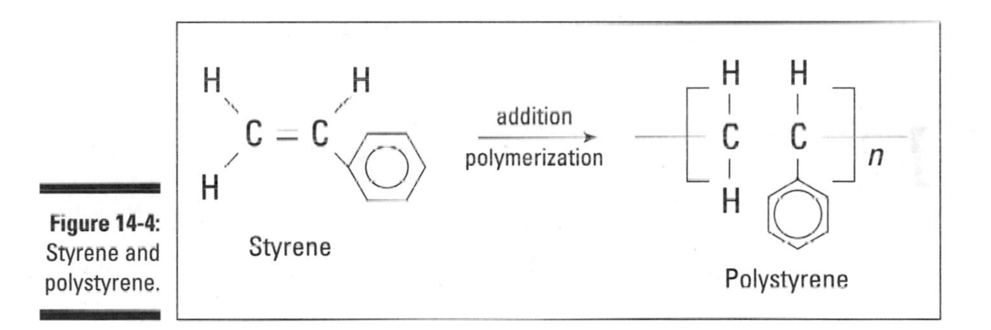

Figure 14-4: Styrene and polystyrene.

Polystyrene (Styrofoam) is a rigid polymer that is used for making foam drink cups, egg cartons, clear rigid drinking glasses, insulating materials, and packing materials. Environmentalists have criticized its use because recycling it is more difficult than other plastics and because it's so widely used.

Polyvinyl chloride: Pipes and upholstery

Substituting a chloride for one of the hydrogen atoms on ethylene creates the vinyl chloride monomer that can polymerize to polyvinyl chloride (PVC), as shown in Figure 14-5.

PVC is a tough polymer. It's used extensively in rigid pipes of all types, flooring, garden hoses, and toys. Thin sheets of PVC used as simulated leather crack easily, so a plasticizer is added. *Plasticizers* are liquids that are mixed with plastics to soften them and allow them to more closely resemble leather. However, after many years, the plasticizers can evaporate from the plastic, making it brittle and allowing it to crack.

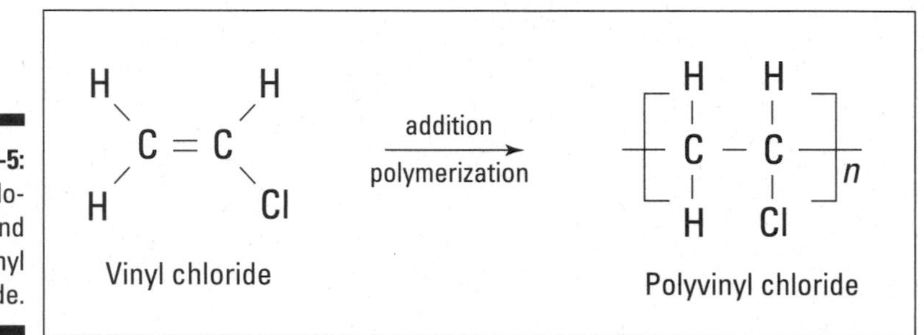

Figure 14-5: Vinyl chloride and polyvinyl chloride.

Polytetrafluoroethylene: Slick stuff

Replace all of the hydrogen atoms on ethylene with fluorine atoms and you have tetrafluoroethylene. The tetrafluoroethylene can be polymerized to polytetrafluoroethylene, as shown in Figure 14-6.

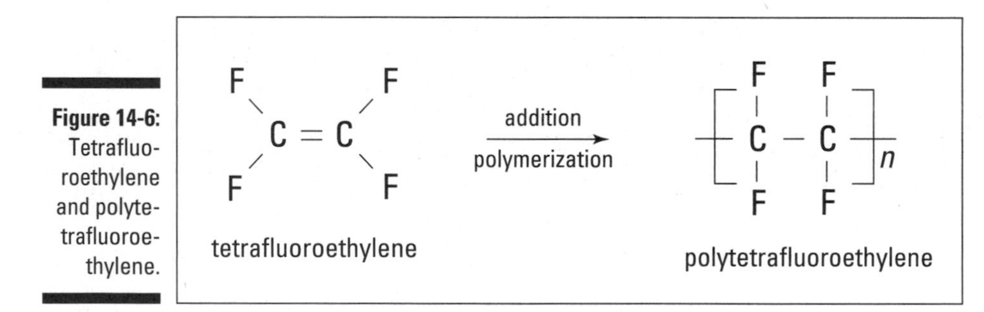

Figure 14-6: Tetrafluoroethylene and polytetrafluoroethylene.

Polytetrafluoroethylene is a material that is hard, heat resistant, and extremely slick. This material is used as bearings, valve seats, and (most importantly to me) a nonstick coating for pots and pans. You can find some other addition polymers in Table 14-1.

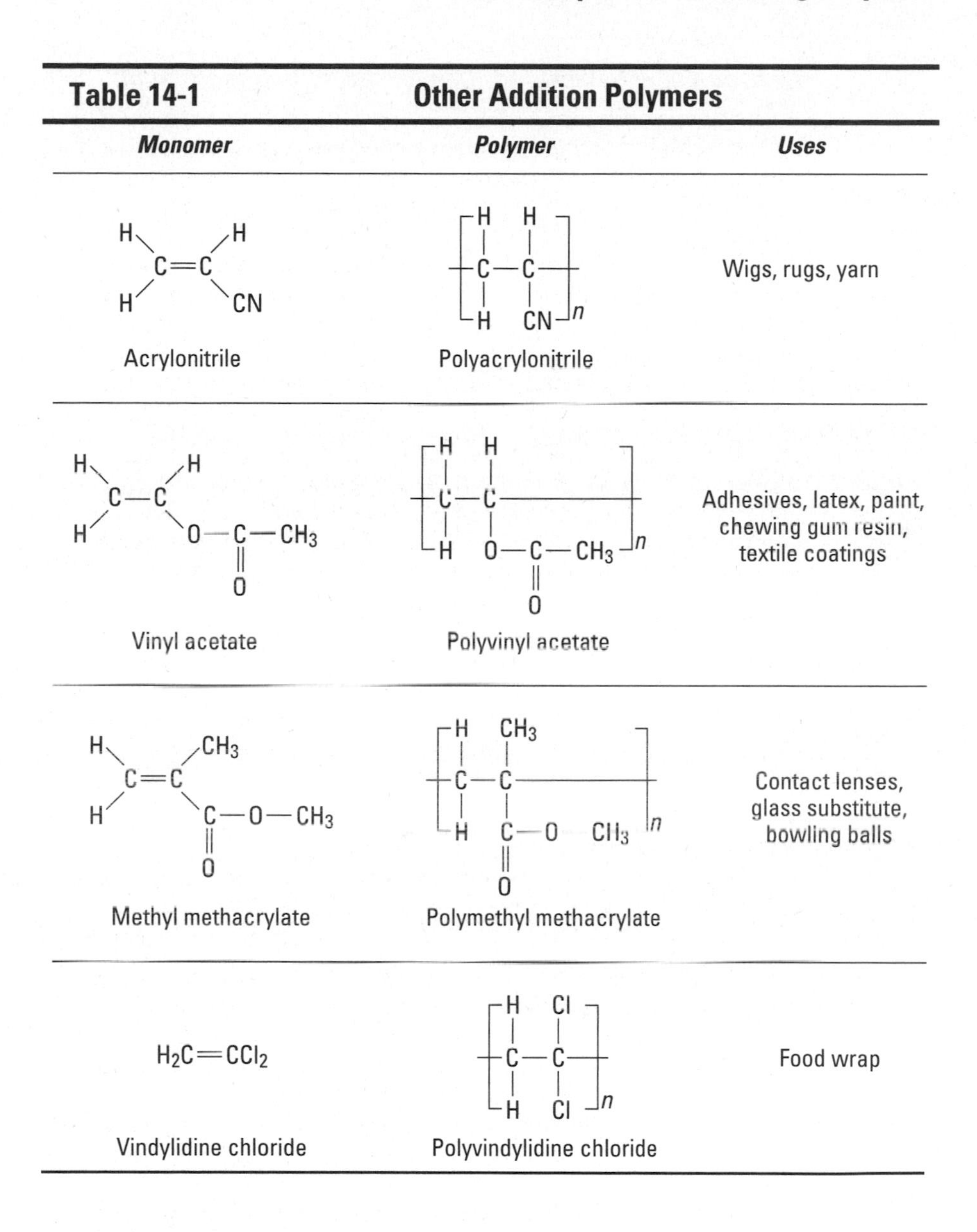

Table 14-1 Other Addition Polymers

Monomer	Polymer	Uses
Acrylonitrile	Polyacrylonitrile	Wigs, rugs, yarn
Vinyl acetate	Polyvinyl acetate	Adhesives, latex, paint, chewing gum resin, textile coatings
Methyl methacrylate	Polymethyl methacrylate	Contact lenses, glass substitute, bowling balls
$H_2C{=}CCl_2$ Vindylidine chloride	Polyvindylidine chloride	Food wrap

Getting rid of something: Condensation polymerization

A reaction in which two chemical species combine with each other by eliminating a small molecule is called a *condensation reaction.* Polymers formed in this fashion are known as *condensation polymers.* No double bond is needed like was required in addition polymerization.

A small molecule, normally water, is eliminated. Commonly, one molecule that is reacting will be an organic acid, and the other molecule will be an alcohol. These two molecules react, splitting off water and forming an organic compound called an *ester.* If a polymer chain grows, it forms a polyester. The following sections describe the common polyesters.

Polyester: Leisure suits and soft drink bottles

If you take ethylene glycol, with its alcohol functional groups on both carbons, and react it with terephthalic acid, with its two organic acid functional groups, you can eliminate water and form the condensation polymer polyethylene terephthalate (PET), a polyester. Figure 14-7 shows the synthesis of PET.

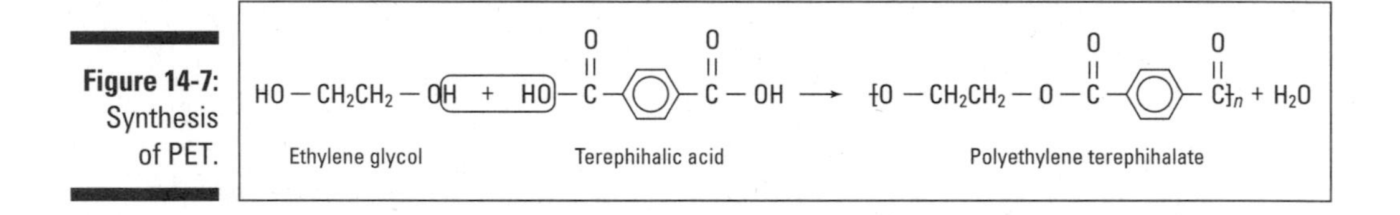

Figure 14-7: Synthesis of PET.

You find this polyester in clothing, automotive tire cord, substitute blood vessels, camera film, and soft drink bottles.

Polyamides: Sheer enough for a woman, strong enough for a (police)man

If you react an organic acid with an amine, you split off water and form an amide. If you use an organic acid that contains two acid ends and an amine that has two amine ends (a *diamine*), then you can polymerize a polyamide. The polyamide is commonly referred to as *nylon.* Figure 14-8 shows the reaction between 1,6-hexanediamine and adipic acid to form Nylon 66. (The 66 indicates that both the anime and the organic acid have six carbon atoms.)

The synthesis of nylon in 1935 had a major impact on the textile industry. Nylon stockings first went on sale in 1939, and nylon was used in parachutes extensively during World War II. Make a slight substitution in one of the carbon backbones, and you have a material strong enough for a bullet proof vest.

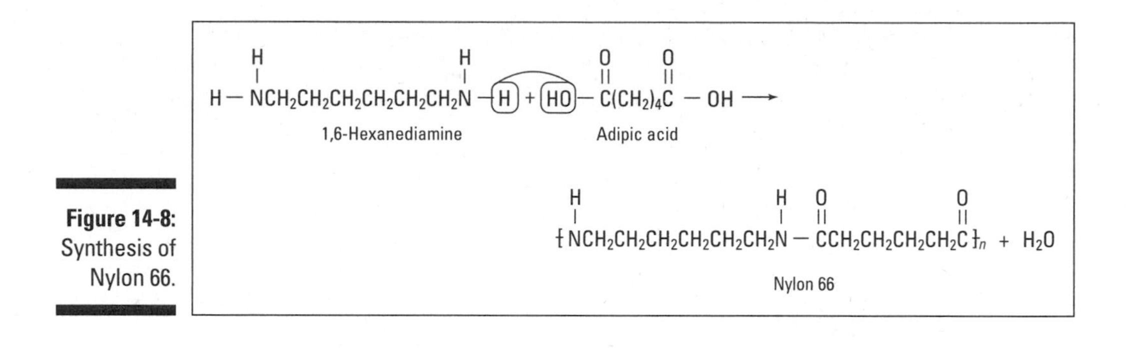

Figure 14-8: Synthesis of Nylon 66.

Silicones: Bigger and better

Because silicon is in the same family as oxygen, chemists can produce a class of polymers that contains silicon in its structure. These polymers are known as silicones. Figure 14-9 shows the synthesis of a typical silicone.

The strong silicon-oxygen bond holds the silicone polymers together; these polymers can have molecular weights in the millions. They're used as gaskets and seals, and they are found in waxes, polishes, and surgical implants. The press has given the most attention to their use as surgical implants.

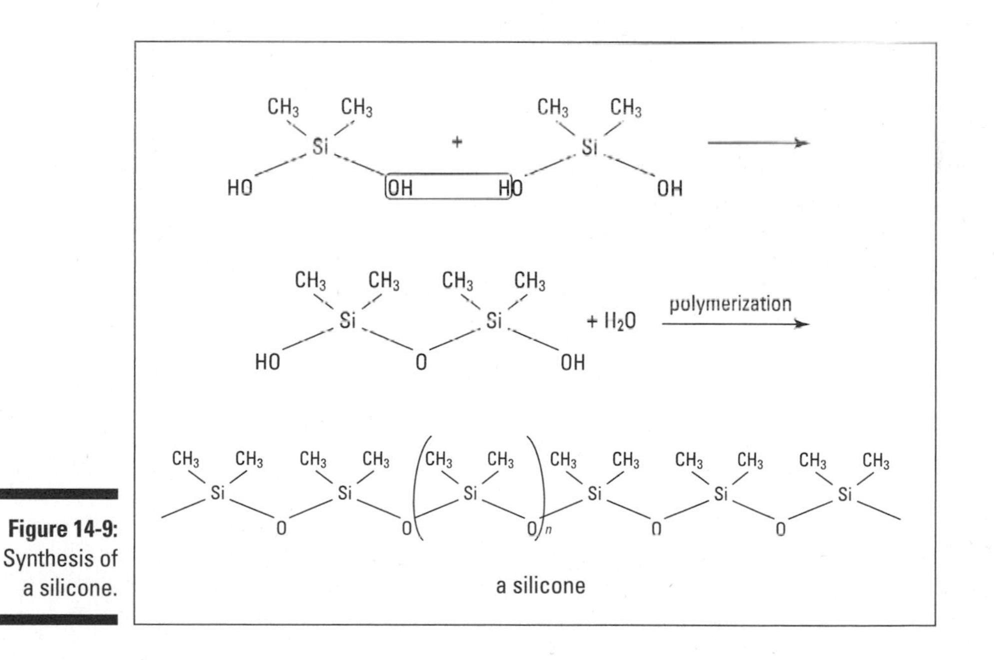

Figure 14-9: Synthesis of a silicone.

Silicone-based implants and prostheses have been used for years. They have been used as shunts, ear prostheses, finger joints, and, of course, breast implants. The implants themselves are filled with silicone oil. Occasionally, an implant leaks and this silicone oil escapes into the body. In 1992, some evidence found that silicone oil may trigger an autoimmune response. Although studies didn't establish a cause-and-effect relationship, many implants were removed, and silicone oil is no longer used in the United States.

Polymers have reshaped society as well as many people's figures. They're useful in a very wide variety of ways, they're relatively inexpensive, and they're durable.

Reducing, Reusing, Recycling Plastics

Because polymers are so durable, figuring out how to dispose of them is a major problem. Plastics have basically an infinite lifetime. Nature has not had a long enough time to develop microorganisms that will do a good job at degrading plastics. If you bury that plastic plate, Styrofoam cup, or disposable diaper in a landfill and then dig it up ten years later, you will see no change. You could even dig it up a hundred years later and get the same results. Waste that contains plastics will be around for a long time.

Some plastics can be burned as fuels. They produce a lot of energy, but they often produce gases that are toxic or corrosive. Society can reduce its reliance on plastics to a certain degree by using cardboard hamburger boxes and cellulose shipping packing instead of Styrofoam, but the best answers so far have been recycling and the engineering of biodegradable polymers.

Thermoplastic polymers can be melted down and reformed. But in order to do so, the plastics must be separated into their various components. Most plastic containers contain a symbol on the bottom that indicates what type of plastic it is made of. Recyclers can use these symbols to separate the plastics into various categories to make recycling easier. Figure 14-10 shows the recycling symbols for plastics and indicates what type of plastic each symbol represents.

PET bottles and HDPE milk cartons are probably the plastics that are recycled the most. But the major problem is not the chemistry involved in the recycling process: Encouraging individuals, families, and businesses to recycle, and developing an easy means to collect and sort the plastics for recycling are the major problems. These polymers are too valuable a resource to simply be buried in some landfill.

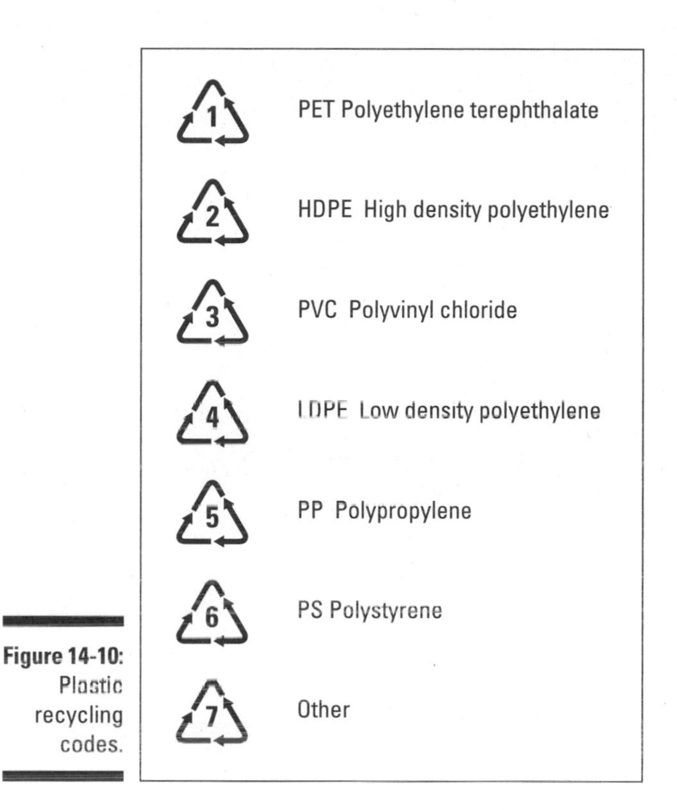

Figure 14-10: Plastic recycling codes.

The second solution, engineering polymers that are biodegradable, has made a great deal of progress during the last few years. These biodegradable plastics, called *bioplastics*, are made from *biomass*, naturally occurring organic materials, such as vegetable oil, corn starch, and other organic material. The first bioplastic that I remember seeing was packing peanuts made from a starch-based bioplastic called *thermoplastic starch*. Starch is mixed with certain chemicals that allow it to be formed like a normal polymer, but it still retains the ability to absorb moisture and thus be degraded in the environment. These packing peanuts could be tilled into the garden soil to help hold the moisture.

Polylactic acid (PLA) is another bioplastic that is transparent. It's produced from cane sugar or glucose. It has the advantage of needing no special equipment to process it. It is used in cups and bottles, and mulch film used in commercial plant production.

With the exception of thermoplastic-starch, the bioplastic industry is still in its infancy. New bioplastic products are just beginning to hit the marketplace. These include biodegradable garbage bags, shampoo bottles, plastic wrap, tea bags, and plastic utensils. The energy cost to produce the bioplastics is high, but with additional research and improved manufacturing, bioplastics can eventually replace many of the petroleum-based, non-biodegradable polymers.

Chapter 15

Bringing Biology into the Lab: Biochemistry

In This Chapter

- Considering carbohydrates
- Tackling amino acids
- Picking proteins
- Analyzing lips
- Comprehending what nucleic acids are

Biochemistry is the study of the chemical process occurring in living organisms. Many of these processes involve the same functional groups seen in Chapter 13 on organic chemistry. In some cases however, the name of the functional group is different in biochemistry, but the properties remain the same. This chapter takes an introductory look at some important classes of compound in biochemistry that you may touch upon in your Chem II class. If you want to delve into biochemistry in more detail, I highly recommend my book with Richard H. Langley *Biochemistry For Dummies* (John Wiley & Sons, Inc.).

Understanding the Basic Elements Associated with Biochemistry

The chemical processes occurring in living organisms involve a wide variety of elements. Having a firm grasp of the following elements can help you when studying biochemistry so you can see the chemical makeup of all living beings.

- **Carbon:** The central element in biochemistry, as in organic chemistry, is carbon. The factors responsible for the usefulness of carbon to biochemistry are much the same as those responsible for the wide variety of organic compounds. These factors include the ability of carbon to catenate and its ability to form strong, stable bonds to most nonmetals.

- **Hydrogen:** Hydrogen is also useful because it's a good inert substituent on carbon. As an *inert* substituent, it occupies a position on the carbon but doesn't complicate reactions as other elements would. In addition, hydrogen ions are necessary for some reactions and for pH control. The behavior of many biologically important molecules, such as DNA, relies on the ability of hydrogen participation in hydrogen bonds.
- **Elements necessary for the development systems:** Even though carbon is the key to all living organisms, a wide variety of other elements is also necessary for life. The elements necessary for the development systems in biochemistry in most cases are the more common elements, such as oxygen, nitrogen, phosphorus, and so on. A comparison of the twenty most abundant elements in the Earth's crust to the biologically important elements results in two very similar lists. One main exception — iodine — does exist. Iodine is a relatively rare element in the environment; however, iodine is extremely important to living organisms, including humans.

The different elements have many uses. Some elements are used in creating the biological structures. Most structures involve *covalent bonds* (the sharing of one or more electron pairs; see Chapter 3 for more information on covalent bonds). Carbon and the other nonmetals are important because they can readily form stable covalent bonds. However, there are structures involving ionic bonds. These structures include the exoskeletons of corals (calcium carbonate, $CaCO_3$), and the teeth and bones of mammals (calcium hydroxyphosphate, $Ca_5(PO_4)_3OH$).

Many elements, especially sodium, potassium, magnesium, calcium, and chlorine, are important as ions in solutions. The metal *cations* (positively charged ions) are very important partially because many of the biologically important compounds are *anionic* (are negatively charged ions). Cations must be present to provide electrical neutrality. See Chapter 3 for a discussion of cations and anions. These ions are also important in maintaining the ionic strength of biological fluids. The concentration of the ions is an important means of regulating the cell's *osmotic pressure* (the pressure needed to just stop osmosis). If the osmotic pressure in a cell is too low, the cell will collapse. If the osmotic pressure in a cell is too high, the cell will rupture. The relative concentrations of specific ions, such as potassium, inside and outside the cell, are important in the regulation of many biologically important functions, such as in the transmission of nerve impulses.

Biological processes, such as photosynthesis and respiration, are oxidation-reduction processes. Many oxidation-reduction processes in biochemistry involve transition metal ions. The transition metals, such as iron, exhibit more than one oxidation state that is stable under biological conditions. Simple oxidation or reduction allows the metal to change from one oxidation state to another.

Contemplating Carbohydrates

Carbohydrates are a fundamental building block of living organisms. A *carbohydrate* is a compound with the general formula $C_x(H_2O)_y$. This group includes the sugars, starches, glycogen, and cellulose. The compounds serve as energy sources, energy storage, and, in some cases, structure. These sections identify the different forms carbohydrates take.

Sticking with one sugar: Monosaccharides

A *monosaccharide* is the simplest type of carbohydrate. The functional groups in monosaccharides are a carbonyl group (as part of either an aldehyde or ketone), and one or more alcohol groups. Each of the carbon atoms present will have at least one oxygen atom attached to it.

The most common monosaccharides are the pentose and hexose sugars. A *pentose* has five carbon atoms (*pent* means five), while a hexose has six carbon atoms (*hex* means six). Additionally, you can classify monosaccharides by the carbonyl group present. If the carbonyl group is part of an aldehyde group, then the sugar is an *aldose*. If the carbonyl group is part of a ketone group, then the sugar is a *ketose*. Figure 15-1 shows these structures. Combining these two systems to produce names such as an aldopentose or a ketohexose is possible.

By convention, when drawing these structures, arrange the carbon atoms vertically with the carbonyl group as near the top of the structure as possible. The placement of the H's and OH's is very important. Switching any pair on a carbon atom will yield a different monosaccharide.

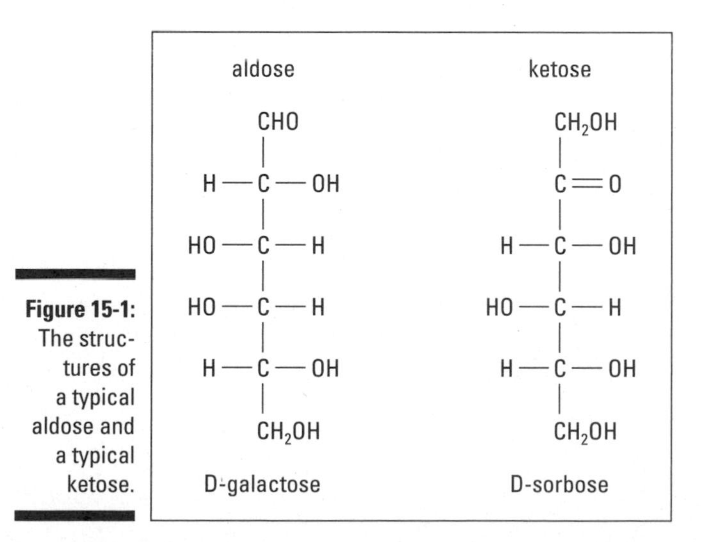

Figure 15-1: The structures of a typical aldose and a typical ketose.

The "D" in the names of these two monosaccharides refers to the effect these molecules have on light. Molecules that rotate light to either the left or right are optically active. A beam of light will rotate to the right when passing through a solution containing any "D" compound. (If there is a shift to the left, the designation "L" replaces the "D".)

The presence of an asymmetric or chiral carbon makes a compound optically active. A *chiral carbon* is a carbon atom with four different groups attached to it. In D-galactose and D-sorbose all the carbon atoms except the top and bottom one in each case and the second from the top in D-sorbose are chiral. *Fructose,* also referred to in layman terms as fruit sugar, is an example of a ketose that forms a ring structure. Figure 15-2 shows the ring structure of D-fructose.

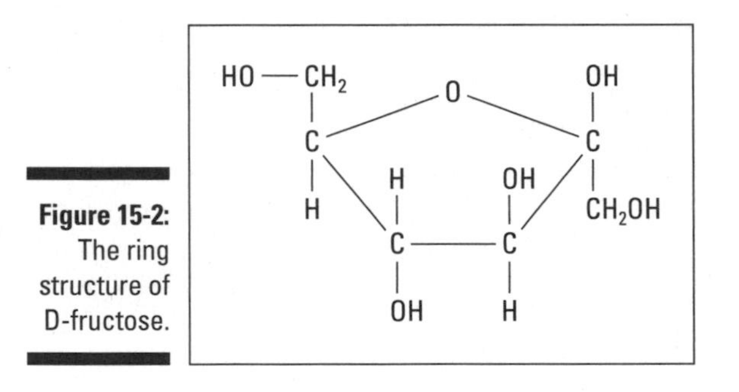

Figure 15-2: The ring structure of D-fructose.

The most common monosaccharide is *glucose,* more commonly known as blood sugar. Figure 15-3 shows the ring structures for D-glucose. (See if you see the difference in two forms of the glucose molecules.)

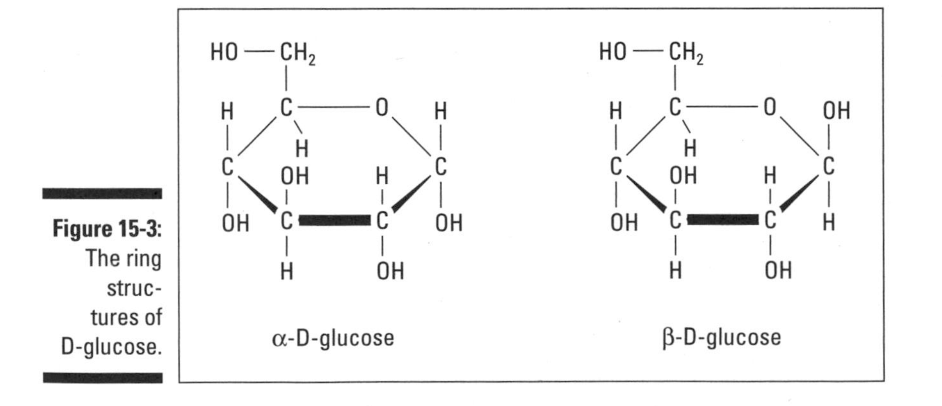

Figure 15-3: The ring structures of D-glucose.

Joining monosaccharides: Disaccharides and polysaccharides

Two monosaccharide molecules may join through a condensation reaction to produce a disaccharide. A *disaccharide* is a carbohydrate containing two monosaccharides joined by a glycoside linkage. A biochemical glycoside linkage is similar to the ether group in organic chemistry.

The joining of a large number of monosaccharide molecules by a process similar to the formation of a disaccharide results in a polymer known as a *polysaccharide*. Three common polysaccharides are

- **Starch:** The linking of numerous α-D-glucose molecules creates a polymeric chain known as starch. Starch is important for energy storage in plants. Branches may form when condensation reactions happen between different OH groups on the rings.
- **Glycogen:** Glycogen is a highly branched polysaccharide used for energy storage in many animals.
- **Cellulose:** A polysaccharide may form by linking numerous β-D-glucose molecules through condensation reactions. This polymer is cellulose. Cellulose is an important structural material for plants. The slight change in the orientation of the linkages results in a molecule that isn't susceptible to digestion by enzymes in animal stomachs.

Getting the Lowdown on Amino Acids

Amino acids are one of the important building blocks in biological systems. I define an *amino acid* (α-amino acid) as a molecule with an amino and a carboxylic acid group connected to the same carbon (the α-carbon). The amino acids are the monomers used to construct the biological polymers known as proteins. Figure 15-4 shows the general structure of an amino acid.

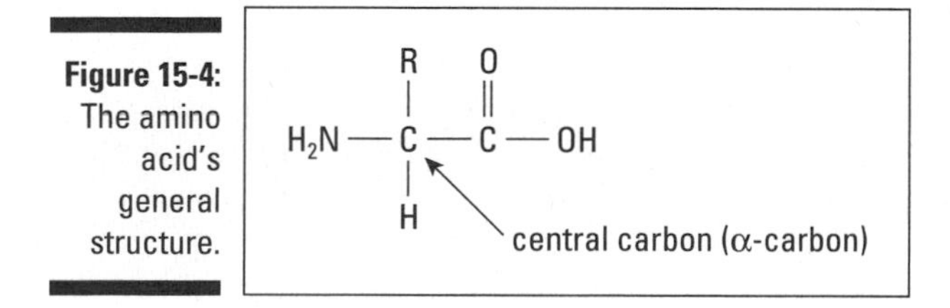

Figure 15-4: The amino acid's general structure.

The R group determines the amino acid's identity. The R-groups are important to the behavior of the amino acids and the properties of the resultant proteins because of the different ways they interact with other molecules or with other parts of the same molecule. The central carbon atom has all the groups attached to it, making it in almost all cases chiral. The natural amino acids are the L-isomers.

This molecule has both an acid group and a base group present. These two groups react to produce a molecule with the same composition but with charges as Figure 15-5 demonstrates.

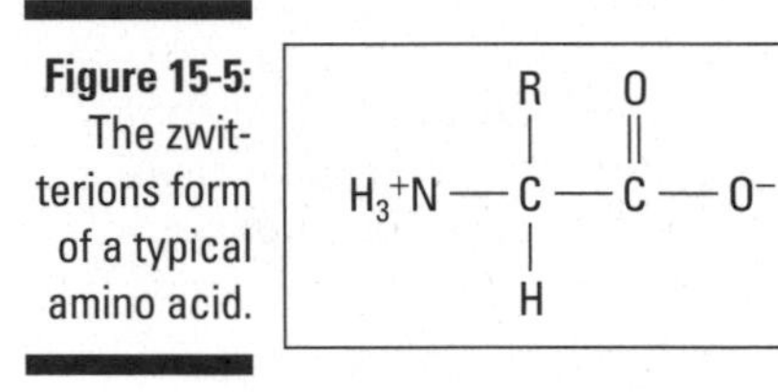

Figure 15-5: The zwitterions form of a typical amino acid.

This figure is an example of a zwitterion. The *zwitterion* form predominates at a pH known as the isoelectric point. At a pH lower than the isoelectric point, some of the O^- gains an H^+ to reform the carboxylic acid group. At a pH higher than the isoelectric point, some of the H_3N^+ loses an H^+ to produce the amine group, H_2N.

The amino acids can link together by condensation reactions to produce proteins. Figure 15-6 displays an example.

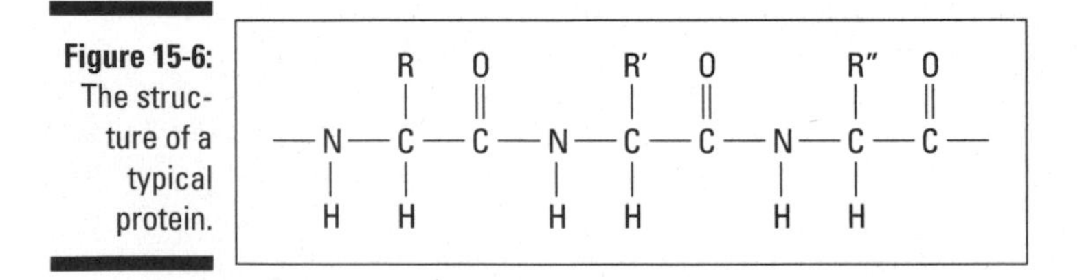

Figure 15-6: The structure of a typical protein.

In biological systems, chemists call the amide groups linking amino acids a peptide bond. A *peptide bond* is an amide functional group joining two amino acids. These three amino acids form a *tripeptide*.

Piecing Together Proteins

A *protein* is a polymer of amino acids. The amino acid groups are the monomers used in the formation of this condensation copolymer. Each amino acid is known as a *residue*. Another name for a protein is a *polypeptide*. This name alludes to the fact that the amino acids are linked by peptide bonds. Replacing the "poly" in polypeptide with a multiplying prefix, such as *tri*, is possible to indicate the number of amino acids present.

The behavior of a protein depends upon the protein's overall structure. Four levels of structure — primary, secondary, teritiary, and quaternary — are possible, which I discuss in the following sections. A protein may utilize one or more of these structures. In order to function properly, a protein must have the correct primary, secondary, tertiary, and possibly quaternary structure.

Forming the sequence: Primary structure

A protein's *primary structure* is the general sequence of amino acids. This structure is always present; all higher levels of structure come from this sequence. If you use the following tripeptide as an example of a small region of a protein, you can get some insight on working with primary structures. This tripeptide has, in order, the amino acids serine, alanine, and glycine.

Figure 15-7: The primary structure of a protein, the sequence of amino acids.

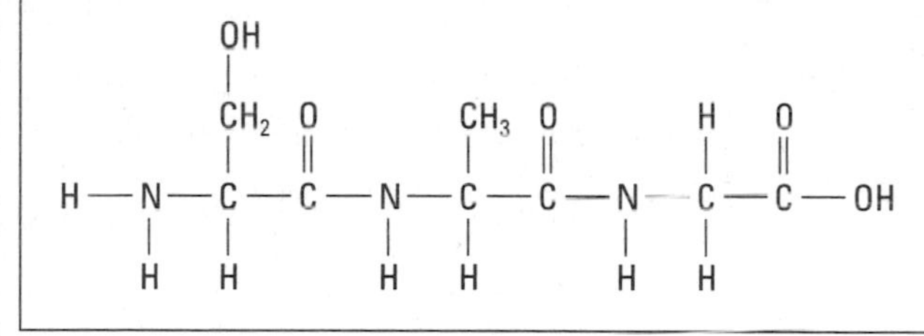

In order to avoid drawing large structures for proteins, replacing the amino acid residues with an abbreviation is possible. The protein structure simply becomes a listing of the abbreviations of the residues from the nitrogen-terminal end to the carbon-terminal end. For our tripeptide example, the abbreviations give *ser-ala-gly*.

You can use an alternate abbreviation method with a single letter for each of the amino acids. This method uses these abbreviations: serine = S, alanine = A, and glycine = G.

Bonding hydrogen: Secondary structure

The *secondary structure* is the part of the protein structure that results from hydrogen bonding between peptide bonds. Figure 15-8 shows this type of interaction.

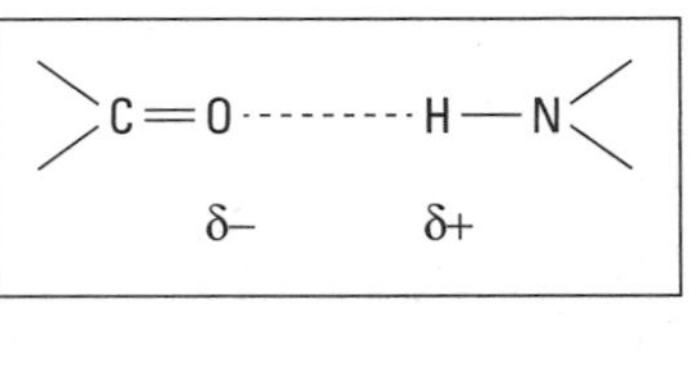

Figure 15-8: The secondary structure of a protein due to hydrogen bonding.

This figure shows an unusually strong hydrogen bond. The presence of the carbonyl group adjacent to the amide nitrogen makes N-H bond more polar than usual, and the presence of the amide nitrogen near the carbonyl oxygen makes the oxygen more polar than normal. Resonance in the peptide bonds holds the group flat and enhances the charge distribution even more.

The presence of a number of hydrogen bonds between the peptide bonds leads to an overall structure present in many proteins. The hydrogen bonds may hold the primary in a helix or coil like a spring. This secondary structure is known as an α-helix.

An alternative secondary structure produced by hydrogen bonds between peptide bonds is the β-keratin structure. This is a *pleated* sheet structure similar in cross section to the central corrugated sheet in a three-layer piece of cardboard.

Interacting with side chains: Tertiary structure

The amino acid side chains may also interact. Such interactions give rise to the tertiary structure of proteins. The *tertiary structure* is that part of the structure of a protein structure that results from any type of interactions between the side chains. These interactions include hydrogen bonding, dipole-dipole forces, London dispersion forces (momentary charge separation, which are hydrophobic interactions — see Chapter 5 for a discussion of London forces), ionic bonds, and covalent bonds.

The formation of a covalent bond occurs when two cysteine side chains interact under oxidizing conditions. Reversing this process under reducing conditions is feasible.

Cystine units are present in many proteins. One occurrence is in the protein in hair. This tertiary structure increases the incidence of curly hair. The application of a mild reducing agent to hair breaks disulfide linkages. After it's broken, reshaping the hair and applying a mild oxidizing agent to form new linkages is possible. Beauty salons follow this process when someone wishes to have his or her hair permed.

Having more than one primary: Quaternary structure

A few proteins have a quaternary structure. A *quaternary structure* is the protein structure when more than one primary structure is present. The different protein chains (primary structures) interact by any of the methods utilized in forming the tertiary structure.

Hemoglobin is an example of a protein with a quaternary structure. This molecule has four separate protein chains. This molecule, like many proteins, contains a nonprotein portion. The *heme* group is a nonprotein group containing an iron(II) ion. The iron ion is responsible for hemoglobin's ability to transport oxygen.

Looking at Lipids

Lipids are a diverse class of biochemicals. They incorporate a variety of functional groups. The property they have in common is their insolubility in water. Biological molecules that are insoluble in water are *hydrophobic* molecules. (Molecules that do dissolve in water are *hydrophyllic* molecules.) The two important classes of lipids that I briefly discuss in the following sections are the steroids and the fats.

Influencing bodily functions: Steroids

Steroids are a group of compounds that contain the same basic structure and are important lipids in the body regulating functions, aiding digestion, and so on (refer to Figure 15-9 for an illustration of this structure).

Different steroids have different functional groups attached to this basic structure, as shown in Figure 15-10.

The ins and outs of cholesterol

The body requires cholesterol to synthesize the sex hormones, adrenal hormones, and vitamin D. Excess cholesterol beyond what is necessary to synthesize other compounds is a problem. A link exists between high levels of cholesterol in the blood and medical problems such as heart disease. Due to its low solubility in water (blood plasma), cholesterol can form deposits in the blood vessels. The precipitated compound clogs the blood vessels and impedes blood flow, which leads to a condition known as hardening of the arteries. A diet low in foods containing cholesterol or that may generate cholesterol may help alleviate this problem. However, if the body detects that a person's diet has too little cholesterol, the body will synthesize the needed cholesterol for survival.

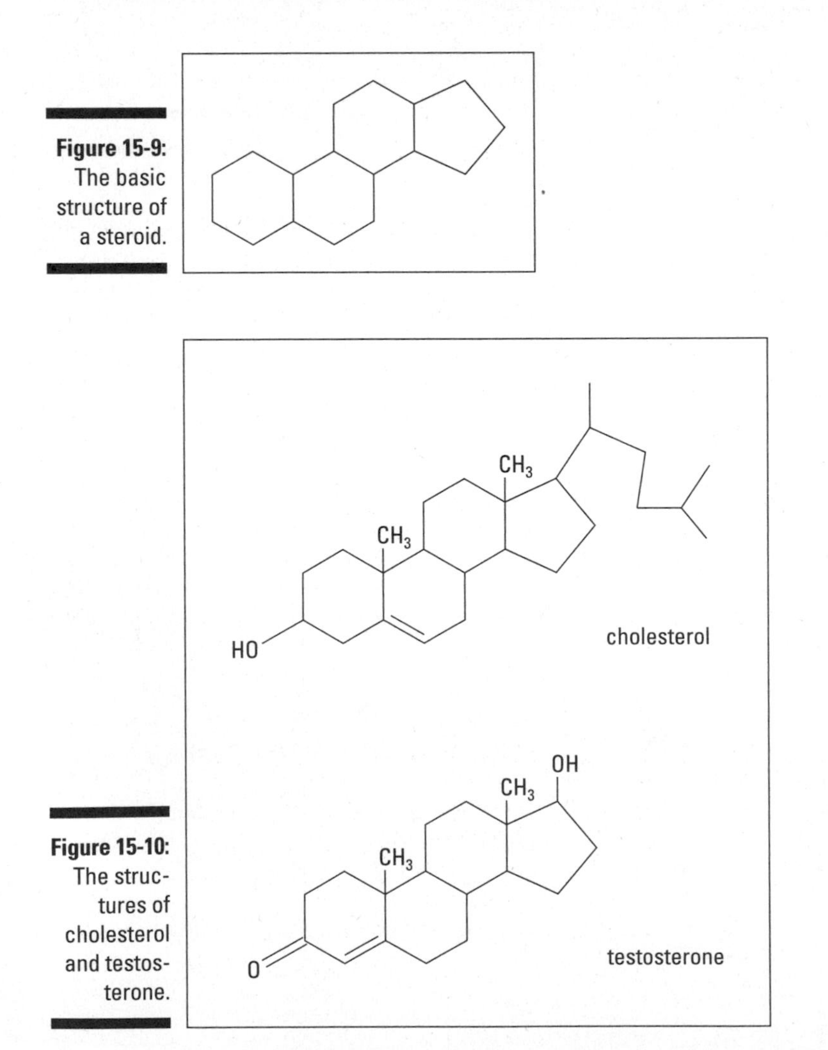

Figure 15-9: The basic structure of a steroid.

Figure 15-10: The structures of cholesterol and testosterone.

Some steroids influence certain bodily functions. These are the steroid hormones. Other steroids have useful medical applications, though abuse is possible. The best-known single steroid is cholesterol. (Check out the nearby sidebar for a quick overview of cholesterol.)

Storing energy: Fats

Fats are esters of the alcohol glycerol and long-chain carboxylic acids known as fatty acids. A fatty acid is a carboxylic acid containing ten or more carbon atoms in a chain. Fats, also referred to as *triglycerides,* are important energy storage molecules in biological systems.

Fats are formed from the reaction of a fatty acid with glycerol. Glycerol, $C_3H_8O_3$, has an alcohol group on each of the carbon atoms. In a fat, each of these three alcohol groups forms an ester. Therefore, a fat is a compound with three ester groups. Figure 15-11 shows the general structure of a fat.

One or more of the fatty acid chains in the fat may be *unsaturated* (contain one or more carbon-carbon double bonds). The presence of one carbon-carbon double bond results in an unsaturated fat, while the presence of more than one carbon-carbon double bond makes the fat polyunsaturated. The more carbon-carbon double bonds present, the lower is the melting point of the fat. A fat with a very low melting point may be a liquid at room temperature. The term *oil* refers to a fat that is a liquid at room temperature. The carbon-carbon double bonds in natural fats are always the cis-isomer, while in manmade fats, both the cis- and trans-isomers may be present. All natural fatty acids have an even number of carbon atoms.

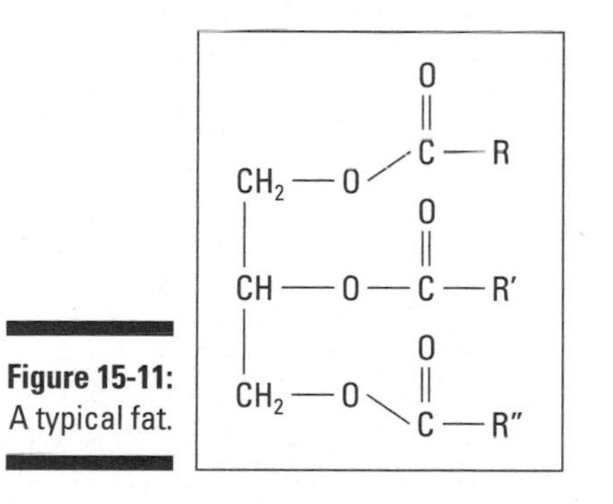

Figure 15-11: A typical fat.

Fats can undergo different reactions that affect their use. The following are those reactions:

- **Hydrogenation:** In the presence of a catalyst, the carbon-carbon double bonds in a fat may undergo *hydrogenation,* an addition reaction. By controlling the amount of hydrogen gas available, controlling the number of bonds undergoing hydrogenation is possible. The partial hydrogenation of a liquid polyunsaturated fat produces a fat with a higher melting point. This process is useful in converting oils into solid fats such as margarine.
- **Hydrolysis:** Fats, due to the presence of ester groups, may undergo *hydrolysis*, which is a reaction with water. Either acids or bases are necessary as hydrolysis catalysts. During digestion, a small amount of fat may undergo hydrolysis under the acid conditions in the stomach; however, most of the hydrolysis of fats occurs under the basic conditions in the small intestine. After hydrolysis, absorption of the hydrolysis products by the body is possible.
- **Micelle formation:** The ability of a soap to function as a cleansing agent is a result of *micelle formation.* This soap can work as a cleansing agent due to the carboxylate ion containing a long nonpolar hydrocarbon chain and an ionic group. The nonpolar portion is soluble in other nonpolar materials such as grease and oil. The ionic group is soluble in polar solvents such as water. When a soap solution encounters an oil particle, the long hydrocarbon chain dissolves in the oil, while the ionic end remains outside. As more soap units dissolve, the outside of the particle takes on a negative charge. This soap/oil combination is a *micelle*. The negative charge leads to each soap/oil particle repelling similar particles and keeps an oil particle from coming back together. To the solvent, water, the charged particle, appears as one large ion. Because it is an ion, the particle is water-soluble.
- **Synthetic soaps:** Detergents have structures similar to soaps; however, a phosphoric acid or sulfuric acid replaces the carboxylic acid group. Soaps tend to be less efficient in hard water because the present metal ions form precipitates with the carboxylate ions present in soap. (The metals ions normally present in hard water are magnesium, calcium, and iron.) Detergents tend not to form precipitates in hard water.
- **Lipid interactions:** Other lipids perform a function similar to soaps in the formation of a micelle. For these lipids, the long nonpolar hydrocarbon portions of the different lipid molecules *dissolve* into each other to form a layer two molecules thick. The outside of this layer has the polar portions of these lipid molecules. This biological structure is a lipid bilayer. A *lipid bilayer* serves as a barrier to diffusion of the material into a cell or out of a cell.

Observing Nucleic Acids

An organism must absorb materials to survive. The organism's nucleic acids aid in the absorption. *Nucleic acids* got their name because they were first found in the cell nucleus and are basically responsible for cell replication. The two most famous nucleic acids are RNA and DNA. The nucleic acids, especially DNA, are large polymeric molecules. The molar mass of a DNA molecule may approach 10^9 grams per mole.

Nucleic acids use an organism's diet in several ways. An organism's diet may include complicated molecules like polysaccharides and proteins. During digestion, the organism breaks complicated molecules down into smaller units, such as monosaccharides and amino acids for absorption. After the molecules are absorbed, the organism must build the complicated molecules necessary for survival. Various catalysts, in the form of enzymes, are necessary for the synthesis of these molecules. *Enzymes* are proteins designed to serve a very specific function. The organism must synthesize enzymes and other proteins. The enzymes facilitate the synthesis of other biological important molecules.

The information directly responsible for the synthesis of the enzymes comes from RNA. RNA obtains this information from DNA. These sections take a closer look at what nucleic acids are and how they work in living organisms.

Eyeing the parts of nucleic acids

All nucleic acids have three components. The following list discusses these components and explains how they interact.

- **Five-carbon sugar:** The five-carbon sugar may be either ribose or deoxyribose. Ribose is present in RNA, which is short for ribonucleic acid. Deoxyribose is present in DNA, which is short for deoxyribonucleic acid. The difference between these two sugars is that deoxyribose has one less oxygen atom. Figure 15-12 shows what RNA and DNA looks like.
- **Nitrogen bases:** Five nitrogen bases occur in the nucleic acids. Three of the bases, adenine, guanine, and cytosine, occur in both DNA and RNA. Thymine is only present in DNA, while uracil only occurs in RNA.

  The nitrogen bases are capable of forming hydrogen bonds to each other. The geometry of the bases is such that only certain combinations can efficiently form hydrogen bonds. Adenine can form hydrogen bonds to either thymine or uracil. Cytosine can effectively form hydrogen bonds to guanine. These hydrogen bonds are not only responsible for the double helix structure of DNA, but also for supplying the information necessary to synthesize enzymes and other proteins.

- **Phosphoric acid:** The addition of the phosphate groups increases the potential energy within the molecule. The reverse reaction, the loss of the phosphate groups, releases this energy. The reverse process is a hydrolysis reaction. Cells produce ATP (a nucleic acid), which migrates to other parts of the cell where hydrolysis releases this stored energy. ATP is an important biological for short-term energy transfer.

DNA is normally present in the cell nucleus in the form of a double helix. Two strands of DNA hydrogen bond to each other. The arrangement of the nitrogen bases is such that an adenine from one of the chains hydrogen bonds to a thymine in the other chain, and a cytosine from one of the chains hydrogen bonds to a guanine of the other chain. The two hydrogen-bonded strands wrap into a helix.

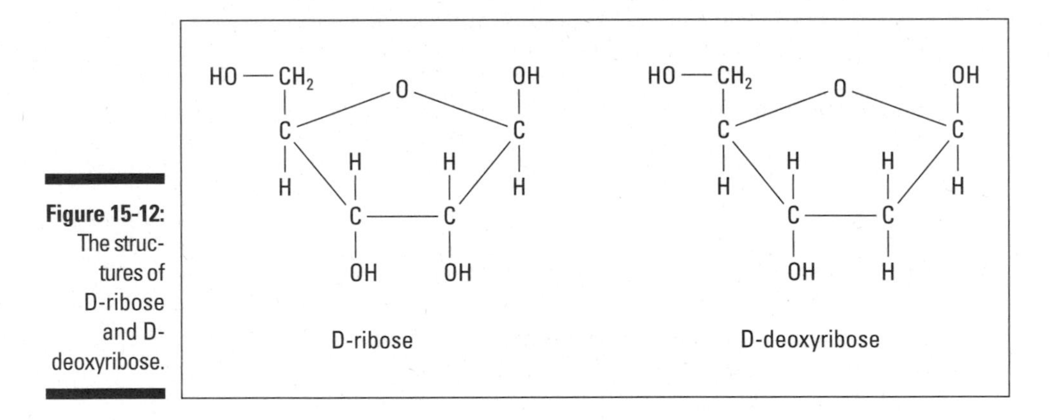

Figure 15-12: The structures of D-ribose and D-deoxyribose.

Seeing what nucleic acids do

Nucleic acids play several important roles in living organisms. They can be broken down into the three following roles:

- **Replication:** *Replication* is the process that generates a new DNA molecule from one DNA to two DNA molecules. When a cell divides, the information contained within the DNA stands passes on to the two cells that form. This process involves the DNA strand unwinding and opening like a zipper. As the two DNA strands separate, new nucleotides enter and hydrogen bonds to the nitrogen bases already present. This process occurs on both strands. Linking the entering nitrogen bases creates a new DNA strand hydrogen bonded to each of the old strands. This process gives two double DNA strands each identical to the original double strand. When the cell divides, one of the DNA double strands goes to the nucleus of each of the forming cells.

- **Transcription:** In a related process, known as transcription, DNA molecules generate RNA molecules. *Transcription* is the process that forms RNA molecules from DNA molecules. In this process, only a portion of the DNA double strand opens. The entering nucleotides are those necessary to generate RNA. These are the same as those necessary to generate DNA, except that uracil replaces thymine, and ribose is the five-carbon sugar. Linking these nucleotides generates RNA.
- **Translation:** The process ends with a process called *translation.* In translation the pattern carried by RNA (specifically messenger-RNA or mRNA) is used to put together specific amino acids into a specific order in order to make a specific protein. The RNA molecules carry information from a small region of the much larger DNA molecule. The RNA molecules diffuse out of the nucleus for the next step in the synthesis of proteins.

Defecting molecules leads to mutations

During all steps in replication, transcription, and translation, hydrogen bonding is important. An error in the translation stage leads to one defective protein molecule, which normally is not a problem. An error in the transcription stage leads to all proteins generated from that particular mRNA being defective. This defection is generally no big deal for the cell, because many other mRNA molecules are present to generate the correct protein. However, an error at the replication stage means that this erroneous information can go to all subsequent generations. Such a change is a *mutation.* A cell has mechanisms to minimize the chance of a mutation or to correct such errors.

A number of potential consequences can result from a mutation. There is a slight change that the new sequence may code for the same amino acid. If the coded amino acid is the same, the protein formed will be the same. The most likely change will be the formation of a different amino acid. The new amino acid will yield a new protein. The new protein may or may not function like the old protein. If it behaves differently, it may be either better or worse than the old one. Improved function can make the cell better able to survive. The effect of the new amino acid will depend upon the identity of the amino acid and its position within the protein.

REMEMBER

The worst type of mutation is where only a portion of the protein forms. This portion won't be able to function properly. If this protein is necessary for the survival of the organism, its absence, in complete form, will lead to the death of the organism.

Designing drugs: Biotechnology bringing biochemistry and chemistry together

Biotechnology is the application of chemistry and biochemistry to make or modify the products of living organisms. This field is an extremely fast-growing area of science and holds a great deal of potential for future advancement. Biotechnology has been involved in genetics in the design of bacteria and yeasts that will metabolize certain substances, such as toxic materials in chemical landfills. Forensic scientists have used biotechnological techniques in a number of ways, including DNA fingerprinting. Agricultural scientists have used it in designing more disease-resistant crops. Pharmacologists have used it in medicine in the design of drugs.

In the past, the design of new drugs was often trial and error, and sometimes serendipity. If scientists identified a natural substance had medical properties, they tested it on animals. If they discovered a positive effect, they isolated the active ingredient and developed synthetic methods in order to produce the drug in larger quantities and more cheaply. A classic example is the discovery and development of aspirin.

Since the invention of aspirin, scientists have discovered several other over-the-counter analgesic (pain-relieving) drugs. One of the considerations in the search for new drugs, such as analgesics, is its biological *half-life,* the amount of time that it takes for one-half of the drug to leave the blood stream. The goal is to find a drug that is effective and with a long-enough half-life so that the dosage will not have to be repeated too often. The shorter the half-life the more often a new dose of the drug will need to be administered. In general, a drug's half-life depends on first-order kinetics.

With society's modern understanding of the action of drugs, bypassing the trial-and-error process and designing drugs that are involved in a condition's or disease's specific biological target is possible. Researchers may use computational methods to design drugs of a specific structure that can interact with the reactive site of the targeted molecule. Chemists can then synthesize and test these molecules. However, serendipity still plays a role in drug development.

Part IV
Describing Descriptive Chemistry

In this part . . .

In the first chapter of this part, I show you the chemistry of petroleum. I sling terms like cracking and reforming around, as well as discuss what that octane rating really means. Maybe I can give you a feel for the processing that goes into the gasoline you use in your automobile. In Chapter 17, I leave an old fuel (petroleum) for a new fuel — nuclear power. I explain the different types of nuclear decay processes, discuss fission and fusion, and show to how to deal with half-life problems.

Chapter 18 delves into the chemistry in the home. General chemistry typically doesn't cover this topic, but I believe that you deserve an opportunity to explore the practical side of chemistry a little. I discuss the chemical nature of soaps and detergents, deodorants and antiperspirants, and drugs, such as Minoxidil and Viagra. I hope it is an hair-raising and uplifting experience.

Chapter 16

Examining the Ins and Outs of Petroleum

In This Chapter

- Discovering how petroleum is refined
- Understanding the gasoline story
- Finding alternatives to gasoline

Today's modern society is based on petroleum. Automobiles run on gasoline, and most homes are heated with petroleum. Petroleum provides the feedstock for the expansive petrochemical industry. Petroleum is used to make plastics, paints, medicines, textiles, herbicides, and pesticides; the list is almost endless. Every year the Unites States consumes more than six billon barrels of petroleum. Wars have been fought over petroleum, and countries rise from poverty to prosperity thanks to petroleum.

In this chapter, I show you how petroleum is refined and converted into useful products. I concentrate on the production of gasoline, because it's one of the most economically important uses of petroleum. I show you some of the problems that have been caused by the reliance on the internal combustion engine.

Understanding How Crude Oil Is Refined

In Chapter 13, I discuss hydrocarbons, those compounds composed of just carbon and hydrogen. Crude oil (sometimes referred to as Black Gold or Texas Tea), as it comes out of the ground, is a complex mixture of hydrocarbons of varying molecular weights. The lighter hydrocarbons are gases dissolved in the liquid mixture, while the heavier hydrocarbons are higher molecular weight solids that are also dissolved in the liquid. This complex organic soup mixture was formed from decaying animal and plant material that was in the earth's crust for a very long time. Because it takes an extremely long time for petroleum to form (many millions of years), it's called a *nonrenewable resource.*

Before the hydrocarbon mixture can really be of much economic value, it must be refined. The refining process separates the mixture into groups of hydrocarbons, and in some cases, the molecular structure of the hydrocarbons is changed. The refining process occurs at a refinery. The refineries produce the refined mixtures and individual compounds that are used for gasoline and feedstock for the vast petrochemical industry. A number of processes occur at the refinery, starting with the fractional distillation of the crude petroleum. The following sections outline these processes.

Separating chemicals: Fractional distillation

Distillation is an important procedure in organic chemistry, and it is the first step in the refining process. The distillation process that is commonly used in the refining industry is fractional distillation.

You may have had the experience of simmering soup in a covered pot on the stove. You remove the lid and find water on the inside of the lid. The heat has caused a little of the water in the soup to evaporate from the liquid, and the vapors have condensed back into a liquid on the inside of the cooler lid. This is the most basic example of a process called distillation. In the laboratory, I can take a mixture of liquids and carefully heat them. The liquid with the lowest boiling point will boil first. I can then condense this vapor back to a liquid and collect it. The substance with the next highest boiling point will then begin to boil, and so on. I can use the process of distillation as a means for separating the components of a mixture and in many cases is the first step in purifying them.

In fractional distillation, the petroleum mixture is heated and different fractions (groups of hydrocarbons with similar boiling points) are collected. Figure 16-1 shows the fractional distillation of crude oil.

The crude oil is brought into the refinery by pipeline and is initially heated and vaporized in a furnace. The hot vapors are then allowed to enter a huge distillation column, called a *fractional distillation tower.* The vapors containing the lowest molecular weight hydrocarbons rise to the top of the tower and are collected. The higher the molecular weight of the hydrocarbons, the lower the level to which they will rise. These fractions, or mixtures of hydrocarbons that have much closer boiling points than the original crude oil, are then collected. Because the compounds in the fractions will be somewhat similar in size and complexity, they can be used for the same purposes in the chemical industry. About six fractions are commonly collected.

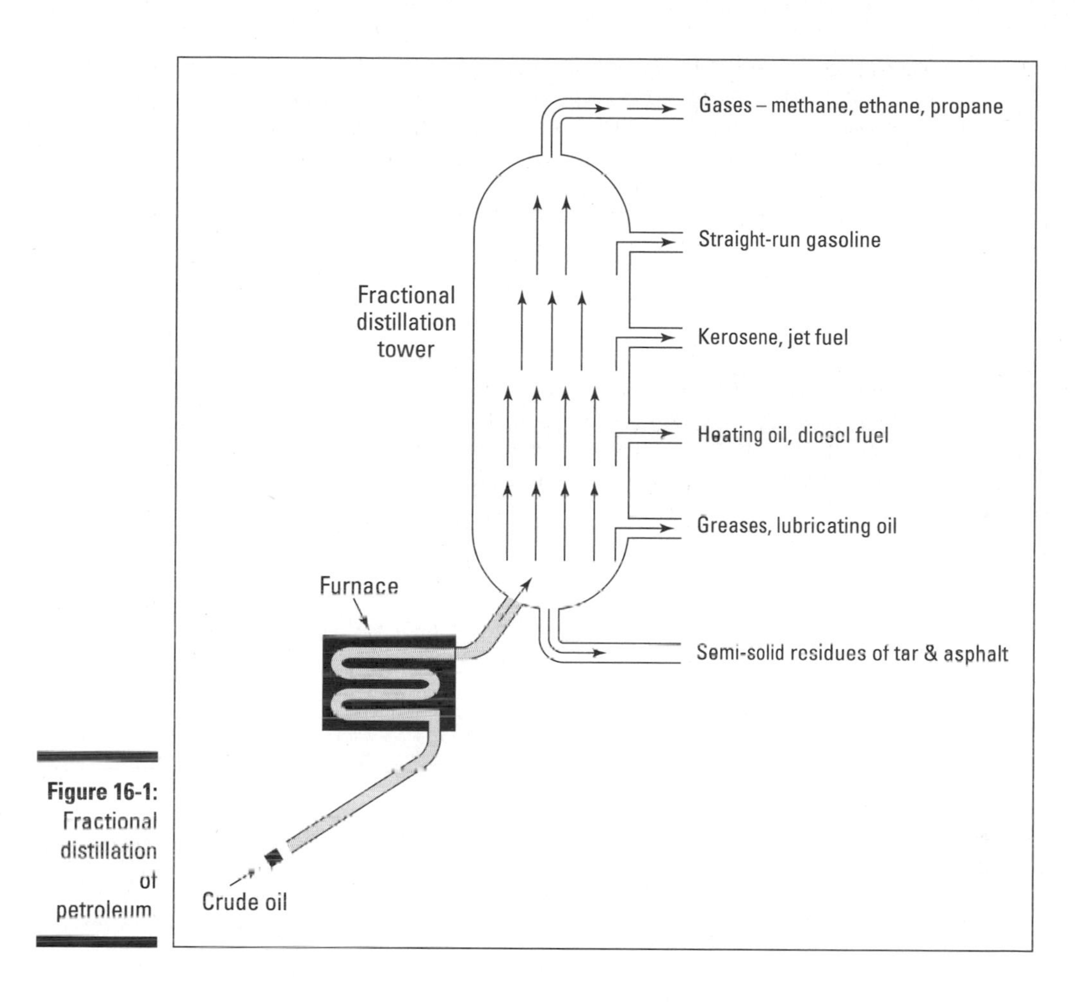

Figure 16-1: Fractional distillation of petroleum

- The first fraction is composed of the lowest molecular weight hydrocarbons, which are gases. A major component of this fraction is *methane* (CH_4), a gas that is sometimes called "marsh gas" because it was first found in marshes. Its primary use is as a fuel, natural gas, because it's a very clean-burning gas. *Propane* (C_3H_8) and *butane* (C_4H_{10}) are also found in this fraction. These two gases are normally collected and put under pressure, a process that causes them to liquefy. A truck can then transport them as liquefied petroleum (LP) gas where they can be used as fuel. The propane is commonly sold for use in gas grills and the butane for lighters. This fraction is also used as starting materials in the synthesis of plastics.

- The second fraction is composed of hydrocarbons of C_5H_{12} (pentane) to $C_{12}H_{26}$ (dodecane), with boiling points below the 150 degree Celsius range. This fraction is commonly called *natural gasoline* or *straight-run gasoline* because it can be used in automobile engines with little additional refining. With each barrel (42 gallons) of crude oil that starts out in the tower, less than a quarter of a barrel of straight-run gasoline is produced.
- The third fraction is composed of hydrocarbons of 12 to 16 carbon atoms in the boiling range of 150 to 275 degrees Celsius. This fraction is used as *kerosene* and *jet fuel*. Later in this chapter, I tell you how this fraction is also used to make additional gasoline.
- The fourth fraction is composed of hydrocarbons in the 12 to 20 carbon-atom chains, with a boiling range of 250 to 400 degrees Celsius. This fraction is used for *heating oil* and *diesel fuel*. Again, it can be used in the production of additional gasoline.
- The fifth fraction is composed of hydrocarbons in the 20 to 36 carbon atom range, with boiling points of 350 to 550 degrees Celsius. They are used as *greases, lubricating oils,* and *paraffin-based waxes.*
- The sixth fraction is composed of the residue of semisolid and solid materials. It is used as *asphalt* and *tar*.

Cracking up: Catalytic cracking

A barrel of crude oil yields a wide variety of products, but they aren't all the same value. Gasoline is the product of petroleum that is in the highest demand. No way can the straight-run gasoline fraction that comes directly from the crude oil keep pace with the demand for gasoline. Somebody got the bright idea that if they took a fraction of higher molecular weight hydrocarbons and broke it down into smaller chains, the lower molecular weight hydrocarbons could be used for gasoline. The idea of *catalytic cracking* was born!

In a catalytic cracking plant ("cat crackers," as they are called in Texas), fractions in the C_{12} to C_{20} range are heated in the absence of air with a catalyst. This process causes the long *alkanes* (compounds of carbon and hydrogen with only carbon-to-carbon single bonds) to break apart into smaller alkanes and *alkenes* (hydrocarbons with at least one carbon-to-carbon double bond). For example, suppose that you take $C_{20}H_{42}$ and "crack it":

$$CH_3{-}(CH_2)_{18}{-}CH_3 \rightarrow CH_3{-}(CH_2)_8{-}CH_3 + CH_2 = CH{-}(CH_2)_7{-}CH_3$$

This process yields hydrocarbons that are useful in the production of gasoline. In fact, the double bonds actually give it a higher octane rating, as I show you later in this chapter.

In order to produce more gasoline, the fraction that is used for heating oil is broken down into smaller hydrocarbons that are suitable for gasoline. Using this fraction can present a problem if a severe winter hits and the demand for heating oil skyrockets. Oil companies watch the long-range weather forecasts closely. In the summer, when the demand for gasoline is high, fractions that can be used for heating oil are converted to gasoline to meet the demand. Then, as fall arrives, the refineries shift their production schedule somewhat. They reduce the amount of gasoline they produce and increase the amount of heating oil so that the winter demand for heating oil can be met. But the refineries don't want to overproduce heating oil and have to store large amounts, so they try to second-guess the weather to develop a supply that will meet the demand. If the winter is unexpectedly severe, however, the demand will outstrip the supply, and people will run out of heating oil and perhaps freeze. It's a real balancing act.

Moving molecular parts around: Catalytic reforming

As the internal combustion engine gained popularity as a mode of transportation, chemists noted that if the gasoline contained only straight-chained hydrocarbons, it would not burn properly; it would have a tendency to "knock" or "ping." Hydrocarbons with branched structures burned much better. In order to increase the amount of branching in the petroleum hydrocarbon fraction being used for gasoline, a process called *catalytic reforming* was developed. In this process, the hydrocarbon vapors are passed over a metal catalyst such as platinum, and the molecule is rearranged into one with a branched structure or even a cyclic structure. Figure 16-2 shows the catalytic reforming of n-hexane to 2-methylpentane and to cyclohexane.

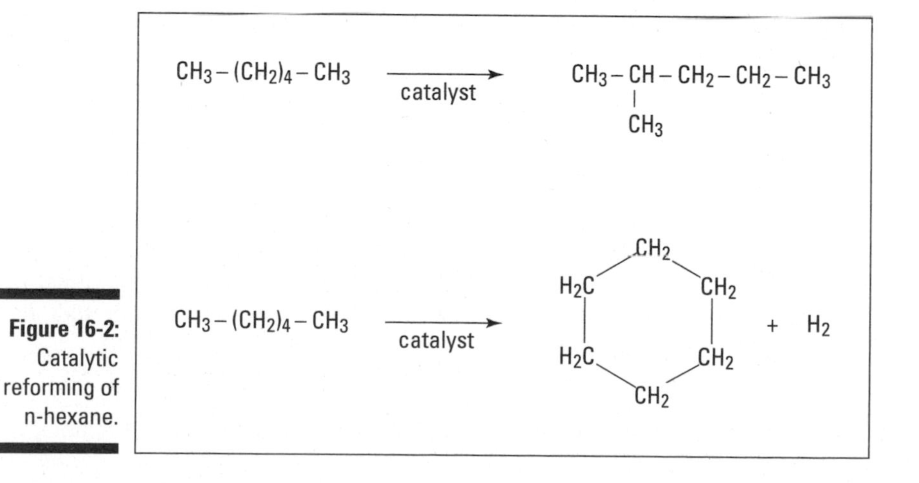

Figure 16-2: Catalytic reforming of n-hexane.

This same process is used extensively to produce benzene and other aromatic compounds for use in the manufacture of plastics, medicines, and synthetic materials. (For a discussion of aromatic compounds, see Chapter 13.)

Telling the Gasoline Story

In order for you to better understand the properties of gasoline, you need a basic understanding of how gasoline burns and reacts in an internal combustion engine. The gasoline is mixed with air (a mixture of nitrogen, oxygen, and so on) and injected into the cylinder as the piston moves to the bottom of the cylinder. The piston then begins to move upward, compressing the gasoline-air mixture. At just the right moment, the spark plug fires, igniting the mixture. The hydrocarbons react with the oxygen in the cylinder, producing water vapor, carbon dioxide, and, unfortunately, large amounts of carbon monoxide. This reaction is an example of converting the potential energy contained in the hydrocarbon bonds to the kinetic energy of the hot gas molecules. The increase in the number of gas molecules boosts the pressure tremendously, shoving the piston down. The linear motion is then converted to a rotary motion, which powers the wheels. Off you go!

The efficient burning of the gasoline in an automobile engine is a complex matter, highly dependent on the properties of the hydrocarbon mixture. The following sections explore some of these properties — the ins and outs of gasoline.

Converting gasoline into a vapor: Volatility

The gasoline-air mixture must ignite at exactly the right moment in order for the engine to operate properly. This process is largely a property of the gasoline and not the engine itself (assuming the timing is set correctly, the spark plugs are good, the compression ratio is okay, an so on). The *volatility* (how easily it is converted into a vapor) of the hydrocarbon fuel is important. Volatility is related to the boiling point of the hydrocarbon. In fact, manufacturers blend (adjust the hydrocarbon mixture) their gasoline to match the climate.

Winter gas is more volatile than summer gas. Some fuels are prone to produce knocking or pinging in an engine. This propensity to cause knocking or pinging may be a result of *preignition,* where the igniting of the gasoline occurs before the compression of the fuel-air mixture is complete, or *spotty ignition,* where combustion starts taking place at a number of sites in the cylinder instead of right around the spark plug electrode. Again, this is a property of the gasoline and not the engine. The energy content of the fuel is important, but how efficiently it burns in the cylinder is just as important.

Determining how good gas is: Octane ratings

The *octane rating scale* was developed to rate the burning characteristics of a gasoline. In the early stages of the development of the internal combustion engine, scientists and engineers found that certain hydrocarbons burned well in an internal combustion engine. They also found that certain hydrocarbons did not burn well in these engines. One hydrocarbon that didn't burn well was n-heptane (straight-chained heptane). However, 2,2,4-trimethylpentane (commonly called *isooctane*) had excellent burning characteristics. These two compounds were chosen to define the *octane rating scale.* The hydrocarbon n-heptane was assigned an octane rating of zero, while isooctane was given a value of 100. Blends of gasoline are then burned in a standard engine, compared to the scale and are rated. If a particular gasoline blend burns 90 percent as well as isooctane, then it is assigned an octane value of 90, and so on. Figure 16-3 shows the octane scale and the octane values of certain pure compounds.

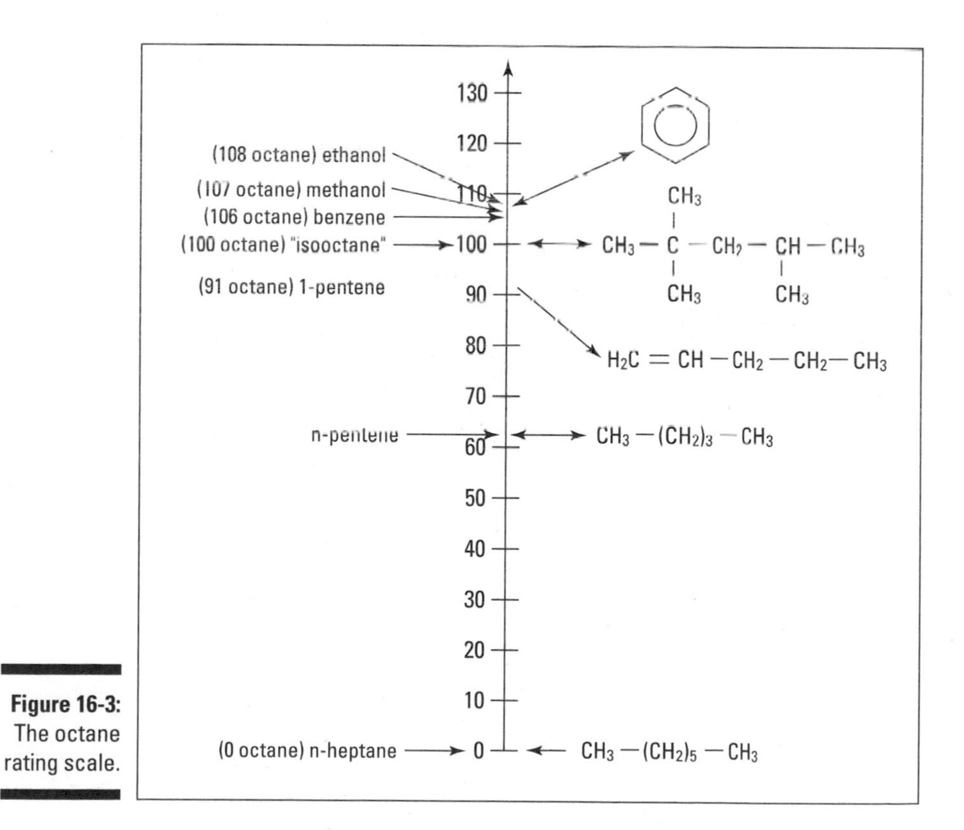

Figure 16-3: The octane rating scale.

Figure 16-3 highlights a couple of things that are useful to note in terms of octane rating and chemical structure. The n-pentane has an octane value of 62. Its octane value can be increased to 91 by introducing a double bond (making it 1-pentene) and making it an unsaturated compound. The octane value increases by almost 30-point with the introduction of the double bond. Recall that the process of catalytic reforming introduces chains, and cracking introduces double bonds. Not only do these two processes increase the amount of gasoline that is produced, but they also improve the quality of the gasoline's burning characteristics.

Also notice that benzene, an aromatic compound, has an octane value of 106. Its burning characteristics are better than isooctane. Other substituted aromatic compounds have octane ratings of almost 120. However, benzene and some related compounds are health hazards, so they aren't used.

The octane rating that is posted on gas pumps is really an average of two kinds of ratings. The *Research octane rating (R)* relates to the burning characteristics of the fuel in a cold engine. The *Motoring octane value (M)* refers to how the fuel behaves while you are cruising down the interstate. If you average *R* and *M,* — $(R+M) \div 2$ — you get the posted octane rating.

Put the lead in, get the lead out: Additives

The first gasoline engines had a compression ratio that was much lower than today's automobile engines and required lower octane gas. However, as engines became more powerful, gasoline with a higher octane rating was required. Cracking and reforming added significant cost to the gasoline. The search was on for something cheap that could be added to the gasoline to effectively increase the octane rating. The substance TEL was found.

Adding a little bit of tetraethyllead (1 milliliter per liter of gasoline), or TEL, to gasoline increases its octane rating by 10 to 15 points. *Tetraethyllead* is basically a lead atom with four ethyl groups attached to it. Figure 16-4 shows the structure of TEL.

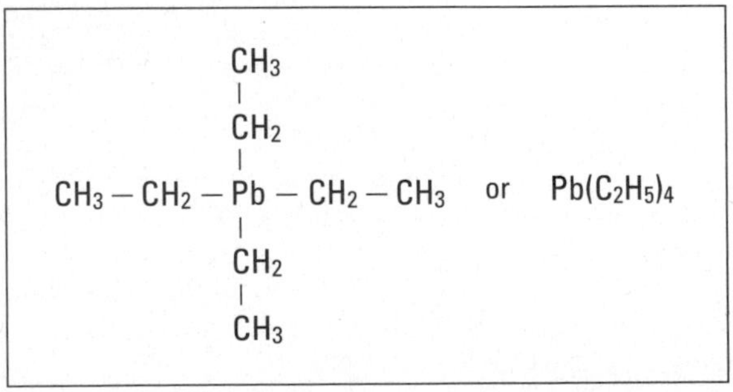

$$
\begin{array}{c}
CH_3 \\
| \\
CH_2 \\
| \\
CH_3-CH_2-Pb-CH_2-CH_3 \\
| \\
CH_2 \\
| \\
CH_3
\end{array}
\quad \text{or} \quad Pb(C_2H_5)_4
$$

Figure 16-4: The composition of tetraethyllead.

TEL was quite effective as an additive to increase the octane rating and prevent engine knocking. It was used for many years. However, the Clean Air Act of 1970 indirectly did it in.

Oops! Automobile exhausts are bad for the air

Earlier in this chapter, I discuss how hydrocarbon fuel burns in the cylinders of internal combustion engines. During this process, not all of the hydrocarbon molecules are converted to water and CO/CO_2. Unburned hydrocarbons and oxides of both sulfur and nitrogen were being released into the environment from automobiles (along with lead from the TEL that was later discovered to be very toxic). These gaseous pollutants dramatically increased the amount and severity of air pollution and gave rise to health hazards like photochemical smog.

The catalytic converter to the rescue

In the United States, the Clean Air Act of 1970 mandated the reduction of automotive pollutant emissions. The most effective way to accomplish the reduction of emissions was through the use of the *catalytic converter.* A catalytic converter is shaped like a muffler and is connected to an automobile's exhaust system. It has a solid catalyst, either palladium or platinum, inside. When the exhaust gases pass over the catalyst, the catalytic converter helps to complete the oxidation of the hydrocarbons and carbon monoxide to carbon dioxide and water.

Lose the lead

The catalytic converter worked well at reducing the automotive emissions as long as no lead was in the fuel. If leaded gasoline was used, the lead vapor in the exhaust gases coated the catalyst, rendering it useless. So the government and environmental groups pushed for getting the lead out. Now finding leaded gasoline in the United States is very difficult, although it is still available in some foreign countries.

With TEL no longer available as an octane booster, chemists tried to find other compounds to replace it. Aromatic compounds were effective in enhancing the octane value, but they were discovered to be serious health hazards. Recently, methyl alcohol, tert-butyl alcohol, and methyl tert-butyl ether (MTBE) have been used as octane boosters. Although none of these are as effective as TEL, the partial redesign of the internal combustion engine has allowed the use of slightly lower octane fuels.

MTBE has been used extensively as a gasoline additive. Not only does it boost the octane rating but it also acts as an *oxygenate,* a compound containing oxygen that increases the efficiency of the complete hydrocarbon combustion. However, MTBE was removed from gasoline due to increasing evidence that it was related to respiratory illnesses and possible cancers in humans. It is unlikely that it will be reintroduced.

Eyeing Alternatives to the Internal Combustion Engine

With the obvious problems associated with the internal combustion engine, why hasn't it simply been banned? The simple answer is that no economical, environmentally friendly, and safe alternative to the internal combustion engine is available that matches society's accustomed style of transportation. Check out some of the proposed alternatives:

- **Electric vehicles:** This type of vehicle has gained a little in popularity and use in recent years. Batteries supply all of the energy for the car's operation. Electric vehicles do not produce pollution at the vehicle site. If, however, the electricity that will be used to recharge the batteries is produced from the combustion of fossil fuels, electric vehicles simply switch the pollution burden from one site to another. These vehicles also have short driving ranges and long recharge times. They may, however, be well suited for urban commuting.
- **Hybrid electric vehicles (HEVs):** This type of vehicle has both a small gasoline engine and a battery/electric motor combination. In one configuration, the electric motor is used in the acceleration, stop-and-go, and idling portions of transportation — the portions that have historically contributed the most to pollution — while the internal combustion engine is used when the vehicle is cruising. Fuel efficiency is increased as emission levels drop. In another configuration, the gasoline engine is simply used to recharge the batteries. Several HEVs are now on the market.
- **Methanol or ethanol:** These fuels, as well as combination methanol-gasoline fuels, are being used on a limited basis in the United States. They have the advantage of being cleaner burning, and they have a higher octane rating than gasoline. The disadvantages include the need for larger gas tanks, higher fuel costs, and the corrosive nature of the fuel.
- **Propane-powered vehicles:** These vehicles are already being used worldwide. The emissions are much less than those of a conventional gasoline engine, but the high flammability and limited supplies of propane available limit its increased use.
- **Compressed natural gas (CNG):** This fuel has possibilities for the future because of the relatively large supply and low cost of natural gas. It is a much cleaner burning fuel, but it requires strong, heavy fuel tanks. Compressed natural gas vehicles also must be refueled much more often than gasoline powered vehicles.

Other potential fuels are looming on the horizon. For the present, the most practical solution may be to reduce the reliance on private transportation and shift the burden to public transportation. This solution may or may not be practical in certain parts of the world.

Chapter 17

Feeling the Power of Nuclear Chemistry

In This Chapter

- Comprehending radioactivity and transmutation
- Investigating natural radioactive decay
- Finding out about half-lives and radioactive dating
- Understanding nuclear fission
- Finding out about nuclear fusion

Most of chemistry deals, in one way or the other, with how *valence electrons* (the electrons in the outermost energy levels of atoms) are lost, gained, or shared. The nucleus of the atom, to a very large degree, isn't involved in typical chemical reactions.

However, in this chapter, I talk about the nucleus and the changes it can undergo. I discuss about radioactivity and the ways an atom can decay. I discuss half-lives and show you why they're important in the storage of nuclear waste products. I also discuss nuclear fission in terms of bombs, power plants, and the hope that nuclear fusion holds for mankind.

I'm a child of the Atomic Age. I grew up with open air testing of nuclear weapons. I remember being warned not to eat snow because it might contain fallout and friends building fallout shelters, A-bomb drills at school, X-ray machines in shoe stores, radioactive Fiesta stoneware, and radium watch hands. Atomic energy was new, exciting, and scary all those years ago, and it still is.

In Chem II you relate the half-lives of radioisotopes to the half-lives discussed in reaction kinetics. You can see that many of the same equations apply to both situations.

Tracing Radioactivity and Transmutation

Before you can completely grasp what nuclear chemistry is, you need to know the basics of atomic structure. Chapter 3 gives you a quick review of atomic structure, if you're interested. This section just provides a quick overview of what nuclear chemistry is.

The *nucleus,* that dense central core of the atom, contains both protons and neutrons. Electrons are outside the nucleus in energy levels. Protons have a positive charge, neutrons have no charge, and electrons have a negative charge. A neutral atom contains equal numbers of protons and electrons. But the number of neutrons within an atom of a particular element can vary. Atoms of the same element that have differing numbers of neutrons are called *isotopes.* Figure 17-1 shows the symbolization chemists use to represent a specific isotope of an element.

In the figure, X represents the symbol of the element found on the periodic table, Z represents the *atomic number* (the number of protons in the nucleus), and A represents the *mass number* (the sum of the protons and neutrons in that particular isotope). If you subtract the atomic number from the mass number (A – Z), you get the number of neutrons in that particular isotope. A short way to show the same information is to simply use the element symbol (X) and the mass number (A) — for example, Pb-208.

Certain isotopes can undergo a couple of processes that fall into the radioactivity category — radioactive decay (radioactivity) and transmutation (radioactive processes that can be initiated by an outside source). Let me show you those two processes.

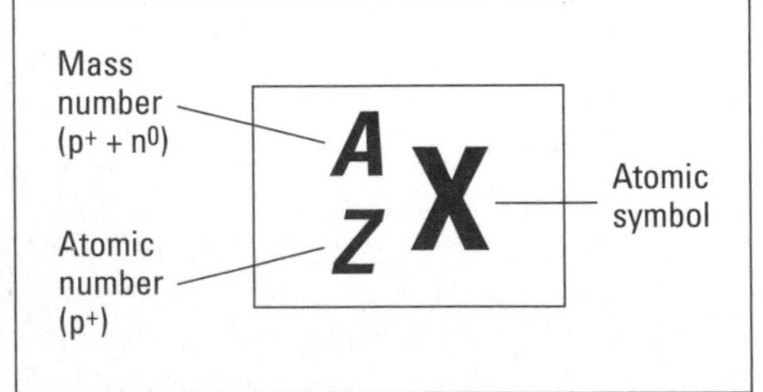

Figure 17-1: Representing a specific isotope.

Radioactivity: An unstable nucleus decays

Radioactivity is defined as the spontaneous decay of an unstable nucleus. An unstable nucleus may break apart into two or more other particles with the release of some energy. This breaking apart can occur in a number of ways, depending on the particular atom that's decaying.

You can often predict one of the particles of a radioactive decay by knowing the other particle. Doing so involves something called *balancing the nuclear reaction.* (A *nuclear reaction* is any reaction involving a change in nuclear structure.)

Balancing a nuclear reaction is a fairly simple process. You should know how to represent a reaction:

Reactants → Products

Reactants are the substances you start with, and *products* are the new substances being formed. The arrow, called a *reaction arrow,* indicates that a reaction has taken place.

For a nuclear reaction to be balanced, the sum of all the atomic numbers on the left-hand side of the reaction arrow must equal the sum of all the atomic numbers on the right-hand side of the arrow. The same is true for the sums of the mass numbers.

Transmutation: Under humanity's control

The radioactive process in the previous section occurs spontaneously. The unstable nucleus breaks apart into more stable ones. Man has no control over this process. But in *transmutation,* scientists can initiate a radioactive process and control it to produce a specific product. This is how artificial (manmade) elements are formed.

The first artificial isotope, was created in 1933 by bombarding aluminum foil with alpha particles. (Work with me here. I'm just trying to get to a point.) You observe that a neutron is created along with another isotope, and you want to figure out what the other isotope is.

The equation for this example is:

$${}^{27}_{13}\text{Al} + {}^{4}_{2}\text{He} \rightarrow ? + {}^{1}_{0}\text{n}$$

To figure out the unknown isotope (represented by ?), do the following steps:

1. **Balance the atomic numbers in the equation.**

   The sum of the atomic numbers on the left is 15 (13 + 2), so you want the sum of the atomic numbers on the right to also equal 15. Right now, you have an atomic number of 0 on the right; 15 – 0 is 15, so that's the atomic number of the unknown isotope. This atomic number identifies the element as phosphorus (15).

2. **Balance the mass numbers in the equation.**

   The sum of the mass numbers on the left is 31 (27 + 4), and you want the sum of the mass numbers on the right to equal 31, too. Right now, you have a mass number of 1 on the right; 31 – 1 is 30, so that's the mass number of the unknown isotope.

   Now you know that the unknown isotope is a sulfur isotope (P-30). And here's what the balanced nuclear equation looks like:

   $${}^{27}_{13}\mathrm{Al} + {}^{4}_{2}\mathrm{He} \rightarrow {}^{30}_{15}\mathrm{P} + {}^{1}_{0}\mathrm{n}$$

REMEMBER

This equation represents a nuclear transmutation, the conversion of one element into another. Nuclear transmutation is a process human beings control. P-30 is an isotope of phosphorus that doesn't exist in nature. It's a *man-made isotope.* Alchemists, those ancient predecessors of chemists, dreamed of converting one element into another (usually lead into gold), but they were never able to master the process. Chemists are now able, sometimes, to convert one element into another.

Examining How Nature Does It: Natural Radioactive Decay

Many isotopes are perfectly stable. Carbon-12, for example, is a stable isotope of carbon. However, other isotopes may be unstable and eventually undergo nuclear decay. Caron-14 is a radioactive isotope of carbon. If an isotope is unstable, the nucleus undergoes nuclear decay and breaks apart. Sometimes the product of that nuclear decay is unstable itself and undergoes nuclear decay, too. For example, when Radium-230 (a radioactive isotope of radium) initially decays, it produces Rn-228, which decays to Po-224. This decay continues until, finally, Pb-206 is produced. Pb-206 is stable, and the decay sequence stops.

Before I show you *how* radioactive isotopes decay, I want to briefly explain *why* a particular isotope decays. The nucleus has all those positively charged protons shoved together in an extremely small volume of space. All those protons are repelling each other. The forces that normally overpower the repelling protons and hold the nucleus together, the "nuclear glue," sometimes can't do the job, and so the nucleus breaks apart, undergoing nuclear decay. The following sections walk you through whether an element is stable or not and then the different ways radioactive decay happens.

Identifying stability

Stability can sometimes be correlated with the number of protons and neutrons. If an isotope of an element has a magic number of protons or neutrons, which are 2, 8, 20, 28, 50, 82, and 126, then it tends to be more stable than if it doesn't have one of these numbers. If both the protons and neutrons are a magic number, then it tends to be especially stable and many times plentiful.

All elements with 84 or more protons are unstable; they eventually undergo decay. Other isotopes with fewer protons in their nucleus are also radioactive. The radioactivity corresponds to the neutron/proton ratio in the atom. If the neutron/proton ratio is too high (caused by too many neutrons or too few protons), the isotope is said to be *neutron rich* and is, therefore, unstable. Likewise, if the neutron/proton ratio is too low (it has too few neutrons or too many protons), the isotope is unstable. The neutron/proton ratio for a certain element must fall within a certain range for the element to be stable, which is why some isotopes of an element are stable and others are radioactive. The *band of stability* shown in Figure 17-2 is the best way to represent whether an element is stable or not.

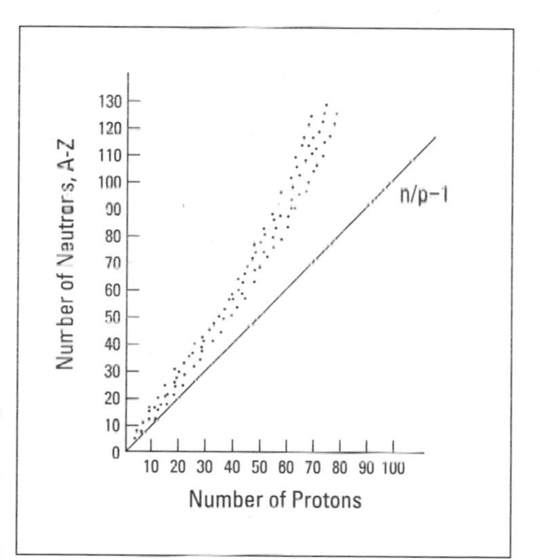

Figure 17-2: The band of stability.

In the belt of stability the dots represent the stable isotopes at various numbers of neutrons and protons. To use the table, find the isotope's number of protons and its number of neutrons. If that combination doesn't lie on the dotted area, then that isotope isn't stable and will, sooner or later, undergo nuclear decay. Depending on whether the isotope lies above or below the belt of stability, gives you clues as to the type of decay mode you may observe with that isotope. Notice that the neutron/proton ratio starts out at about a 1:1 ratio for low atomic numbers, but at high atomic numbers it is about a 1.5:1 ratio.

Decaying in a natural way

Naturally occurring radioactive isotopes decay in three primary ways. The following sections examine these ways in more depth.

Alpha emission

An *alpha particle* is defined as a positively charged particle of a helium nucleus. An alpha particle is composed of two protons and two neutrons, so it can be represented as a helium-4 atom. As an alpha particle breaks away from the nucleus of a radioactive atom, it has no electrons, so it has a 2+ charge. Therefore, it's a *cation,* a positively charged ion.

But electrons are basically free — easy to lose and easy to gain. So normally, an alpha particle is shown with no charge because it very rapidly picks up two electrons and becomes a neutral helium atom instead of a cation.

Large, heavy elements, such as uranium and thorium, tend to undergo alpha emission. This decay mode relieves the nucleus of two units of positive charge (two protons) and four units of mass (two protons + two neutrons). What a process! Each time an alpha particle is emitted, four units of mass are lost. I wish I could find a diet that would allow me to lose four pounds at a time!

Thorium-230 (Th-230) is an alpha particle emitter, as shown in the following equation:

$$^{230}_{90}\text{Th} \rightarrow {}^{226}_{88}\text{Ra} + {}^{4}_{2}\text{He}$$

Here thorium-230 undergoes nuclear decay with the release of an alpha particle. The other remaining isotope must have a mass number of 226 (230 – 4) and an atomic number of 88 (90 – 2), which identifies the element as radium (Ra). (If this subtraction confuses you, check out how to balance equations in the section "Transmutation: Under humanity's control," earlier in this chapter.)

Beta emission

A *beta particle* is essentially an electron that's emitted from the nucleus. (Now I know what you're thinking — electrons aren't in the nucleus. Keep on reading to find out how they can be formed in this nuclear reaction.) Nickel-63 (Ni-63) is a beta particle emitter:

$$^{63}_{28}\text{Ni} \rightarrow {}^{63}_{29}\text{Cu} + {}^{0}_{-1}\text{e}$$

Here, the nickel-63 gives off a beta particle (an electron), leaving an isotope with a mass number of 63 (63 – 0) and an atomic number of 29 (28 – (–1). An atomic number of 29 identifies the element as copper (Cu).

Notice that the mass number doesn't change in going from Nickel-63 to Cu-63, but the atomic number increases by one. In the nickel nucleus, a neutron was converted (decayed) into a proton and an electron, and the electron was emitted from the nucleus as a beta particle.

$$^{1}_{0}\text{n} \rightarrow {}^{1}_{+1}\text{p} + {}^{0}_{-1}\text{e}$$

Isotopes with a high neutron/proton ratio often undergo beta emission, because this decay mode allows the number of neutrons to be decreased by one and the number of protons to be increased by one, thus lowering the neutron/proton ratio.

Gamma radiation emission

Because no mass change is associated with gamma emission, I refer to gamma emission as *gamma radiation emission*. Gamma radiation is similar to X-rays — high energy, short wavelength radiation. Gamma radiation commonly accompanies both alpha and beta emission, but it's usually not shown in a balanced nuclear reaction. Some isotopes, such as cobalt-60 (Co-60), give off large amounts of gamma radiation. Co-60 is used in the radiation treatment of cancer. The medical personnel focus the gamma rays on the tumor, thus destroying it.

Decaying in less common ways

In addition, a couple of less common types of radioactive decay also happen. The following sections outline these ways.

Positron emission

Although positron emission doesn't occur with naturally occurring radioactive isotopes, it does occur naturally in a few man-made ones. A *positron* is essentially an electron that has a positive charge instead of a negative charge. A positron is formed when a proton in the nucleus decays into a neutron and a positively charged electron. The positron is then emitted from the nucleus:

$$^{1}_{+1}\mathrm{p} \rightarrow {}^{1}_{0}\mathrm{n} + {}^{0}_{+1}\mathrm{e}$$

This process occurs in a few isotopes, such as carbon-11 (C-11), as shown in the following equation:

$$^{11}_{6}\mathrm{C} \rightarrow {}^{11}_{5}\mathrm{B} + {}^{0}_{+1}\mathrm{e}$$

The C-11 emits the positron, leaving an element with a mass number of 11 (11 – 0) and an atomic number of 5 (6 – 1). An isotope of boron (B), B-11, has been formed.

If you watch *Star Trek,* you may have heard about antimatter. The positron is a tiny bit of antimatter. When it comes in contact with an electron, both particles are destroyed with the release of energy. Luckily, not many positrons are produced: If a lot of them were produced, you'd probably have to spend a lot of time ducking explosions.

Electron capture

Electron capture is a rare type of nuclear decay in which an electron from the innermost energy level is captured by the nucleus. This electron combines with a proton to form a neutron. The atomic number decreases by one, but the mass number stays the same:

$$^{1}_{+1}\text{p} + ^{0}_{-1}\text{e} \rightarrow ^{1}_{0}\text{n}$$

The following equation shows the electron capture of mercury-201 (Hg-201):

$$^{201}_{80}\text{Hg} + ^{0}_{-1}\text{e} \rightarrow ^{201}_{79}\text{Au} + \text{X} - \text{rays}$$

The electron combines with a proton in the mercury nucleus, creating an isotope of gold (Au-201). Now that's what I call a great trade — mercury to gold!

The capture of the 1s electron leaves a vacancy in the 1s orbitals. Electrons drop down to fill the vacancy, releasing energy not in the visible part of the electromagnetic spectrum but in the X-ray portion. This emission of X-rays is what reveals that electron capture is taking place.

Determining Half-Lives and Radioactive Dating

If you could watch a single atom of a radioactive isotope, C-14, for example, you wouldn't be able to predict when that particular atom might decay. It might take a millisecond, or it might take a century. You simply have no way to tell.

But if you have a large enough sample — what mathematicians call a *statistically significant sample size* — a pattern begins to emerge. It takes a certain amount of time for half the atoms in a sample to decay. It then takes the same amount of time for half the remaining radioactive atoms to decay, and the same amount of time for half of those remaining radioactive atoms to decay, and so on. The amount of time it takes for one-half of a sample to decay is called the *half-life* of the isotope, and it's given the symbol $t_{1/2}$. Table 17-1 shows this process.

Table 17-1 Half-Life Decay of a Radioactive Isotope

Half-Life	*Percent of Radioactive Isotope Remaining*
0	100.00
1	50.00
2	25.00

Half-Life	Percent of Radioactive Isotope Remaining
3	12.50
4	6.25
5	3.12
6	1.56
7	0.78
8	0.39
9	0.19
10	0.09

The half-life decay of radioactive isotopes isn't linear. For example, you can't find the remaining amount of an isotope as 7.5 half-lives by finding the midpoint between 7 and 8 half-lives. This decay is an example of an exponential decay, shown in Figure 17-3.

If you want a firmer grasp of how to calculate a half-life, the importance of half-lives, and how to use half-lives, keep on reading the following sections.

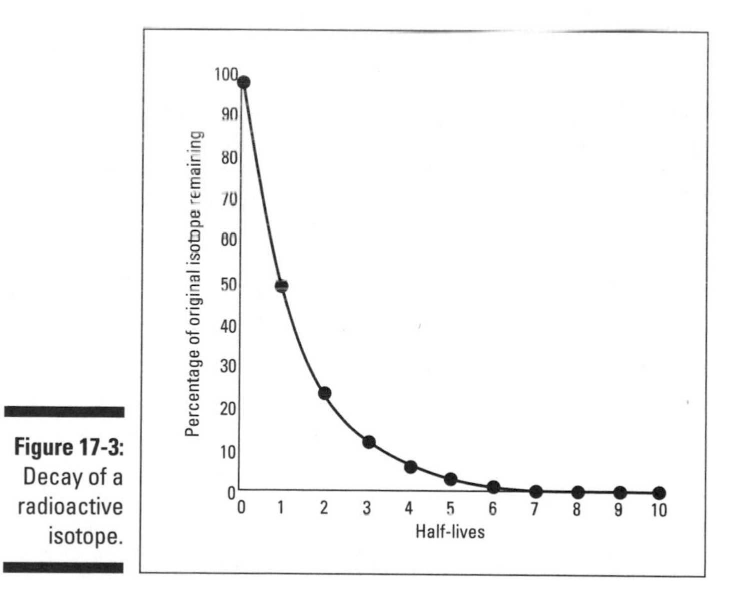

Figure 17-3: Decay of a radioactive isotope.

Figuring out half-lives

Calculations involving simple-multiples of half-lives is straightforward. You can use the table in the previous section as a guide. The amounts don't have to be expressed in percentages — you can use grams, micrograms, or some unit of radioactive emission.

If you want to find times or amounts that aren't associated with a simple multiple of a half-life, you can use these two equations:

$$(1) \qquad t_{\frac{1}{2}} = \frac{0.693}{\mathrm{k}}$$

$$(2) \ln\left(\frac{\mathrm{N_t}}{\mathrm{N_0}}\right) = -\mathrm{kt}$$

In the equation, *ln* stands for the *natural logarithm* (not the base 10 log; it's that *ln* button on your calculator, not the *log* button), N_o is the amount of radioactive isotope that you start with, N_t is the amount of radioisotope left at some time (t), k is a constant, and $t_{1/2}$ is the half-life of the radioisotope. If you know the half-life and the amount of the radioactive isotope that you start with, you can use this equation to calculate the amount remaining radioactive at any time.

Suppose you find a wooden bowl that has 70.0 percent the radioactivity as that of C-14 ($t_{1/2}$ of 5,730 years). You can use the two preceding equations to calculate the age of the handle (when the tree died or was cut down). Because you know the half-life, you can use equation (1) to calculate k:

$$t_{\frac{1}{2}} = \frac{0.693}{\mathrm{k}}$$

$$5730 \text{ yrs} = \frac{0.693}{\mathrm{k}}$$

$$k = \frac{1.21 \times 10^{-4}}{\text{yrs}}$$

You can now substitute k into equation (2) and solve for t:

$$\ln\left(\frac{\mathrm{N_t}}{\mathrm{N_0}}\right) = -\mathrm{k}t$$

$$\ln\left(\frac{60}{100}\right) = -\frac{1.21 \times 10^{-4}}{\text{yrs}}t$$

$$-0.5108 = -\frac{1.21 \times 10^{-4}}{\text{yrs}}t$$

$$t = 2.9 \times 10^{3} \text{yrs}$$

I can also use these same two equations to calculate the half-life if I know the amount (N_t) and the time (t).

Half-lives may be very short or very long. Table 17-2 shows the half-lives of some typical radioactive isotopes.

Table 17-2 Half-Lives of Some Radioactive Isotopes

Radioisotope	*Radiation Emitted*	*Half-Life*
Kr-94	Beta	1.4 seconds
Rn-222	Alpha	3.8 days
I-131	Beta	8 days
Co 60	Gamma	5.2 years
H-3	Beta	12.3 years
C-14	Beta	5,730 years
U-235	Alpha	4.5 billion years
Re-187	Beta	70 billion years

Introducing radioactive dating

A useful application of half-lives is *radioactive dating,* which has to do with figuring out the age of ancient things.

Cosmic radiation in the upper atmosphere produces carbon-14 (C-14), a radioactive isotope of carbon. Carbon dioxide is the primary carbon-containing compound in the atmosphere. A very small amount of carbon dioxide contains C-14. Plants absorb C-14 during photosynthesis, so C-14 is incorporated into the cellular structure of plants. Animals then eat plants, which makes C-14 a part of the cellular structure of all living things.

As long as an organism is alive, the amount of C-14 in its cellular structure remains constant. But when the organism dies, the amount of C-14 begins to decrease. Scientists know the half-life of C-14 (5,730 years, listed in Table 17-2), so they can figure out how long ago the organism died (refer to the problem in "Figuring out half-lives").

Scientists have also used radioactive dating using C-14 to determine the age of skeletons found at archeological sites. Carbon-14 dating can only be used to determine the age of something that was once alive. It can't be used to determine the age of a moon rock or a meteorite. For nonliving substances, scientists use other isotopes, such as potassium-40.

Clarifying Nuclear Fission

In the 1930s, scientists discovered that some nuclear reactions can be initiated and controlled. They usually accomplished this task by bombarding a large isotope with a second, smaller one — commonly a neutron. The collision caused the larger isotope to break apart into two or more elements, which is called *nuclear fission.* The nuclear fission of uranium-235 is shown in the following equation:

$$^{235}_{92}\text{U} + ^{1}_{0}\text{n} \rightarrow ^{142}_{56}\text{Ba} + ^{91}_{36}\text{Kr} + 3^{1}_{0}\text{n}$$

Reactions of this type also release a lot of energy. Where does the energy come from? Well, if you make an accurate measurement of the masses of all the atoms and subatomic particles you start with and all the atoms and subatomic particles you end up with and then compare the two, you find that some mass is missing. Matter disappears during the nuclear reaction. This loss of matter is called the *mass defect.* The missing matter is converted into energy.

These sections explain just what you need to know in your Chem II class about the ins and outs of nuclear fission.

Dealing with critical mass and chain reactions

You can actually calculate the amount of energy produced during a nuclear reaction with a fairly simple equation developed by Einstein: $E = mc^2$. In this equation, E is the amount of energy produced, m is the missing mass, or the mass defect, and c is the speed of light, which is a rather large number. The speed of light is squared, making that part of the equation a *very* large number that, even when multiplied by a small amount of mass, yields a *large* amount of energy. For example when a mole of U-235 decays to Th-234, the mass defect is 5×10^{-6} kg, which when converted to energy amounts to 5×10^{11} joules!

Notice that the initial reaction used one neutron, but three neutrons were produced. These three neutrons, encountering other U-235 atoms, can initiate other fission reactions, producing even more neutrons. It's the old domino effect. In terms of nuclear chemistry, it's a continuing cascade of nuclear fissions called a *chain reaction.* The chain reaction of U-235 is shown in Figure 17-4.

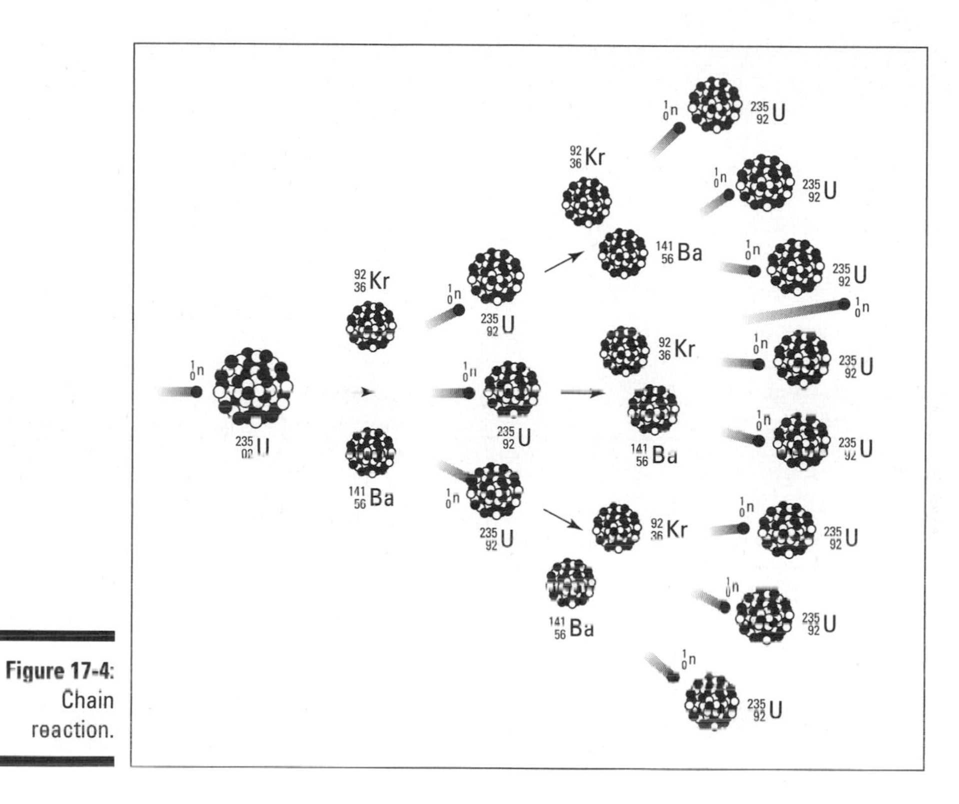

Figure 17-4: Chain reaction.

Whether or not a certain nuclear decay produces a chain-reaction depends on whether or not there was more neutrons produced than were used during the nuclear reaction. If you were to write the equation for the nuclear fission of U-238, the more abundant isotope of uranium, you'd use one neutron and but only get one back out. You can't have a chain reaction with U-238. But isotopes that produce an *excess* of neutrons in their fission support a chain reaction. This type of isotope is said to be *fissionable,* and there are only two main fissionable isotopes used during nuclear reactions — uranium-235 and plutonium-239.

A certain minimum amount of fissionable matter is needed to support a self-sustaining chain reaction, and it's related to those neutrons. If the sample is small, then the neutrons are likely to shoot out of the sample before hitting a U-235 nucleus. If they don't hit a U-235 nucleus, no extra electrons and no energy are released. The reaction just fizzles. The minimum amount of fissionable material needed to ensure that a chain reaction occurs is called the *critical mass.* Anything less than this amount is called *subcritical.*

Nuclear energy to kill — Atomic bombs

Because of the tremendous amount of energy released in a fission chain reaction, the military implications of nuclear reactions were immediately realized. The first atomic bomb was dropped on Hiroshima, Japan, on August 6, 1945.

In an atomic bomb, two pieces of a fissionable isotope are kept apart. Each piece, by itself, is subcritical. When it's time for the bomb to explode, conventional explosives force the two pieces together to cause a critical mass. The chain reaction is uncontrolled, releasing a tremendous amount of energy almost instantaneously.

The real trick, however, is to control the chain reaction, releasing its energy slowly so that ends other than destruction can be achieved.

Powering civilization: Power plants

The secret to controlling a chain reaction is to control the neutrons. If the neutrons can be controlled, then the energy can be released in a controlled way. That's what scientists have done with nuclear power plants.

In many respects, a nuclear power plant is similar to a conventional fossil fuel power plant. In this type of plant, a fossil fuel (coal, oil, natural gas) is burned, and the heat is used to boil water, which, in turn, is used to make steam. The steam is then used to turn a turbine that is attached to a generator that produces electricity.

The big difference between a conventional power plant and a nuclear power plant is that the nuclear power plant produces heat through nuclear fission chain reactions.

Making electricity

Most people believe that the concepts behind nuclear power plants are tremendously complex. That's really not the case. Nuclear power plants are very similar to conventional fossil fuel plants.

Fuel rods in the reactor core contain the fissionable isotope All the fuel rods together comprise the critical mass. Control rods, commonly made of boron or cadmium, are in the core, and they act like neutron sponges to control the rate of radioactive decay. Operators can stop a chain reaction completely by pushing the control rods all the way into the reactor core, where they absorb all the neutrons. The operators can then pull out the control rods a little at a time to produce the desired amount of heat.

A liquid (water or, sometimes, liquid sodium) is circulated through the reactor core, and the heat generated by the fission reaction is absorbed. The liquid then flows into a steam generator, where steam is produced as the heat is absorbed by water. This steam is then piped through a steam turbine that's connected to an electric generator. The steam is condensed and recycled through the steam generator. This forms a closed system; that is, no water or steam escapes — it's all recycled.

The liquid that circulates through the reactor core is also part of a closed system. This closed system helps ensure that no contamination of the air or water takes place. But sometimes problems do arise.

Eyeing the good and bad of nuclear power plants

The United States has 104 nuclear reactors, which produce a little less than 20 percent of the country's electricity. In France, almost 80 percent of the country's electricity is generated through nuclear fission. Nuclear power plants have certain advantages. No fossil fuels are burned (saving fossil-fuel resources for producing plastics and medicines), and the process uses no combustion products, such as carbon dioxide, sulfur dioxide, and so on, to pollute the air and water. But significant problems are associated with nuclear power plants, including the following.

Cost: They require lots of money

Nuclear power plants are expensive to build and operate. The electricity that's generated by nuclear power costs about twice as much as electricity generated through fossil fuel or hydroelectric plants. Another problem is that the supply of fissionable uranium-235 is limited. Of all the naturally occurring uranium, only about 0.75 percent is U-235. A vast majority is nonfissionable U-238. At current usage levels, nuclear power plants will be out of naturally occurring U-235 in fewer than 100 years. A little bit more time can be gained through the use of breeder reactors. But there's a limit to the amount of nuclear fuel available in the earth, just as there's a limit to the amount of fossil fuels.

Safety: Three Mile Island and Chernobyl

Although nuclear power reactors really do have a good safety record, the distrust and fear associated with radiation make most people sensitive to safety issues and accidents. The most serious accident to occur in the United States happened in 1979 at the Three Mile Island Plant in Pennsylvania. A combination of operator error and equipment failure caused a loss of reactor core coolant. The loss of coolant led to a partial meltdown and the release of a small amount of radioactive gas. Fortunately no lives were lost and none of the plant personnel or general population was injured.

Unfortunately that wasn't the case at Chernobyl, Ukraine, in 1986. Human error, along with poor reactor design and engineering, contributed to a tremendous overheating of the reactor core, causing it to rupture. Two explosions and a fire resulted, blowing apart the core and scattering nuclear material into the atmosphere. A small amount of this material made its way to Asia and much of Europe. Hundreds of people died. Many others felt the effect of radiation poisoning. Instances of thyroid cancer, possibly caused by the release of I-13, have risen dramatically in the towns surrounding Chernobyl. Scientists won't know the full effects of this disaster for many years.

Disposal: Throwing it away

The fission process produces large amounts of radioactive isotopes. If you refer to Table 18-2, you notice that some of the half-lives of radioactive isotopes are rather long. Those isotopes are safe after ten half-lives. The length of ten half-lives presents a problem when dealing with the waste products of a fission reactor.

Eventually, all reactors must have their nuclear fuel replenished. And as governments disarm nuclear weapons, they must deal with what to do with their radioactive material. Many of these waste products have long half-lives. How can the isotopes be safely stored until their residual radioactivity has dropped to safe limits (ten half-lives)? How can the environment and society, and future generations to come, be protected from this waste? These questions are undoubtedly the most serious problem associated with the peaceful use of nuclear power.

Nuclear waste is divided into low-level and high-level material, based on the amount of radioactivity being emitted.

- **Low-level waste:** In the United States, low-level wastes are stored at the site of generation or at special storage facilities. The wastes are basically buried and guarded at the sites.
- **High-level wastes:** High-level wastes pose a much larger problem. They're temporarily being stored at the site of generation, with plans to eventually seal the material in glass and then in drums. The material will then be stored underground in Nevada. At any rate, the waste must be kept safe and undisturbed for at least 10,000 years. Other countries face the same problems. Some nuclear material has been dumped into deep trenches in the sea, but many nations have discouraged this practice.

Breeder reactors: Making more nuclear stuff

Only the U-235 isotope of uranium is fissionable, because it's the only isotope of uranium that produces the excess of neutrons needed to maintain a chain-reaction. The far more plentiful U-238 isotope doesn't produce those extra neutrons.

The other commonly used fissionable isotope, plutonium-239 (Pu-239), is very rare in nature. But scientists can make Pu-239 from U-238 in a special fission reactor called a *breeder reactor.* Uranium-238 is first bombarded with a neutron to produce U-239, which decays to Pu-239. The process is shown in Figure 17-5.

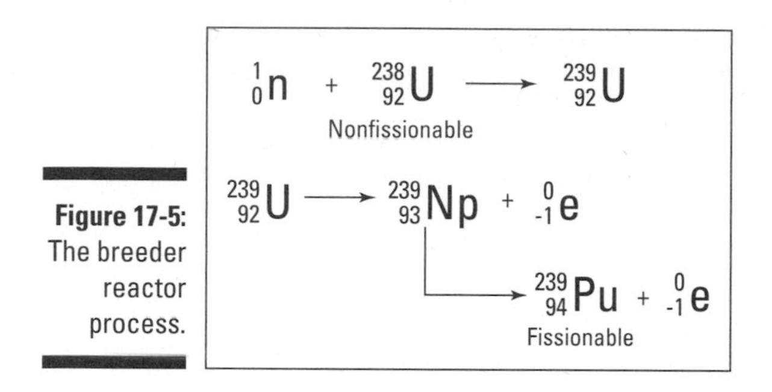

Figure 17-5: The breeder reactor process.

Breeder reactors can extend the supply of fissionable fuels for many years, and France is currently using them. But the United States is moving *slowly* with the construction of breeder reactors because of several problems associated with them. First, they're extremely expensive to build. Second, they produce large amounts of nuclear wastes. And, finally, the plutonium that's produced is much more hazardous to handle than uranium and can easily be used in an atomic bomb.

Using the Sun's Power: Nuclear Fusion

Soon after the fission process was discovered, another process, called *fusion,* was discovered. Fusion is essentially the opposite of fission. In fission, a heavy nucleus is split into smaller nuclei. With fusion, lighter nuclei are fused into a heavier nucleus.

The fusion process is the reaction that powers the sun. On the sun, in a series of nuclear reactions, four isotopes of hydrogen-1 are fused into a helium-4 with the release of a tremendous amount of energy. Here on earth, two other isotopes of hydrogen are used: H-2, called deuterium, and H-3, called tritium. Deuterium is a minor isotope of hydrogen, but it's still relatively abundant. Tritium occurs naturally in minute amounts, but it can easily be produced by bombarding deuterium with a neutron. The fusion reaction is shown in the following equation:

$$^{3}_{1}H + ^{2}_{1}H \rightarrow ^{4}_{2}He + ^{1}_{0}n$$

The first demonstration of nuclear fusion — the hydrogen bomb — was conducted by the military. A hydrogen bomb is approximately 1,000 times as powerful as an ordinary atomic bomb.

The isotopes of hydrogen needed for the hydrogen bomb fusion reaction were placed around an ordinary fission bomb. The explosion of the fission bomb released the energy needed to provide the *activation energy* (the energy necessary to initiate, or start, the reaction) for the fusion process.

Nuclear fusion as used in a hydrogen bomb may be fine for warfare, but in order to use the fusion process as an energy source for everyday life, scientists have to harness the power, much the same way as they did for nuclear fission. However, developing a fusion power plant has proven to be much more difficult than developing a fission one. These sections look at some challenges facing nuclear scientists and what the future holds.

Figuring out whether a fusion reactor is possible

The goal of scientists for the last 50 years has been the controlled release of energy from a fusion reaction. If the energy from a fusion reaction can be released slowly, it can be used to produce electricity. It will provide an unlimited supply of energy that has no wastes to deal with or contaminants to harm the atmosphere — simply nonpolluting helium. But achieving this goal requires overcoming these three problems:

Temperature

The fusion process requires an extremely high activation energy. Heat is used to provide the energy, but it takes a *lot* of heat to start the reaction. Scientists estimate that the sample of hydrogen isotopes must be heated to approximately 40,000,000 K. (K represents the Kelvin temperature scale. To get the Kelvin temperature, you add 273 to the Celsius temperature. Chapter 1 explains all about Kelvin and his pals Celsius and Fahrenheit.)

But 40,000,000 K is hotter than the sun! At this temperature, the electrons have long since left the building; all that's left is a positively-charged *plasma,* bare nuclei heated to a tremendously high temperature. Presently, scientists are trying to heat samples to this high temperature through two ways — magnetic fields and lasers. Neither one has yet achieved the necessary temperature.

Time

Time is the second problem scientists must overcome to achieve the controlled release of energy from fusion reactions. The charged nuclei must be held together close enough and long enough for the fusion reaction to start. Scientists estimate that the plasma needs to be held together at 40,000,000 K for about one second.

Containment

Containment is the third major problem facing fusion research. At 40,000,000 K, everything is a gas. The best ceramics developed for the space program would vaporize when exposed to this temperature. Because the plasma has a charge, magnetic fields can be used to contain it — like a magnetic bottle. But if the bottle leaks, the reaction won't take place. And scientists have yet to create a magnetic field that won't allow the plasma to leak. Using lasers to zap the hydrogen isotope mixture and provide the necessary energy bypasses the containment problem. But scientists haven't figured out how to protect the lasers themselves from the fusion reaction.

Considering fusion's future

The latest estimates indicate that science is a few years away from showing that fusion can work by reaching the so-called *break-even point,* where the reaction produces more energy than is put in. It will then be another 20 to 30 years before a functioning fusion reactor is developed. But scientists are optimistic that controlled fusion power will be achieved. The rewards are great — an unlimited source of nonpolluting energy.

An interesting by-product of fusion research is the *fusion torch* concept. With this idea, the fusion plasma, which must be cooled in order to produce steam, is used to incinerate garbage and solid wastes. Then the individual atoms and small molecules that are produced are collected and used as raw materials for industry. It seems like an ideal way to close the loop between waste and raw materials. Time will tell if this concept will eventually make it into practice.

Radon: What a gas

Radon is a radioactive isotope that's received a lot of publicity. Radon-222 is formed naturally as part of the decay of uranium. It's an unreactive noble gas, so it escapes from the ground into the air. Because it's heavier than air, it can accumulate in basements.

Radon itself has a short half-life of 3.8 days, but it decays to Polonium-218, a solid. So if radon is inhaled, solid Po-218 can accumulate in the lungs. Po-218 is an alpha emitter, and even though this type of radiation isn't very penetrating, it has been linked to increased instances of lung cancer. In many parts of the United States, radon testing is performed before selling a house. Commercial test kits can be opened, left in the basement area for a specified amount of time, and then sent to a lab for analysis. The question of whether radon represents a serious problem is still being investigated and debated.

Chapter 18

Chemistry in the Home

In This Chapter

- Handling chemistry with laundry
- Clarifying chemistry in the kitchen
- Seeing how chemistry pops up in the bathroom
- Understanding chemistry in the medicine cabinet

You may be surprised to discover that you probably come into direct contact with more chemicals and chemistry in your own home than anywhere else. The kitchen is filled with cleaners, soaps, and detergents, most of which are contained in plastics bottles. Your kitchen has a myriad of chemical reactions that take place while cooking. No wonder consumer chemistry is sometimes called *kitchen chemistry*. My wife is glad to have her own private chemist handy, especially for cleaning silver and finding a solvent to remove an adhesive. Furthermore, your bathroom is filled with medicines, soaps, toothpaste, and cosmetics.

In this chapter, I cover a few topics from consumer product chemistry. I show you the chemistry behinds soaps, detergents, and cleaners. I talk a little bit about medicines and drugs, and then I show you some things about personal care products, permanents, tanning products, and perfumes. I hope that you'll gain an appreciation for chemistry and what it has done to make your life better and easier.

Coping with Chemistry in the Laundry Room

The laundry room contains some of the most heavy-duty chemistry in your home. No matter whether you're using soaps, detergents, or other cleaners to remove stains and odors, chemistry plays a pivotal role. The following sections discuss the many faces chemistry shows up in laundry room.

Reducing the water tension: Surfactants

Have you ever got distracted and forgot to put the laundry detergent in the washer? I doubt that the clothes came out very clean. You may have gotten some surface dirt off, but the grease and oil stayed right where it was. You failed to add a *surfactant,* also called a surface active agent, to the wash. A surfactant is a substance that can bridge the gap between the grease and oils and the water.

Without the surfactant, the grease and oil stayed on the clothes because "like dissolves like." Grease and oils are nonpolar materials, and water is a polar substance, so water isn't going to dissolve the grease and oil. Surfactants reduce the surface tension of water, allowing it to "wet" nonpolar substances like grease and oil. Surfactants work because they have two ends:

- **A nonpolar end:** The nonpolar end is called the *hydrophobic* (water-fearing) end. This end is normally composed of a long hydrocarbon chain. The nonpolar end dissolves in the nonpolar grease and oil.
- **A polar end:** The other part of the surfactant molecule is called the *hydrophilic* (water-loving) end. This end is normally an ionic end with a negative charge *(anionic)*, a positive charge *(cationic),* or both *(amphoteric).*

  Some surfactants have no charge *(nonionic).* A vast majority of the surfactants on the market are anionic surfactants, because they're cheaper to produce. Figure 18-1 shows a typical anionic surfactant.

When a surfactant is added to water, the hydrophobic end dissolves in the oil and grease, while the hydrophilic end becomes attracted to the polar water molecules. The grease and oil are broken into very tiny droplets called *micelles,* with the hydrophobic (hydrocarbon) end of the surfactant sticking into the droplet and the hydrophilic end sticking out into the water. This structure with the hydrophobic end in the grease and the hydrophilic end in the water gives the droplet a charge (a negative charge in the case of an anionic surfactant). These charged droplets repel each other and keep the oil and grease droplets from joining together. These micelles remain dispersed. They eventually go down the drain with the used wash water. The two general types of surfactants that are used in the cleaning of clothes are soaps and detergents.

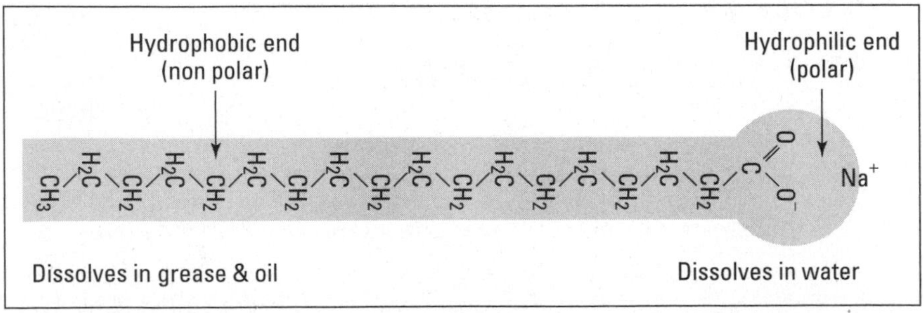

Hydrophobic end (non polar)
Hydrophilic end (polar)
Dissolves in grease & oil
Dissolves in water

Figure 18-1: A typical anionic surfactant.

Keep it clean: Soap

Soaps, which date back almost 5,000 years, are certainly the oldest and most well-known surfactant for cleaning. The specific type of organic reaction that is involved in the production of soap, commonly called *saponification,* is a hydrolysis reaction of fats or oils in a basic solution. The products of this reaction are glycerol and the salt of the fatty acid. Figure 18-2 shows the hydrolysis of tristearin to sodium stearate, a soap. (This is the same soap, or surfactant, shown in Figure 18-1.)

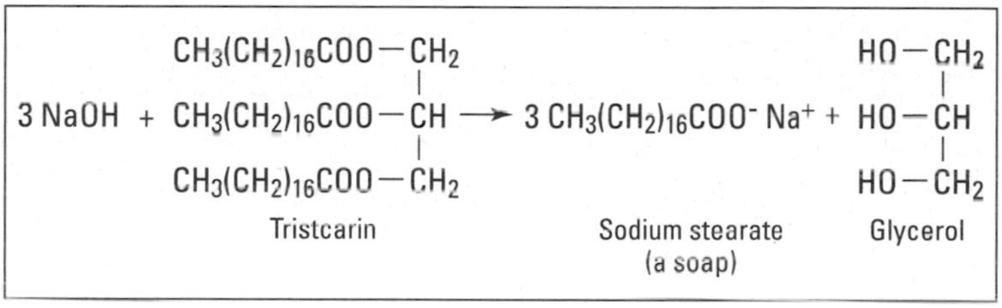

Figure 18-2: Production of a soap by saponification.

$$
\begin{array}{lclcl}
 & & CH_3(CH_2)_{16}COO-CH_2 & & HO-CH_2 \\
 & & \quad\quad\quad\quad\quad\quad\quad\quad | & & \quad\quad | \\
3\,NaOH & + & CH_3(CH_2)_{16}COO-CH & \longrightarrow\ 3\,CH_3(CH_2)_{16}COO^-\,Na^+ & +\ HO-CH \\
 & & \quad\quad\quad\quad\quad\quad\quad\quad | & & \quad\quad | \\
 & & CH_3(CH_2)_{16}COO-CH_2 & & HO-CH_2 \\
 & & \text{Tristearin} & \text{Sodium stearate (a soap)} & \text{Glycerol}
\end{array}
$$

Grandma made her soap by taking animal fat, adding it to water and lye (sodium hydroxide, NaOH), and boiling it in a huge iron kettle. The lye came from wood ashes. After cooking for hours, the soap rose to the top. It was then skimmed off and pressed into bars. However, Grandma didn't know much about reaction stoichiometry. She usually had an excess of lye, so her soap was very alkaline.

Today, soap is made a little differently. The hydrolysis is generally accomplished without the use of lye. Coconut oil, palm oil, and cottonseed oil are used in addition to animal tallow. For bar soaps, an abrasive, such as pumice, is occasionally added to aid in the removal of tough grease and oil from your skin. In addition, perfumes may be added, and air may be mixed with the soap to get it to float.

Soap, however, has a couple of big disadvantages:

- If soap is used with acidic water, the soap will be converted to fatty acids and lose its cleaning ability.
- If soap is used with hard water (water containing calcium, magnesium, or iron ions), a greasy insoluble precipitate (solid) will form. This greasy deposit is commonly called *bathtub ring.* Not only does this deposit form in your bathtub, but it also appears on your clothes, dishes, and so on.

  You can avoid the deposit in a couple of ways:

  - You can use a whole-house water softener (see "Make it soft: Water softeners," later in this chapter).
  - You can buy a synthetic soap that doesn't precipitate with hard-water ions. These synthetic soaps are called *detergents.*

Replacing the choking builders

The builder that was used in early laundry detergents was sodium tripolyphosphate. It was cheap and safe. It was, however, an excellent nutrient for water plants. It caused an increase in the growth of algae in lakes and streams, choking out fish and other aquatic life. States began banning the use of phosphates in detergents in order to control this problem. Sodium carbonate and *zeolites* (complex aluminosilicates — compounds of aluminum, oxygen, and silicon) have been used as replacements for the polyphosphates, but both are less than ideal. There really hasn't been an effective, cheap, and non-toxic replacement for the polyphosphate builders. This is an area of research that is still quite active.

Another type of surfactant: Detergents

Detergents have the same basic structure as the soap. Their hydrophobic end, composed of a long nonpolar hydrocarbon chain that will dissolve in the grease and oil, is the same, but their hydrophilic (ionic) end is different. Instead of carboxylate ($-COO^-$), the hydrophilic end may have a sulfate ($-O\text{-}SO_3^-$), a hydroxyl (OH^-), or some other polar group that won't precipitate with hard water.

Laundry detergents contain a number of other compounds in addition to the detergent surfactant. The compounds in laundry detergents are

- **Builders:** These compounds increase the surfactant's efficiency by softening the water (removing the hard water ions) and making it alkaline. (See the nearby sidebar for how builders have evolved.)
- **Fillers:** Compounds such as sodium sulfate (Na_2SO_4) are added to give the detergent bulk and to keep it free flowing.
- **Enzymes:** These biological catalysts are sometimes added to help remove protein-based stains such as blood and grass.
- **Sodium perborate:** $NaBO_3$ is sometimes added as a solid bleach to help remove stains. It works by generating hydrogen peroxide in water. It's much gentler on textiles than chlorine bleach. However, it's most effective in hot water, which can present a problem if you like to wash in cold water.
- **Suspension agents:** These compounds are added to help keep the dirt in solution.
- **Corrosion inhibitors:** These compounds coat washer parts to help prevent rust.
- **Optical brighteners:** These compounds make white clothes appear extra clean and bright. These very complex organic compounds deposit themselves as a thin coating on the clothes. They absorb ultraviolet light and re-emit it as a blue light in the visible part of the spectra. Figure 18-3 shows this process.
- **Coloring agents and perfumes:** Detergents also include these additives.

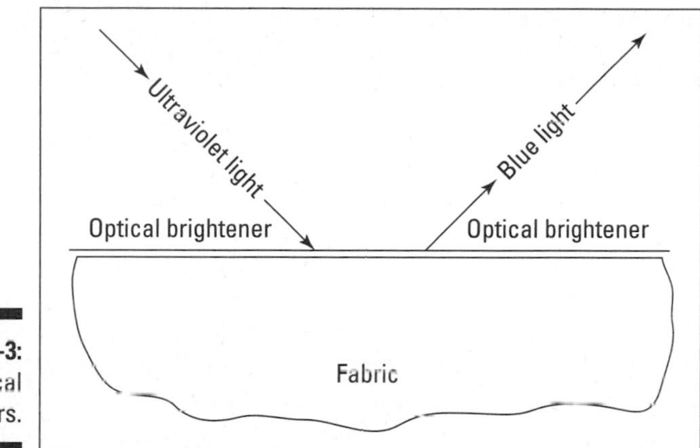

Figure 18-3: Optical brighteners.

Make it soft: Water softeners

Another way chemistry pops up in the laundry room is in your water softener. A home water softener is a way to prevent hard water and bathtub ring by simply removing the cations responsible for the hard water before they reach the house. Figure 18-4 shows a water-softening unit.

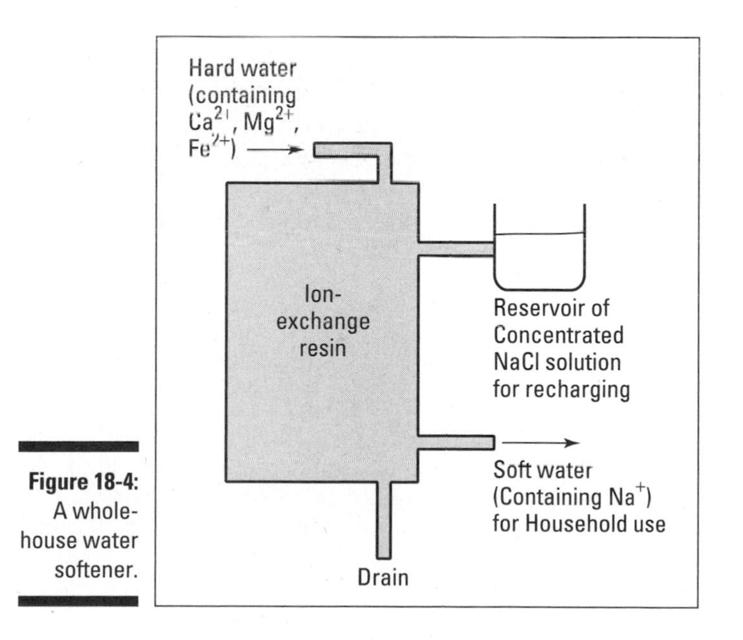

Figure 18-4: A whole-house water softener.

The water softener consists of a large tank containing an ion-exchange resin. The resin is charged when a concentrated sodium chloride solution runs through it. The sodium ions are held to the polymer material of the resin. The hard water passes through the polymer, and the calcium, magnesium, and iron ions are exchanged for the sodium ions on the resin (that's where the term *ion-exchange* resin comes from). The softened water contains sodium ions, but the hard-water ions remain in the resin. After a while, the resin must be recharged with more sodium chloride from the reservoir. The wastewater that contains the Ca^{2+}, Mg^{2+}, and Fe^{2+} is drained from the resin tank.

If you limit your sodium intake because of high blood pressure, you should avoid drinking the softened water, because it will have a high sodium ion concentration.

Make it whiter: Bleach

Bleaches are another obvious way that chemistry shows up in the laundry room. Bleaches use redox reactions to remove color from material (see Chapter 12 on electrochemistry for a discussion of redox reactions). Most bleaches are oxidizing agents. The most common bleach used in the home is a 5 percent solution of sodium hypochlorite. This type of bleach is produced by bubbling chlorine gas through a sodium hydroxide solution shown here:

$$2\ NaOH(aq) + Cl_2(g) \rightarrow NaOCl(aq) + NaCl(aq) + H_2O(l)$$

The chlorine released by hypochlorite bleaches can damage fabrics. Also, these types of bleaches don't work well on polyester fabrics. Some bleaches containing sodium perborate are now on the market that are gentler on fabrics. This type of bleach generates hydrogen peroxide, which in turn decomposes liberating oxygen, shown here:

$$2\ H_2O_2 \rightarrow 2\ H_2O(l) + O_2(g)$$

Cooking Up Chemistry in the Kitchen

Look under your kitchen sink and you can count the different chemical products. Even more, these products are stored in plastic bottles that are made with, you guessed it, chemicals. The following sections point out the different ways that chemicals show up in your kitchen.

Clean it all: Multipurpose cleaners

Most multipurpose cleaners are composed of some surfactant and disinfectant. Ammonia is commonly used because of its ability to react with grease and the fact that it leaves no residue. Pine oil, a solution of compounds called *terpenes,* is used for its pleasant odor, its ability to dissolve grease, and its antibacterial nature.

Be careful when mixing household cleaning products, especially bleach with ammonia or muratic acid (HCl). This solution can generate toxic gases that can be quite dangerous.

Wash those pots: Dishwashing products

Dishwashing detergent is much simpler than laundry detergent. It has some surfactant (normally a nonionic one), a little colorant, and something to make your hands feel soft. Dishwashing detergent isn't nearly as alkaline as laundry detergent. However, automatic dishwasher detergents are highly alkaline and contain only a little surfactant. They use the high pH to saponify the fats (like the process used to make soap), and high water temperature and agitation to clean the dishes. They're composed mostly of sodium metasilicate (Na_2SiO_3), for its alkalinity, sodium tripolyphosphate ($Na_5P_3O_{10}$), which acts as a detergent, and a little chlorine bleach.

Checking out Chemistry in the Bathroom

A lot of chemistry goes on in your bathroom. You probably have all those skin and hair care products that make you look, smell, and even taste good. These sections take a closer look at how chemistry plays a role in your bathroom.

Detergent for the mouth: Toothpaste

Walk down any toothpaste isle and you will see a wide variety of toothpastes with different colors, flavors, and so on. Although they may look different, they all contain the same basic ingredients. The two primary ingredients in toothpaste are surfactant (detergent) and abrasive. The abrasive is for scraping the film off the teeth without damaging the tooth itself. Common abrasives are chalk ($CaCO_3$), titanium dioxide (TiO_2), and calcium hydrogen phosphate ($CaHPO_4$). Other additives are added to give the toothpaste color, flavoring, and so on. Table 18-1 gives the general formula for toothpaste. The percentages and specific chemical compounds may vary from toothpaste to toothpaste.

Table 18-1 Typical Formulation for Toothpaste

Function	*Possible Ingredient*	*Percentage*
Solvent and filler	Water	30–40%
Detergent	Sodium lauryl sulfate, soap	4%
Abrasive	Calcium carbonate, calcium hydrogen phosphate, titanium dioxide, sodium metaphosphate, silicia, aluminia	30–50%
Sweetener	Glycerine, saccharin, sorbitol	15–20%
Thickener	Gum cellulose, carrageenan	1%
Fluoride	Stannous or sodium fluoride	1%
Flavoring	Oil of wintergreen, peppermint, strawberry, lime, and so on	1%

The addition of stannous or sodium fluoride is effective in the prevention of dental cavities, because the fluoride ion actually becomes part of the tooth enamel, making the enamel stronger and more resistant to the attack of acids.

Phew! Deodorants and antiperspirants

Sweating helps your body regulate its internal temperature. Sweat contains *amines,* low molecular weight fatty acids, and proteins, in addition to sodium chloride and other inorganic compounds. Some of these organic compounds have a disagreeable odor. Bacterial action can certainly make the odor worse. Deodorants and antiperspirants can control the socially unacceptable odor. Here is how these two work:

- *Deodorants* contain fragrances to cover up the odor and an antibacterial agent to destroy the odor causing bacteria. They may also contain substances such as zinc peroxide that oxidize the amines and fatty acids to less odorous compounds.
- *Antiperspirants* inhibit or stop perspiration. They act as an *astringent,* constricting the sweat gland ducts. The most commonly used antiperspirants are compounds of aluminum — aluminum chlorohydrates ($Al_2(OH)_5Cl$, $Al_2(OH)_4Cl_2$, and so on), hydrated aluminum chloride ($AlCl_3{\cdot}6H_2O$), and others.

Skin care chemistry: Keeping it soft and pretty

Chemistry also plays an important role for keeping you soft and pretty. You may be surprised at what's in some of the stuff you put on your skin.

Creams and lotions

The skin is a complex organ composed primarily of protein and naturally occurring *macromolecules,* complex, compounds containing many, many atoms. Some synthetic macromolecules are called plastics. Healthy skin contains about 10 percent moisture. Creams and lotions work to soften and moisturize the skin.

Emollients are skin softeners. Petroleum jelly (mixture of alkanes, with 20+ carbons, isolated from crude oil), lanolin (mixture of esters isolated from sheep wool fat), and coco butter (mixture of esters isolated from the cacao bean) are excellent skin softeners.

Skin creams are normally made of oil-in-water or water-in-oil emulsions. An *emulsion* is a colloidal dispersion of one liquid in another (see my book *Chemistry For Dummies,* 2nd edition for a discussion of colloids). It tries to soften and moisturize the skin at the same time. Cold creams remove makeup and moisturize, while vanishing creams make the skin appear younger by filling in wrinkles. Typical formulations for cold cream and vanishing cream are

Cold Cream Formulation	***Vanishing Cream Formulation***
20–50% water	70% water
30–60% mineral oil	10% glycerin
12–15% beeswax	20% stearic acid/sodium stearate
	5–15% lanolin or whale wax
	1% borax
	trace of perfume

Body and face powders

Body and face powders are used to dry and smooth the skin. The main ingredient in both types of powder is talc ($Mg_3(Si_2O_5)_2(OH)_2$), a mineral that can absorb both oil and water. Astringents are added to reduce sweating, and binders are added to help the powders stick to the skin better. Face powders often contain dyes to give color to the skin. Table 18-2 shows a typical formulation for body powder, while Table 18-3 shows a typical formulation for face powder.

Table 18-2 Typical Formulation for Body Powder

Ingredient	*Function*	*Percentage*
Talc	Absorbent, bulk	50–60%
Chalk ($CaCO_3$)	Absorbent	10–15%
Zinc oxide (ZnO)	Astringent	15–25%
Zinc stearate	Binder	5–10%
Perfume, dye	Odor, color	trace

Table 18-3 Typical Formulation for Face Powder

Ingredient	*Function*	*Percentage*
Talc	Absorbent, bulk	60–70%
Zinc oxide	Astringent	10–15%
Kaolin (Al_2SiO_5)	Absorbent	10–15%
Magnesium and zinc stearates	Texture	5–15%
Cetyl alcohol	Binder	1%
Mineral oil	Emollient	2%
Lanolin, perfume, dyes	Softening, odor, color	2%

Making up those eyes

Eye shadow and mascara are composed primarily of emollients, lanolin, beeswax, and colorants. Mascara darkens the eyelashes, making them appear longer. Typical formulations for eye shadow and mascara are

Eye Shadow

55–60% petroleum jelly

5–15% fats and waxes

5–10% lanolin

15–25% zinc oxide

1–5% dyes

Mascara

45–50% soap

35–40% wax and paraffin

5–10% lanolin

1–5% dyes

Kissable lips: Lipstick

Lipstick keeps the lips soft and protects them from drying out, while adding a desirable color. It's composed mostly of wax and oil. These ingredients must be balanced carefully so that the lipstick can go on easily without running and come off easily, but not too easily, when the wearer is ready to remove it. The color normally comes from a precipitate (solid) of some metal ion with an organic dye, commonly called a *lake.* The metal ion tends to intensify the color of the dye. Table 18-4 shows a typical lipstick formulation.

Table 18-4 A Typical Lipstick Formulation

Ingredient	*Function*	*Percentage*
Castor oil, mineral oil, fats	Dye solvent	40–50%
Lanolin	Emollient	20–30%
Carnauba wax or beeswax	Stiffener	15–25%
Dye	Color	5–10%
Perfume and flavoring	Odor and taste	trace

Beautiful nails: Nail polish

Nail polish is a synthetic lacquer that owes its flexibility to a polymer and a plasticizer. The polymer is normally nitrocellulose. The solvents used in the polish are acetone and ethyl acetate, the same substances used for nail polish removers.

Smelling good: Perfumes, colognes, and aftershaves

The major difference between perfume, cologne, and aftershave is the amount of fragrance used. Perfumes are commonly composed of 10 to 25 percent fragrance, while colognes use 1 to 3 percent, and aftershaves use less than 1 percent. These fragrances are usually organic esters, alcohols, ketones, and aldehydes. Perfumes also contain *fixatives,* compounds that help keep the other fragrances from evaporating too rapidly. Several of these fixatives have disagreeable odors or histories themselves: Civetone comes from the glands of the skunk-like civet cat, ambergris is sperm-whale vomit, and indole is isolated from feces.

Perfumes are usually mixtures of *notes,* fragrances with similar aromas but different volatilities. The most volatile is called the *top note,* which is what you initially smell. The *middle note* is the most noticeable, while the *end note* fragrances are responsible for the lingering odor of the perfume. Figure 18-5 shows the chemical structure of several of the fragrances commonly used in perfumes.

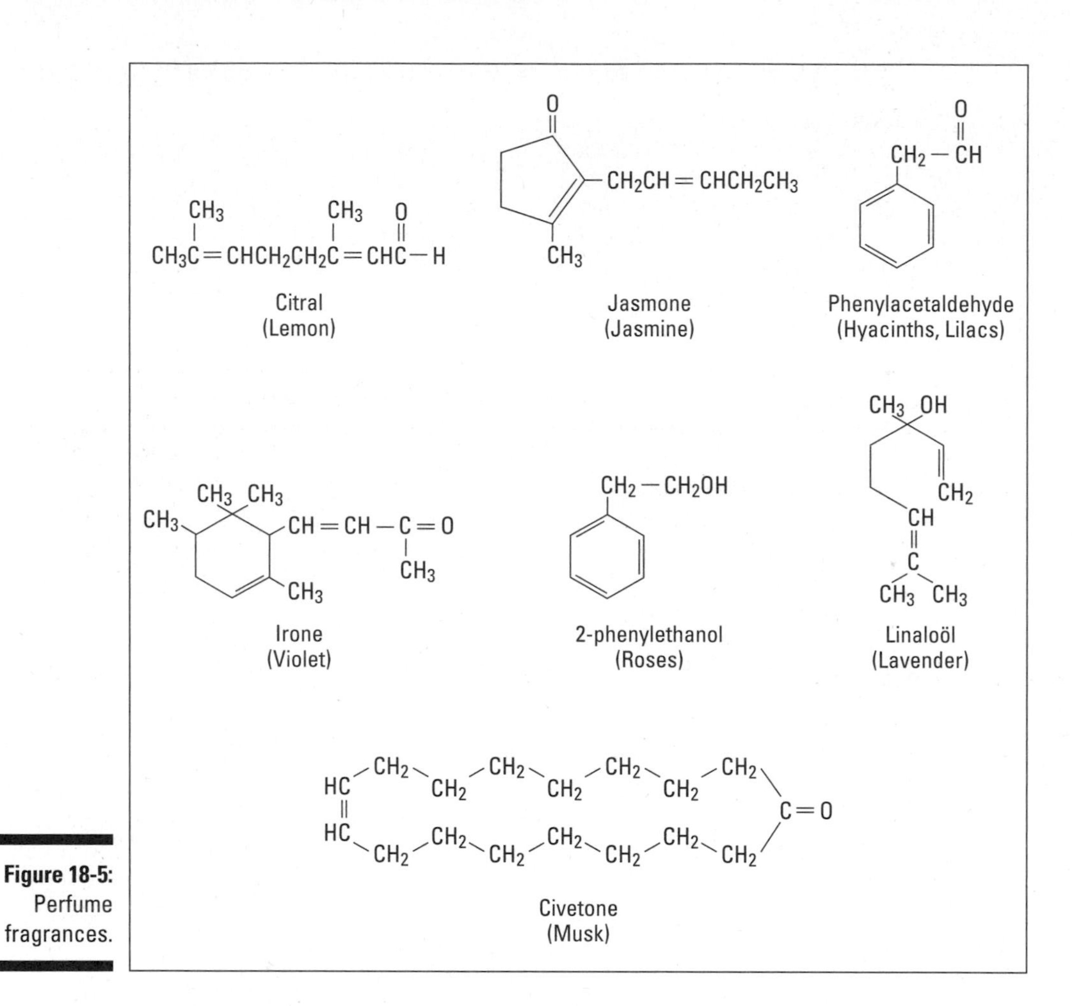

Figure 18-5: Perfume fragrances.

Suntan lotion and sunscreen: Brown is beautiful

A suntan is nature's way of protecting your body against the sun's harmful UV rays. The UV spectrum is composed of two regions:

- **UV-A region:** The UV-A region is of slightly longer wavelengths and tends to produce a tan rather than a burn.
- **UV-B region:** The UV-B radiation is what is responsible for quick sunburns. Repeated exposure to these harmful UV rays, especially the UV-B rays, is related to an increase in the occurrences of skin cancers such as melanoma.

Suntan lotion and sunscreens allow an individual to be exposed to the sun for longer periods of time without burning. They give the body a chance to produce *melanin,* a dark pigment that acts as a natural shield for the sun's UV rays, and that desirable brown skin tone. These products are given a Sun Protection Factor (SPF) rating. The *SPF value* is a ratio of the amount of time required to tan (or burn) with the product versus without it. An SPF value of 10 would then indicate that the individual can be exposed to the sun 10 times as long without burning when using the product.

Suntan lotions and sunscreens protect the skin by partially or totally blocking the sun's radiation in the UV range. An opaque cream of zinc oxide and titanium dioxide is the most effective type of sunscreen. This cream is effective at blocking out all the sun's rays. It is generally applied to the nose and the tops of the ears. Some suntan lotions and sunscreens block both the UV-A and UV-B regions; other types selectively block the UV-B regions, allowing the UV-A rays through to increase tanning rates.

A number of chemical substances are effective at blocking UV radiation. Para-aminobenzoic acid (PABA), benzophenone, and cinnamates are commonly used to block UV radiation. Figure 18-6 shows the structures of several compounds that are used in suntan and sunscreen products.

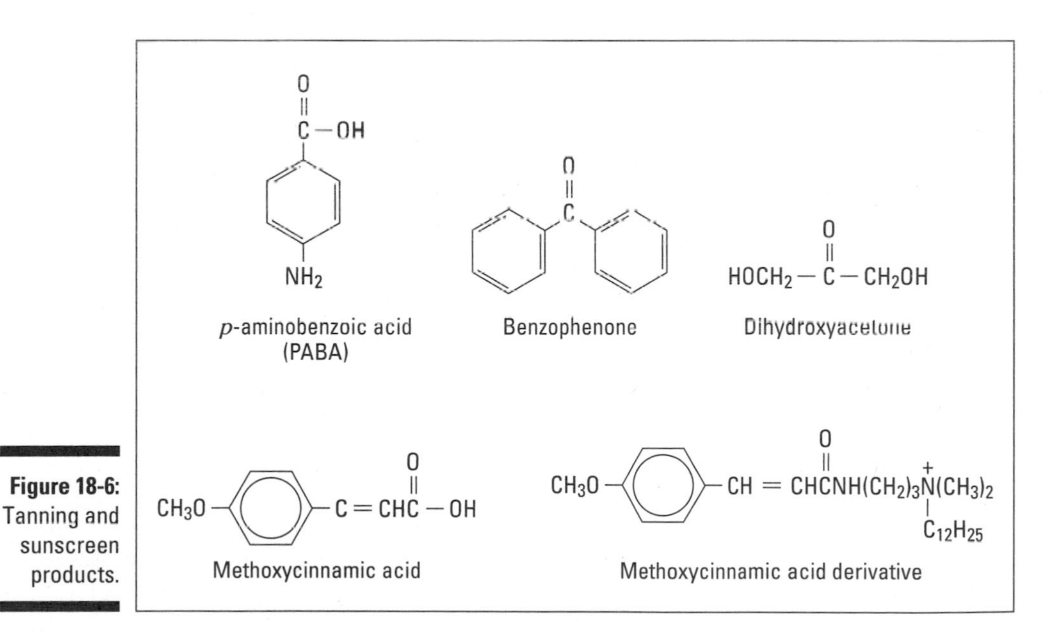

Figure 18-6: Tanning and sunscreen products.

The dihydroxyacetone shown in Figure 18-6 produce a tan without exposure to the sun. It reacts with the skin to produce a brown pigment.

Recently there has been a move away from the use of PABA. It is somewhat toxic, and a significant number of individuals are allergic to it.

Clean it, color it, curl it: Hair care chemistry

Hair is composed of a protein called *keratin.* The protein chains in the hair strand are connected to each other by what is called a *disulfide bond,* a sulfur-to-sulfur bond from the *cystine* (an amino acid component of hair) on one protein chain to another cystine on another protein chain. Figure 18-7 shows a portion of hair and disulfide crosslinks joining two protein chains.

These crosslinks give hair its strength. I say more about this disulfide bond when I tell you about permanents a little later in this section.

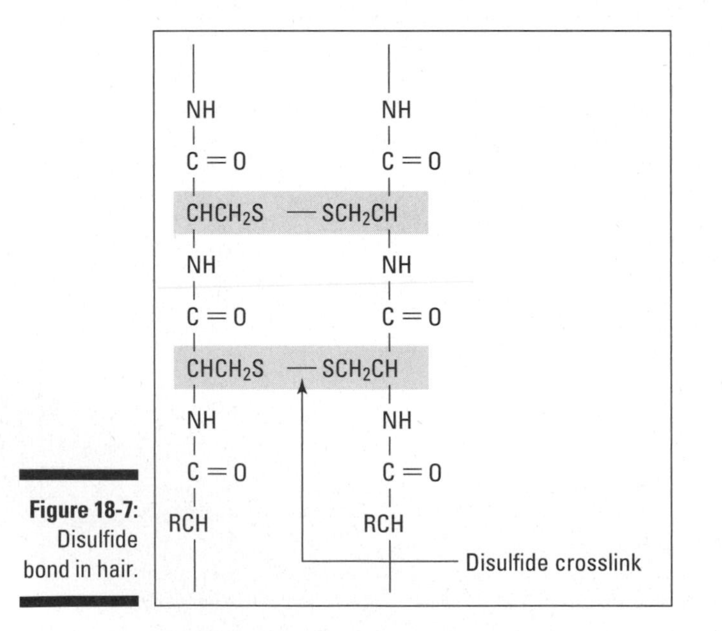

Figure 18-7: Disulfide bond in hair.

Shampoos: Detergents for the hair

Modern shampoos are simple surfactants, such as sodium lauryl sulfate and sodium dodecyl sulfate. There are other ingredients, however, that combine with the metal ions in hard water in order to help prevent the soap from precipitating with these metal ions (bathtub ring in the hair). Other ingredients give a pleasant odor, replace some of the natural lubricants in the hair (conditioners), and adjust the hair's pH. Hair and skin are slightly acidic. A very alkaline (basic) shampoo will damage hair, so the pH is commonly adjusted to the 5 to 8 pH range. A higher pH may also make the scales on the hair cuticles fan out and reflect the light poorly, making the hair look flat and dull. A protein is sometimes added to the shampoo to help glue damaged splint-ends together. Colorants and preservatives are also commonly added.

Dyes and bleaches: Changing your hair color

Hair contains two pigments — melanin and phaeomelanin. Melanin has a dark-brown color, and phaeomelanin has a reddish-brown color. The natural color of the hair is determined by the relative amounts of these pigments. Redheads have much less melanin; brunettes have much more. Blondes have very little of either.

You can *bleach* hair by using hydrogen peroxide to oxidize these colored pigments to their colorless forms. However, bleached hair becomes weaker and more brittle, because the hair protein is broken down into lower molecular weight compounds. Perborate compounds, which tend to be more expensive than bleach, and chlorine-based bleaches are also sometimes used to bleach hair.

You can change the color of your hair temporarily by using dyes that simply coat the hair strands. These compounds are composed of complex organic molecules. They're too large to penetrate the hair strand, so they simply accumulate on the surface. Semi-permanent color can be added by using dyes with smaller molecules that can penetrate into the hair. These dyes frequently contain complexes of chromium or cobalt. The dyes withstand repeated washing, but because the molecules contained in the dye were small enough to penetrate the hair initially, they eventually migrate out.

Permanent dyes are actually formed inside the hair. Small molecules are forced into the hair and then are oxidized, normally by hydrogen peroxide, into colored complexes that are too large to migrate out of the hair. The color then becomes permanent on the portion of the hair that was treated. To maintain the color, the process has to be repeated as new hair grows out. This maintenance program keeps hairdressers in business.

Another type of hair coloring is made to change color gradually, over a period of weeks, so that the change will go unnoticed. A solution of lead acetate is applied to the hair. The lead ions react with the sulfur atoms in the hair protein, forming lead (II) sulfide (PbS), which is black and very insoluble. Instead of losing its color in the sunlight like other dyes, PbS-treated hair actually darkens.

Take it off, all off! Depilatories

Depilatories remove hair by chemical reaction. They contain a substance, usually sodium sulfide, calcium sulfide, or calcium thioglycolate, that disrupts the disulfide linkages in the hair and dissolves it. The formulations commonly contain a base such as calcium hydroxide to raise the pH and enhance the action of the depilatory. A detergent and a skin conditioner such as mineral oil are also generally added to depilatories.

Permanents — that aren't

The disulfide bonds that I mention earlier in this chapter are responsible for the shape of your hair, whether it's straight or curly. In order to affect the shape of hair, those disulfide bonds must be broken and reformed into a new orientation. Suppose that you want to make your straight hair curly (referred to as a *permanent*). You initially treat your hair with some reducing agent that breaks the disulfide bonds; thioglycolic acid (HS-CH_2-COOH) is commonly used. Then you change the orientation of the protein chains of the hair by using curlers. Finally, you treat the hair with an oxidizing agent such as hydrogen peroxide to reform the disulfide bonds in their new locations. Water-soluble polymers are used to thicken the solutions, ammonia is used to adjust the pH to a basic level, and a conditioner is used to complete the formulation. Figure 18-8 shows this process.

Hair is straightened in exactly the same fashion, except it's stretched straight instead of curled. Obviously, as new hair grows in, you need to repeat the process. I guess *permanent* refers to the fact that trips to the beauty parlor become a permanent part of your life.

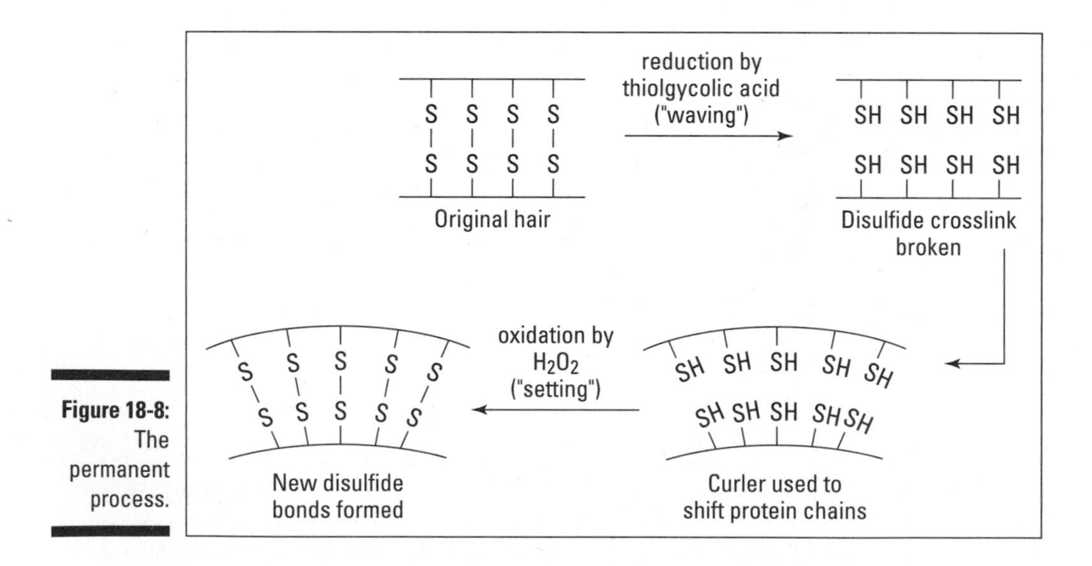

Figure 18-8: The permanent process.

Chemistry in the Medicine Cabinet

Take a quick peek in the medicine cabinet. You probably see a lot of drugs and medicines. I could write a book talking about the chemistry of their reactions and interactions, but I want to stay brief, so the following sections are dedicated to chemistry in your medicine cabinet.

Not tonight, I have a headache: The aspirin story

As early as the fifth century B.C., it was known that chewing willow bark could relieve pain. But it wasn't until 1860 that the chemical compound responsible for the analgesic effect, salicylic acid, was isolated. It had a very sour taste and caused irritation of the stomach. In 1875, chemists created sodium salicylate. It caused stomach irritation, but it was less bitter than the salicylic acid. Finally, in 1899, the German Bayer Company began marketing acetylsalicylic acid, made by reacting salicylic acid with acetic anhydride, under the trade name of aspirin. Figure 18-9 traces the history of aspirin.

Aspirin is the most widely used drug in the world. More than 55 billion aspirin tablets are sold annually in the United States.

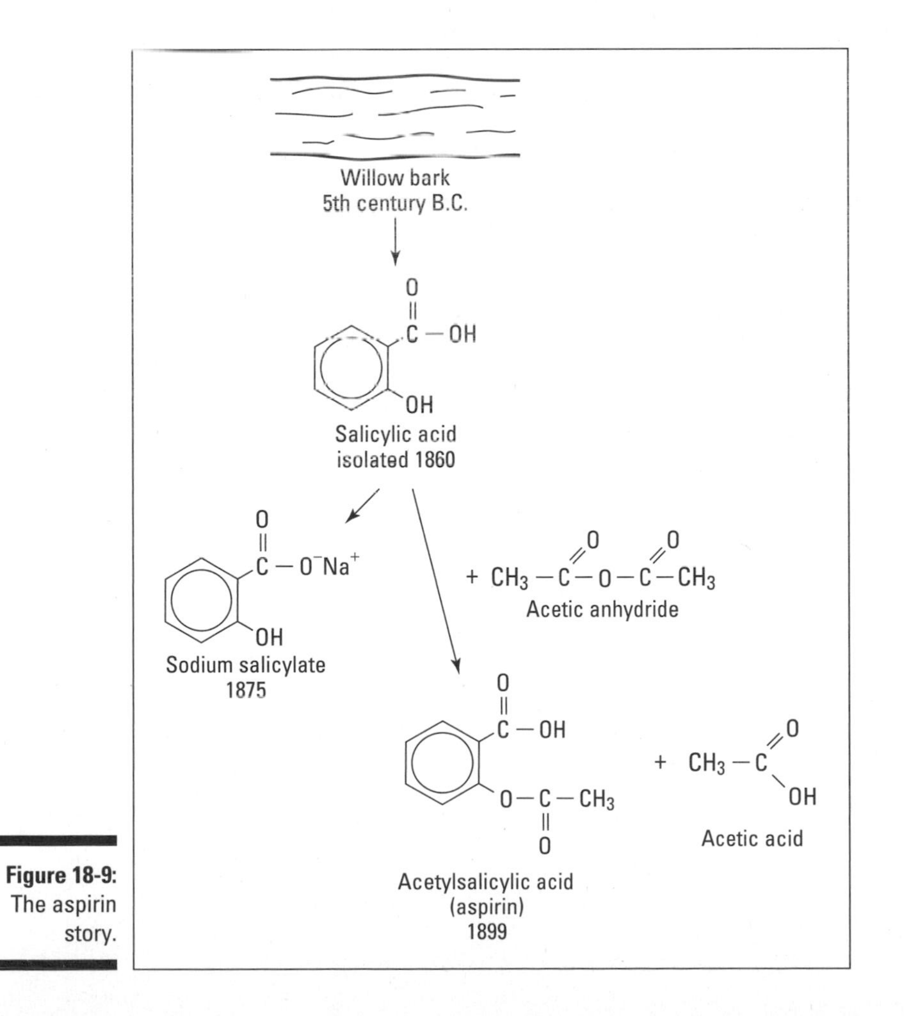

Figure 18-9: The aspirin story.

Introduction to serendipity: Minoxidil and Viagra

Science proceeds by hard work, training, intuition, hunches, and luck. That luck is sometimes called serendipity, accidental discoveries. Or, as I like to say, "finding something you didn't know you were looking for." Because I'm in the medicine cabinet anyway, I may as well mention a couple of serendipitous discoveries right now.

Male pattern baldness affects many millions of men and women in the world. Minoxidil, the current over-the-counter treatment for baldness, was discovered quite by accident. It was being used as an oral treatment for high blood pressure, when patients reported hair growth. Now it is usually applied topically instead of orally.

The much-publicized properties of Viagra were discovered in much the same fashion. It was also being used as a treatment for high blood pressure, as well as angina, when its side effect was reported. In fact, male patients refused to return the unused portion of their medications during clinical trials. Both serendipitous discoveries have spawned multimillion-dollar industries and made countless men and women very happy. Between the two, things are growing at both ends.

Part V
The Part of Tens

In this part . . .

Just like every *For Dummies* book, this part has some fun and snappy chapters. Here I have three chapters that you can read and re-read to help you when you need them. Chapter 19 shows you my top ten tips for passing Chem II. Studying chemistry is hard work for most people; it certainly was for me. I hope these tips can help make studying my favorite subject a bit easier and more enjoyable.

Chapter 20 contains what I consider the ten most important mathematical relationships that you will need in Chem II. These deal with kinetics, equilibriums, thermodynamics, and electrochemistry.

Finally, in Chapter 21 I show you ten typical careers that chemists can pursue. These range from patent attorney to chemistry teacher. Imagine, doing what you love and getting paid for it!

Chapter 19

Ten Terrific Tips for Passing Chem II

In This Chapter

- Figuring out techniques that can help your exam scores
- Finding ways to make your studying more efficient

You're taking or getting ready to take a Chemistry II course and you're interested in maximizing your course grade along with your knowledge. In my book *Chemistry For Dummies,* 2nd Edition (John Wiley & Sons, Inc.), I present ten tips for passing Chem I. If you used them, I hope they helped. The techniques for studying Chem II are really no different than Chem I, so I keep the same general tips, this time focused specifically to Chem II. In this chapter I lay out helpful tips to help you excel in your Chem II course.

Developing and Sticking to a Regular Study Schedule

Studying chemistry is different than studying history. Cramming the night before an exam won't result in a good chemistry grade. Chemistry is a subject that you have to work at in a steady fashion, giving yourself time to think about and process the material.

Try to budget some time every day to your study of chemistry and then stick to a regular schedule. If you study some every day, you won't have to cram the night before the test. How much time should you devote? The general rule of two hours studying for every hour in class is a pretty good one. Sometimes you may have to spend more, sometimes less. And put away your phone; texting while studying isn't an effective way to study.

Striving For Understanding — Don't Just Memorize

When I was in school, many of my classmates tried to simply memorize their way through their chemistry classes. And it didn't work for them, especially when they hit the final exam and they had to try to recall an entire semester's worth of chemistry. I, on the other hand, had such a bad memory that I concentrated on actually learning the material, and I had a much easier and more productive time. Granted, you do have to memorize some things, such as formulas, but I urge that you strive for understanding. Don't simply memorize the fact that the size of atoms decrease in going from left to right in a period, but master the concept why. That way, when faced with a similar situation, you can reason your way through.

Let me give you another example specific to Chem II. In this book, I devote four chapters to equilibriums, but the basic concepts are all the same. After you truly understand these basic concepts, you can apply them to every equilibrium situation. The less you have to memorize, the less to forget in the middle of an exam.

Doing the Homework

Chemistry isn't a spectator sport. During my forty-plus years of teaching, I've had many students tell me that those problems looked so easy when I worked them, but they couldn't do them on the exam. They weren't successful because they never practiced. Have you ever watched a professional athlete or musician perform? They make it look easy. But behind that performance was practice, practice, and more practice. The same is true in chemistry; to be good, you must practice. Homework is practice. The more homework problems you do, the better will be your performance on an exam. Find additional problem sets and do them.

Using Additional Resources

When I'm preparing a lesson on an unfamiliar topic, I may use several different resources to help me reach a level of understanding that allows me to explain the concept to my students. I use numerous different books to see the approach that different authors use. I use the Internet to find current information or additional problem sets. If I have colleagues that are familiar with the topic, I talk to them. Preparing a lecture takes several hours. You may need to do the same.

No matter how good your textbook's author is, some text will be less than crystal clear. Try another author's approach. I know that you probably don't have an extensive chemistry library (probably this guide and your textbook), so go to the library and work with some other text. Do an online search for topics that are troubling you. Look at several sources from several angles. Sure, doing so takes time, but a lot less time (and money) than repeating the course.

Skimming the Material before Going to Class

Skimming the text before going to class can help set the stage for the lecture. I tell my students to spend about 15 minutes reading ahead in their text before coming to class. Just read for general understanding. You don't work problems until after class. Just try to get an overview of the topic that your teacher will be covering. Doing so really can make the lecture more meaningful. After class, you can reread in depth; before class, you're looking for the big picture.

Taking Good Notes

I've been tempted to buy the T-shirt that says, "If my lips are moving, you should be writing!" Theoretically, you should be able to just read the textbook, work the exercises, and ace the exams. It never seems to work that way. Your instructor (hopefully) won't simply be reading from the text. She will be explaining things in a different way than the textbook author does, which is valuable. When she's explaining, you should be taking good notes. Taking good notes allows you to capture those explanations and techniques. Your notes can help jog your memory when studying; the more complete the notes, the more useful they become.

Recopying Your Lecture Notes

Lecture notes tend to be messy. In a chemistry class using a laptop to take notes is difficult because of all the special symbols, structures, formulas, and so on. Therefore, most students take handwritten notes. Your notes will probably be sloppy because you've been writing so fast. Sloppy notes are difficult to study, so I suggest you recopy your notes as you study — within 24 hours if possible after the lecture.

Recopying your notes can make them more useful. Recopying your notes also becomes a reinforcing technique. If something isn't clear, you can use your text and/or other sources to supplement and resolve any misconceptions. When recopying, you can take a little more time drawing those structures and writing those formulas. You can even take the time to word process the notes; drawing may still be an issue, but you can still overcome them.

Asking Questions

The only stupid question is the one you didn't ask. Ask questions in class if you don't understand something. Waiting and not asking won't make the topic clearer. If you're uncomfortable about asking questions during class, go to your instructor's office during office hours. Have him explain it again. Maybe a different approach can help you. Most professors and teachers entered teaching because they enjoy helping people; please give them an opportunity to help. You (or someone else) is paying a great deal of money for your instructor's time, so get your money's worth.

Getting a Good Night's Sleep before Exams

If you've kept up with your studying and homework, then all that you should need to do the night before an exam is a little brief review and then a good night's sleep. Mastering concepts the night before an exam is very difficult. If you haven't learned them by now, you aren't going to by staying up all night. The best thing you can do is to get a good night's sleep so that you will be fresh and relaxed for the exam. You need to be able to think your way through the chemistry test and doing so is difficult if you keep falling asleep. All-nighters are a no-no.

Paying Particular Attention to Charges and Significant Figures on Exams

If you know the material, you can still get a bad grade by making careless errors. Make sure you haven't skipped any problems, you've filled your answer sheet out properly, and so on. Checking to make sure you have shown the correct charges on charged species (ions, subatomic particles, and so on) and that you have reported your answer to the correct number of significant figures can save you a couple of points here and there. These points add up. Ask my student who left off 41 charges on an exam at one-point each. It was a hard lesson for him, but a valuable one.

Chapter 20

Ten Must-Know Formulas for Chem II

In This Chapter

- Figuring out which are the really important formulas for Chem II
- Reviewing how to use these formulas

In your Chemistry II course, you will encounter many different math formulas to help you make different calculations. In this chapter, I give you a brief review of what I consider to be the most important formulas that you'll use in Chem II. As you read these formulas, remember that you must know when to apply them and how to apply them correctly.

Molarity Equation

The most widely used concentration unit in general chemistry is molarity (M). *Molarity (M)* is defined as the moles of solute per liter of solution:

$$M = \frac{\text{Moles of solute}}{\text{L of solution}}$$

Fully labeling the solute in this equation is always a good idea, such as "moles of NaCl" instead of just "moles" so that you don't try to cancel the wrong substance. Check out Chapter 5 for molarity calculations.

Rate Law Equation

The *rate law equation* relates the change in concentration of the reactants to the speed of the reaction. For the general reaction:

$$aA + bB \rightarrow cC + dD$$

the rate law is:

$$\text{Rate} = k[A]^m[B]^n$$

In this equation, the rate is an experimentally determined value. The k is a rate constant, which you need to determine. The [ ] terms refer to the concentrations of the reactant substances involved in the reaction. The exponents, *m* and *n,* are the orders of reaction. The order(s) of reaction give information about what will happen to the reaction rate if the concentration of that particular reactant changes. Chapter 7 on kinetics makes use of this rate law equation.

Homogeneous Equilibrium Constant Expression

The *equilibrium constant* describes the relationship between the amounts of the reactants and the products at a certain temperature. For the general equilibrium:

$$aA + bB \rightleftarrows cC + dD$$

the equilibrium constant expression is:

$$K_c = \frac{[C]^c[D]^d}{[A]^a[B]^b}$$

In the expression, K is the equilibrium constant, the subscript *c* indicates this constant is expressed in terms of concentrations (not pressures, p), the brackets (as usual) stand for molar (moles/L) concentration, the uppercase letters are the reactant and product species, and the lowercase superscripts are the coefficients in the balanced chemical equation. Chapter 8 uses equilibrium constant expressions a lot.

Acid and Base Equilibrium Constant Expressions

The *acid and base equilibrium constant* expressions describe the relationship between the amounts of reactants and products in aqueous acid-base systems. For the following general weak-acid equilibrium:

$$HA(aq) \rightleftarrows H^+(aq) + A^-(aq)$$

the equilibrium constant expression is:

$$K_a = \frac{[H^+][A^-]}{[HA]}$$

For a general weak-base equilibrium:

$$B(aq) + H_2O(aq) \rightleftarrows HB^+(aq) + OH^-(aq)$$

the equilibrium constant expression is

$$K_b = \frac{[HB^+][OH^-]}{[B]}$$

The concentration of water (or any pure liquid or solvent or solid) does appear in the equilibrium constant expression. K is the equilibrium constant, the subscript *b* indicates that this is an equilibrium constant expression for a weak base, and the brackets indicate molar concentrations.

Here is a freebie about the autoionization of water:

$$H_2O(aq) \rightleftarrows H^+(aq) + OH^-(aq)$$

$$K_w = [H^+][OH^-] = 1.00 \times 10^{-14}$$

And also:

$$K_a \times K_b = K_w$$

Acid base constant expressions are used a lot in Chapter 9.

pH/pOH Equations

pH is a measure of the molar concentration of the hydrogen ion in solution while *pOH* is a measure of the molar concentration of the hydroxide ion in solution. The following equations show the interrelationships between these quantities.

$$PH = -\log[H^+] \text{ and } pOH = -\log[OH^-]$$

Another freebie:

$$pH + pOH = 14.00$$

You can find these equations in Chapter 9.

Henderson-Hasselbalch Equation

The *Henderson-Hasselbalch* equation is most commonly used in buffer problems, especially in biochemistry. There are two forms of the Henderson-Hasselbalch equation:

$$pH = pK_a + \log\frac{[CB]}{[CA]}$$

and

$$pOH = pK_b + \log\frac{[CA]}{[CB]}$$

CA stands for the conjugate acid and CB the conjugate base. I use the Henderson-Hasselbalch equation in Chapter 9.

Solubility Product Equation

The *solubility product equation* is used to describe the equilibrium situation when a not-so-soluble salt is dissolving in water. For the general dissociation of a sparingly soluble salt:

$$M_aX_b(s) \rightleftarrows aM^{x+}(aq) + bX^{z-}(aq)$$

In this equation, $x+$ and $z-$ are the magnitude of the positive and negative charge, respectively; the equilibrium constant expression (solubility product expression) is

$$K_{sp} = [M^{x+}]^a[X^{z-}]^b$$

You use this expression in Chapter 17.

Gibbs Free Energy Equation

The *Gibbs Free Energy* is the best indicator about whether a reaction will be spontaneous or nonspontaneous. It has the form:

$$\Delta G^0_{reaction} = \Delta H^0_{reaction} - T\Delta S^0_{reaction}$$

In this equation ΔG° is the Gibbs Free Energy of a reaction under standard conditions of 1 atm (or 1 bar) for gases and 1 M for solutions at 25°C; ΔH° is the enthalpy of the reaction under standard conditions; T is the Kelvin temperature; and ΔS° is the entropy of the reaction under standard conditions.

A spontaneous process has $\Delta G < 0$. A nonspontaneous process has $\Delta G > 0$. When $\Delta G = 0$, the process is at equilibrium.

The Gibbs Free Energy equation appears in Chapter 11.

Nernst Equation

The *Nernst equation* allows for the calculation of the actual cell potential if the electrochemical cell isn't at standard conditions:

$$E_{cell} = E^0_{cell} - \left(\frac{RT}{nF}\right) \ln Q$$

or

$$E_{cell} = E^0_{cell} - \left(\frac{0.0592}{n}\right) \log Q \text{ at } 25^{\circ}\text{ C}$$

R is the ideal gas constant, T is the Kelvin temperature, n is the number of electrons transferred, F is Faraday's constant, and Q is the activity quotient (same form as the equilibrium constant expression). The second form, involving the log Q is the more useful form.

If the electrochemical cell is at equilibrium:

$$\log K_c = \frac{nE^0_{cell}}{0.0592}$$

This equation gives a method for calculating the equilibrium constant from cell potential measurements. I discuss this equation in Chapter12.

Half-Life Equations

The *half-life* ($t_{1/2}$) is the amount of time that it takes for exactly one half of a sample to react (or decay).

For first-order reactions, use this equation:

$$t_{\frac{1}{2}} = \frac{0.693}{k}$$

For second-order reactions, use this equation:

$$t_{\frac{1}{2}} = \frac{1}{k[\mathrm{A}]_0}$$

where $[\mathrm{A}]_0$ is the concentration of A initially, time = 0.

For zero-order reactions, use this equation:

$$t_{\frac{1}{2}} = \frac{[\mathrm{A}]_0}{2k}$$

$[\mathrm{A}]_0$ is the concentration of A initially and k is the rate constant.

You use these half-life equations a lot in Chapter 17.

Chapter 21

Ten Great Chemistry Careers

In This Chapter

- Figuring out what chemists do
- Finding jobs in industry and in teaching

When I advise students who are chemistry majors, sooner or later I get *the* question. What can I do with a degree in chemistry? It's a good question. Most people have never really talked to a chemist. They've talked to store managers, clerks, accountants, lawyers, and so on, but rarely to chemists. A lot of chemists are in the workplace, but maybe chemists have a certain air about them (chemical fumes?) that people detect and avoid asking about their jobs. When people do ask me what I do besides write *For Dummies* books, I tell them that I teach. "Really. What do you teach?" When I say chemistry, I get comments like "I didn't take chemistry in high school. It was too hard!"

Actually, being a scientist is sometimes okay and even cool, especially with television shows like CSI and NCIS. This chapter takes a look at some typical jobs that chemists might hold in case you are thinking about your options.

If you want to be successful as a chemist or in any science area, you must be a good communicator. You need to be able to prepare reports, papers, and summaries that are understandable to your audience. You need to be able to prepare an oral presentation that is intelligible to your audience. Just knowledge of chemistry isn't sufficient to progress in any company, university, or firm. You must be able to communicate well. If you're still in school, pick up a course in technical writing and another in public speaking. Those courses can be as useful to you in your career as organic chemistry.

Patent Attorney

I was recently talking with one of our chemistry graduates who went on to earn his law degree. He ended up as a very successful attorney in patent law and highly recommended the combination of chemistry and law.

Patent attorneys perform patent searches, advise their clients on whether or not their formulation/invention is patentable, provide advice on such topics as product liability and intellectual property, and may even take cases to court for product infringement. A background in chemistry makes them valuable to companies that deal with chemicals or chemical processes. All of these patent law attorneys prepare reports and do presentations to their clients (or the court).

Pharmaceutical/Chemical Sales

Do you like chemistry, but don't like to be in the lab? Then sales may be the job for you. Somebody has to sell all those chemicals or medicines, and buyers like to deal with somebody who knows their science. The salesperson has to answer their customers' questions about the product, toxicity, side effects, and so on. The chemical or pharmaceutical salesperson has to be willing to travel a lot, servicing existing accounts and finding new customers.

This type of salesperson needs to be able to communicate well with clients, their company's management, and the production engineers and chemists. A degree in chemistry is very useful in communicating with the manufacturing and quality control scientists, many of whom are chemists. And you probably figured out that these chemists also prepare a lot of reports for their company and their clients.

Forensic Chemist

You may think that forensic chemists investigate crime scenes, chase killers, and all that you see on CSI; however, most forensic chemists that I've talked to spend the first year or so exclusively in the lab analyzing evidence for drug residues. But some, after years of hard work, do get to work crime scenes. In addition, forensic chemists operate and maintain laboratory instrumentation, analyze biological fluids for DNA matches, analyze for drug residues in foods or biological tissues, analyze gunshot residues, examine hair or other fibers concerning its source, and so on.

Keeping detailed, meticulous records and reports is critical to this profession because they may be called upon to testify in court about their findings.

Biochemistry/Biotechnology

Do you like biology about as well as chemistry? Then you may want to combine your two passions. Biochemistry or biotechnology may be the field for you. Biochemists study the reactions and chemical processes associated with living systems, while biotechnologists work to improve existing biological products (like grains, cattle, and so on) or to develop new biological products (medicines, textiles, and so on).

Biochemists and biotechnologists work in research developing new genetic tests, work in the genetic engineering (cloning) area, and are involved in the development of new drugs. Others work as plant breeders, trying to develop more disease-resistant strains of crops. Others work as biochemical development engineers taking a biochemical process developed in the lab and scaling it through the pilot plant stage to full plant production. Still others become *biostatisticians,* statisticians who work in health-related fields examining how diseases progress and the safety of a new treatment methodology, or even analyzing healthcare costs.

Agricultural Chemist

I was raised on a farm and my dad always said that to be a good farmer you had to be a good chemist. Agricultural chemists, also referred to as *agrochemists,* collect and analyze samples for nutrient levels as well as levels of pesticides, heavy metals, or toxins. They operate and maintain a wide range of instrumentation including spectrophotometers and pH meters and may be called upon to operate farm equipment such as tractors, plows, and sprayers. Some agrochemists specialize in animal feeds; others specialize in the testing of pesticides.

They may do presentations to such diverse groups as corporation CEOs and farmers as well as preparing reports showing their data and conclusions and recommendations.

Material Science

Material scientists study the composition and structure of various materials with the goal of developing new products or improving existing ones. One goal is to lighten and strengthen existing products such as golf clubs, tennis rackets, automobiles, and snow skis. Some material scientists work with ceramics and others with metals. Some analyze failed products to determine the reason for their failure.

Some are involved in quality control, testing raw materials and finished products to make sure that they meet specifications. Some may experiment on new ways to combine different materials while others may actually help in the design of new manufacturing plants. All of them prepare reports and do presentations to management, engineers, and the public.

Food and Flavor Chemist

Food and flavor chemists work in the research and development of new foods as well as ways of keeping foods fresher on the shelves. Some test food additives and preservatives to make sure that these products are in compliance with FDA regulations. Others may work on developing new flavors or finding ways to synthesize flavors found in nature. They may analyze foods for the presence of contaminants and bacteria. Others food chemists may analyze foods to determine their nutrient levels, fat percentage, and so on.

Some food scientists are involved in the various preservation processes such as freezing, canning, blanching, pasteurizing, and so on, to ensure that the food remains as fresh and nutritious as possible. Some food chemists work for the FDA and other governmental agencies as inspectors to ensure that regulations are followed in food processing areas and in the shipping of foods. And, you guessed it, all of them prepare reports and do presentations.

Water Quality Chemist

Water quality chemists hold a wide variety of jobs, but all, in one way or another, help ensure that the public's drinking water is safe. Some may be involved in the design of water or wastewater treatment plants. They may design water runoff collection systems for industries or agriculture. They may analyze water samples for the presence of pesticide residue, heavy metals, or bacteria and make recommendations for appropriate methods of treatment. Others may conduct environmental impact studies for either an industry or for some governmental agency.

They may collect water usage data and use it to predict future water needs. Others may use sophisticated computer programs to predict the movement of water or pollutants in the waste supply or perform mathematical modeling of underground water resources. Some may perform quality control tests to ensure that water treatment plants are removing the harmful pollutants from drinking water. All write technical reports and make presentations to a wide variety of individuals and entities.

Cosmetic Chemists

Nearly everyone uses some type of cosmetic product, whether it be lipstick, lip balm, eyeliner, or shaving cream. Each of those products had to be developed and tested. The cosmetic chemist tests all these products. The cosmetic formulator creates the cosmetic. Then another chemist tests it to be sure that it meets governmental regulations. When full production begins, another chemist ensures that both the raw materials and the finished product meet specifications. These quality control chemists work with analytical instrumentation such as spectrophotometers, gas chromatographs, and mass spectrographs. These chemists are always in high demand in the cosmetics industry.

Some companies buy the raw materials for their cosmetics, while others synthesize their own. If they do create their own raw materials, then a synthesis chemist is involved. And every company needs some type of regulatory scientists to make sure that everything is meeting governmental regulations. And all of the cosmetic chemists prepare reports and do presentations.

Chemistry Teaching

After I earned my master's degree in chemistry I had planned to go to work for a chemical company, probably in quality control because I was an analytical chemist. However, the United States Army had another idea. After my stint in the army, I looked around for a job, but it was one of the few times in history that jobs for chemists were scarce. So, I took a job as a lab coordinator at a regional Texas university with the plan that I would stay there for a couple of years and then move on to industry. And I'm still here after 41 years (and a doctorate) because I found that I love teaching. I love presenting the material that I majored in, and I love helping other people develop an appreciation, and the occasional love, of chemistry.

Teaching jobs in chemistry can range from teaching in the public schools at the middle school or high school level (bachelor's degree required) to junior/community college (master's degree required) to the university level (doctorate required). At the university level you get to do research along with teaching. In fact, having your students do research with you is a great form of teaching. You can even work with teachers and people who want to become teachers, which is enjoyable and rewarding.

Index

• Symbols •

° (degree symbol for standard [state] conditions), 201
Δ (change in state function), 192
ΔG (Gibb's Free Energy), 200–202, 336–337
ΔH (change in enthalpy), 192, 193–195
ΔS (change in entropy), 196–197

• A •

A (ampere), 225
A (frequency factor), 114
abrasives in toothpaste, 315–316
acetic acid
  as carboxylic acid, 243
  K_a equilibrium problems, 160–162
  K_b equilibrium problems, 164–165
  as weak acid, 153
acetone, 245
acetylene (ethyne), 240
acetylsalicylic acid, 325
acid and base equilibrium constant expressions, 334–335
acid ionization constant (K_a)
  defined, 153
  problems, solving, 160–162, 172
acid-base equilibrium problems
  K_a, 160–162
  K_b, 163–165
acid-base reactions, Bronsted-Lowry, 155–156
acidic oxides, 154
acidic water and soaps, 311
acidity, determining, 156–158
acids
  amino, 265–266
  carboxylic, 243
  conjugate, 155
  dilution of concentrated, 68
  macroscopic properties of, 145–146
  microscopic properties of, 146–148
  nucleic, 273–275
  strength of, 148–150
  strong, 151–152
  weak, 153
actinide series, 34–35, 36
activation energy
  catalysts, 121–123
  fusion process, 305
  overview, 114–116
actual yield, 57
addition
  exponential or scientific notation, 19–20
  reporting correct number of significant figures, 25
addition polymerization, 250–255
addition reactions, 239–240
aftershaves, 319–320
agents
  oxidizing and reducing, 211
  suspension, in detergents, 312
agricultural chemists, 341
air pollution, 287
alcohols, 239, 242–243
aldehydes (R-COH), 244
aldose sugar, 263
alkali metals, oxidation numbers of, 211
alkaline dry cells, 227
alkaline metals, oxidation numbers of, 211
alkaloids, 246
alkanes
  cycloalkanes, 238
  halogenated hydrocarbons, 238
  molecular and structural formulas, 233–234
  naming, 234–238
  overview, 232–233
alkenes, 238–240
alkynes, 240
alpha emission, 294
alternators, 228
ambergris, 319
amides (R-CO-NH_2), 245–246
amines (R-NH_2)
  overview, 245–246
  in sweat, 316

amino acids, 265–266
ammonia. *See also* Haber (Haber-Bosch) process
  amine and amide functional groups, 245–246
  as base, 150–151, 154
  multipurpose cleaners, 315
  pH, determining, 163–164
  synthesis of, 126–127, 138
ammonium chloride, 168
amorphous solids, 78
ampere (A), 225
amphoteric substances, 159
analytical chemistry, 12–13
anionic compounds, 39
anionic surfactants, 310
anions, defined, 40
anodes, 217
antilog relationships, determining, 158
antimatter, 295
antiperspirants, 316
applied chemistry, 14
aqueous solutions, 64
aromatic hydrocarbons, 241, 284
Arrhenius equation, 114–115
Arrhenius theory, 146–147
asphalt, 282
aspirin, 325
assumptions, checking validity of, 133, 161
astringents, 316
atmospheres, as units of measurement, 83
atomic bombs, 302
atomic number
  defined, 33
  isotopes, 29, 290
  in periodic table, 36
atomic structure
  modeling electrons, 29–30
  nucleus, 28–29
  overview, 10, 290
  representing electrons, 30–33
  subatomic particles, 27–28
atomic weight, 29, 36
aufbau principle, 32
autoionization of water, 159, 335
automatic dishwasher detergents, 315
automobile batteries, 228
Avogadro's law, 90–91
Avogadro's number, 48

• B •

balancing equations
  redox, 212–215
  stoichiometry problems, 52
balancing nuclear reactions, 291
baldness, male pattern, 326
band of stability, 293
barium chloride, 183
bases
  conjugate, 155
  macroscopic properties of, 146
  microscopic properties of, 146–148
  strength of, 148–149, 150–151
  strong, 152–153
  weak, 154
basic oxides, 154
basic research, 14
bathroom chemistry
  deodorants and antiperspirants, 316
  hair care, 322–324
  skin care, 317–322
  toothpastes, 315–316
bathtub ring, 311
batteries, 216, 226–228
benzene, 241, 284, 286
benzophenone, 321
beta emission, 294–295
bimolecular step, 117
binary acids, 149
binary metal hydride, 211
biochemistry
  amino acids, 265–266
  buffer calculations, 165–166
  carbohydrates, 263–265
  carbon, 261
  careers in, 341
  hydrogen, 262
  lipids, 269–272
  nucleic acids, 273–275
  overview, 13, 261
  proteins, 267–269
Biochemistry II For Dummies (Moore and Langley), 13
biomass, 259–260
bioplastics, 259–260
biostatisticians, 341

biotechnology
careers in, 341
designing drugs, 276
overview, 13
bleaches
hair, 323
laundry, 314
body powders, 317–318
Bohr, Niels (scientist), 29–30
Bohr model, 29–30
boiling-point elevation property, 71–72
bomb calorimeters, 229
bombs
atomic, 302
hydrogen, 305
bond strength, 43
bonding. *See also* covalent bonding
coordinate-covalent, 148
hydrogen, 46, 76, 268
ionic, 39–43, 45, 262
overview, 10
peptide, 266
types of, 38
bonding electrons, 44
Boyle's law, 84–85
branched structure of polymers, 249
break-even point, 307
breeder reactors, 304–305
Bronsted-Lowry theory
acid-base reactions, 155–156
overview, 148
buffer capacity, 171
buffers
defined, 165
Henderson-Hasselbalch equation, 167–171, 336
problems, solving, 165–167
builders in detergents, 312
butane
in crude oil, 281
as cycloalkane, 238

• C •

C (coulomb), 225
calculators, using, 20–21
calorie
defined, 19, 193, 229
nutritional, 230
calorimeters
bomb, 229
measuring enthalpies of reaction with, 194
capillary action, 77–78
carbohydrates
disaccharides and polysaccharides, 265
monosaccharides, 263–264
carbon
biochemistry, 261
covalent bonds, 231
hydrocarbons, 232–241
carbon-11, 295
carbon-14 dating, 299
carboxylic acids (R-COOH), 243
careers in chemistry, 339–343
catalysts
defined, 121
effects on equilibrium, 141
Haber process, 143
heterogeneous, 121–122
homogeneous, 122–123
reaction rates, 101
catalytic converter, 287
catalytic cracking, 282–283, 286
catalytic reforming, 283–284, 286
categories of elements
metals, 36–37
nonmetals and metalloids, 37
periods and families, 38
cathode compartments, 217
cathodes, 217
cations
biochemistry, 262
defined, 39, 208
cell notation, 218–219
cells
alkaline dry, 227
concentration, 224
electrochemistry, 216–222
electrolytic, 225, 228
human, and osmotic pressure, 74, 262
cellulose, 247–248, 265
chain reactions, 300–302
Charles, Jacques (chemist), 86
Charles's law, 86–87
chemical equilibrium. *See also* LeChatelier's Principle
equilibrium constant, 128–129
Law of Mass Action, 127–128

chemical equilibrium *(continued)*
overview, 125–127
problem 1, 131
problem 2, 131–132
problem 3, 132–133
problem 4, 133–137
chemical sales, careers in, 340
chemistry. *See also* biochemistry; consumer product chemistry; electrochemistry; nuclear chemistry; organic chemistry
careers, 339–343
general areas/branches, 12–13
kitchen, 309
macroscopic and microscopic viewpoints, 13–14
pure and applied, 14
Chemistry For Dummies, 2e (Moore) , 9
Chemistry I, basic topics, 10–11
Chemistry II
basic topics, 11–12
tips for passing, 329–332
Chernobyl power plant, 303–304
chiral carbon, 264
chlorine
ionic bonding, 39–40
oxidation number of, 212
oxyacids of, 149–150
cholesterol, 270
cinnamates, 321
civetone, 319
Clean Air Act of 1970, 287
cleaners, multipurpose, 315
CLPE (crosslinked polyethylene), 251
cobalt-60, 295
coco butter, 317
coefficients, 193
cold cream formulation, 317
colligative properties
boiling-point elevation, 71–72
defined, 69, 70
freezing-point depression, 72–73
osmotic pressure, 73–74
vapor-pressure lowering, 70–71
collision theory
endothermic reactions, 113–114
exothermic reactions, 112–113
overview, 111–112
colognes, 319–320
combined gas law, 89–90
combustion reactions
alkanes, 238
overview, 229–230
common-ion effect, 184–185
communication, importance of, 339
completion of reactions, 152
complex ion equilibriums
dissociation, 188
formation constant, calculating, 186–188
overview, 185–186
compounds, ionic, 41–43
compressed natural gas, 288
concentrated solutions, 63
concentration cells, 224
concentration units
molality, 69
molarity, 66–69
parts per million/billion, 69–70
percent composition, 64–66
concentration-based equilibrium constant (K_c), 129
concentrations
altering, and equilibrium, 138–139
calculating, of dissolved ions, 180–181
calculating, with Nernst equation, 224
defined, 148–149
Haber process, 143
condensation polymerization, 256–258
conjugate acid-base pairs, 155
constants. *See also* equilibrium constant; rate constant
acid ionization, 153, 160–162, 172
dissociation, 188
equilibrium, for weak bases, 154, 163–165, 175–176
formation, 186–188
gas, 202
instability, 188
pressure-based equilibrium, 130
solubility product, 180
stability, 186–188
water dissociation, 159
consumer product chemistry
bathroom, 315–324
kitchen, 314–315
laundry room, 309–314
medicine cabinet, 324–326

conversions
grams to moles, 49
grams to particles, 51
length, 17
mass, 17
moles to grams, 49–50
moles to particles, 50–51
particles to grams, 51–52
particles to moles, 50
temperature, 18
volume, 18
coordinate-covalent bonds, 148
corrosion inhibitors in detergents, 312
cosmetic chemists, 343
coulomb (C), 225
counted numbers, 23
covalent bonding
electronegativities, 44–45
overview, 43, 262
polar, 45–46
proteins, 268–269
structural formula, 44
creams for skin, 317
crenation, 74
crisscross rule, 42–43
critical mass, 301
crosslinked polyethylene (CLPE), 251
crosslinked structure of polymers, 249
crude oil, refining
catalytic cracking, 282–283, 286
catalytic reforming, 283–284, 286
fractional distillation, 280–282
overview, 279–280
crystallization, enthalpy of, 198
cycloalkanes, 238
cyclohexane, 238
cystine, 322
cystine units, 269

• D •

Dalton's law, 92–94, 132
Daniell cells, 216–218
degree symbol (°) for standard (state) conditions, 201
delocalized electrons, 241
Δ (change in state function), 192
ΔG (Gibb's Free Energy), 200–202, 336–337
ΔH (change in enthalpy), 192, 193–195
ΔS (change in entropy), 196–197
denaturing process, 242
deodorants, 316
deoxyribose, 273, 274
depilatories, 323–324
detergents
dishwashing, 315
hair, 322–323
laundry, 312–313
toothpaste, 315–316
deuterium, 305
D-fructose, 264
D-galactose, 263, 264
D-glucose, 264
diagramming energy levels, 31–32
diesel fuel, 282
diethyl ether, 245
dihydroxyacetone, 321 322
diluted solutions, 63
dilution process, 67–68
dipole-dipole attraction, 46, 76
dipoles, 45–46
diprotic acids, 151
direct electron reactions, 210–211, 216
disaccharides, 265
dishwashing detergents, 315
dispersion (London) forces, 46, 75
disposal of nuclear waste, 304
dissociation
into compounds, 188
into ions, 152
dissociation constant (K_d), 188
dissolving process, 62–63
distillation process in refining industry, 280–282
disulfide bond in hair, 322
division
exponential or scientific notation, 20
reporting correct number of significant figures, 25
DNA, 273–275
drugs, designing, 276
dry cells, 227
D-sorbose, 263, 264
ductile, defined, 37
dyes for hair, 323
dynamic chemical equilibrium, 126–127

• E •

elastomers, 250
electric vehicles, 288
electricity, making, 302–303
electrochemical reactions, 208
electrochemistry
  batteries, 226–228
  cells and cell potentials, 216–222
  combustion reactions, 229–230
  electrolysis, 224–226
  Nernst equation, 222–224
  overview, 12, 207
  redox reactions, 208–215
electrodes, 216, 218
electrolysis, 224–226
electrolytic cells, 225, 228
electron capture, 296
electron clouds, 30
electron configurations, 10, 32–33
electron-dot formula, 44
electronegativities, 44–45
electrons. *See also* indirect electron transfers
  bonding and nonbonding, 44
  delocalized, 241
  described, 28
  gain of, in reduction, 209
  loss of, in oxidation, 208
  modeling, 29–30
  representing, 30–33
  valence, 33, 289
electroplating, 226
electrostatic attraction, 40
emollients, 317
emulsions, 317
end notes in fragrances, 319
endergonic reactions, 114
endothermic reactions, 113–114, 193
energy
  activation, 114–116, 121–123, 305
  change in, determining, 192–193
  kinetic, 111
  of nuclear fission, 300
  in SI system, 19
energy levels
  defined, 30
  diagramming, 31–32
engines, internal combustion
  alternatives to, 288
  overview, 284
enthalpy change (ΔH)
  overview, 192, 193–195
  predicting spontaneity for, 199–200
entropy (S)
  defined, 195–196
  predicting spontaneity for, 199–200
  spontaneity, determining, 196–197, 198–199
enzymes
  as catalysts, 121, 273
  detergents, 312
equations. *See also* conversions
  acid and base equilibrium constant expressions, 334–335
  Arrhenius, 114–115
  energy produced in nuclear reactions, 300
  Gibb's Free Energy, 200–202, 336–337
  half-life, 337–338
  Henderson-Hasselbalch, 167–171, 174, 336
  homogeneous equilibrium constant expression, 334
  ideal gas, 91–92
  molarity, 333
  most important, 333–338
  Nernst, 222–224, 337
  net-ionic, 147, 210–211
  pH/pOH, 335
  quadratic, solving, 21
  rate law, 333–334
  redox, balancing, 212–215
  simple-multiples of half-lives, 298–299
  solubility product, 336
  thermochemical, 193–194
equilibrium. *See also* acid-base equilibrium problems; chemical equilibrium; equilibrium constant
  complex ion, 185–188
  heterogeneous, 179
  overview, 11–12
  problems, solving, 131–137
  solubility, 179–185
equilibrium constant *(K)*
  acid and base equilibrium constant expressions, 334–335
  calculating with Nernst equation, 223–224

homogeneous equilibrium constant expression, 334
overview, 104
transforming into, 128–130
equilibrium constant for weak bases (K_b)
defined, 154
problems, solving, 163–165, 175–176
equivalence points, 171
esters (R-COOR'), 244, 256
ethanoic acid, 243
ethanol (ethyl or grain alcohol), 242–243, 288
ethene (ethylene), 239, 251, 252
ethers (R-O-R'), 245
ethyne (acetylene), 240
exact numbers, 23
Example icon, 5
excited state of electrons, 30
exergonic reactions, 113
exothermic, defined, 72
exothermic reactions, 112–113, 193
EXP key on calculators, 20–21
exponential notation, 19–21
eye shadow formulation, 318

• F •

F (faraday), 225
face powders, 317–318
factor label method, 10, 22–23
families of elements, 38
faraday (F), 225
fats, 271–272
fertilizer, synthetic, and Haber process, 142
fibers, 250
fillers in detergents, 312
firkin, defined, 22
First Law of Thermodynamics, 192, 197
first order rate law, 107–109, 110
fission process. *See* nuclear fission
fissionable isotopes, 301
fixatives, 319
flashlight cells, 227
fluoride in toothpastes, 316
fluorine, oxidation number of, 212
food and flavor chemists, 342
forensic chemists, 340
formaldehyde, 242, 244
formation constant (K_f), 186–188
formic acid, 243
formulas. *See* equations; structural formulas
fractional distillation, 280–282
freezing point, 199
freezing-point depression property, 72–73
frequency factor *(A)*, 114
fructose, 264
functional groups
alcohols, 242–243
aldehydes and ketones, 244
amines and amides, 245–246
carboxylic acids, 243
esters, 244
ethers, 245
overview, 241–242
fusion
enthalpy of, 198
nuclear, 305–307
fusion torch concept, 307

• G •

galvanic cells
calculating potential of, 221–222
defined, 216
writing cell notation, 218–219
gamma radiation emission, 295
gas constant (R), 202
gaseous diffusion process, 94
gaseous effusion process, 94
gases
Avogadro's law, 90–91
Boyle's law, 84–85
Charles's law, 86–87
combined gas law, 89–90
Dalton's law, 92–94
entropy, 195
Gay-Lussac's law, 87–89
Graham's law, 94–95
ideal, 83
ideal gas equation, 91–92
kinetic molecular theory, 81–83
properties, 11, 83–84
reaction stoichiometry, 95–96
solubility and temperature, 63
solutions of, 62

gasoline
  additives, 286–287
  converting into vapor, 284
  octane ratings, 285–286
Gay-Lussac's law, 87–89
Gibb's Free Energy (ΔG), 200–202, 336–337
glasses, 78
glucose, 264
glycerol, 271
glycogen, 265
Graham's law, 94–95
grain alcohol (ethanol), 242–243, 288
grams
  relationship to mole, 48–50
  relationship to particle, 51–52
greases, 282
ground state of electrons, 30

• H •

Haber (Haber-Bosch) process
  factors, 143–144
  overview, 141–142
  reaction stoichiometry, 52–54
  synthesis of ammonia, 126, 138
  thermodynamics, 203–205
hair care chemistry, 322–324
half-cell reduction potentials, 219–220
half-life
  drugs, 276
  equations, 337–338
  of radioactive isotopes, 296–299
  reaction kinetics, 109–110
half-reaction method of balancing redox equations, 212–215
half-reactions, 208
halogenated hydrocarbons, 238
hard water and soaps, 311, 313–314
HDPE (high-density polyethylene), 251, 258
heat capacity, 77
heating oil, 282, 283
hemolysis, 74
Henderson-Hasselbalch equation, 167–171, 174, 336
Hess's Law, 194
heterogeneous catalysis, 121–122
heterogeneous equilibriums, 179
hexane, 236
hexose sugar, 263
high-density polyethylene (HDPE), 251
homework, doing, 330
homogeneous catalysis, 122–123
homogeneous equilibrium constant expression, 334
homogeneous mixtures, 61
Hund's rule, 32
hybrid electric vehicles, 288
hydration process, 240
hydrocarbons
  alkanes, 232–238
  alkenes, 238–240
  alkynes, 240
  aromatic, 241
  crude oil, 280
hydrogen
  biochemistry, 262
  gain of, in reduction, 210
  loss of, in oxidation, 209
  oxidation state of, 211
  in periodic table, 36, 37
hydrogen bombs, 305
hydrogen bonding
  intermolecular forces, 46, 76
  proteins, 268
hydrogenation process, 239–240, 272
hydrolysis, 164, 272
hydronium ions, 151–152, 153
hydrophilic end of surfactants, 310
hydrophobic end of surfactants, 310
hydrophobic molecules, 269
hydrophyllic molecules, 269
hydroxide ions, 152–153
hypertonic, defined, 74

• I •

ICE (Initial, Change, Equilibrium) tables, 135
icons, explained, 4–5
ideal gas equation, 91–92
ideal gases, 83
implants, silicone-based, 258
indicators, acid-base, 177
indirect electron transfers
  cell notation, 218–219
  Daniell cells, 216–218
  defined, 216
  standard reduction potentials, 219–222

indole, 319
induced dipole forces, 75
inert electrodes, 218
inert substituents, 262
Initial, Change, Equilibrium (ICE) tables, 135
inorganic chemistry, 13
insoluble (sparingly soluble), 179, 181–182
inspection method of balancing chemical equations, 212
instability constant (K_{instab}), 188
integrated rate laws, 107–110
intermediates, defined, 112, 120
intermolecular forces
- dipole-dipole, 46, 76
- hydrogen bonding, 46, 76, 268
- induced dipole, 75
- ion-dipole, 76–77
- London (dispersion), 46, 75
- overview, 74–75
- properties of liquids, 77–79
- types of, 46

internal combustion engines
- alternatives to, 288
- overview, 284

International Union of Pure and Applied Chemistry, 234
Internet resources for studying, 330–331
iodine, 262
ion-dipole forces, 76–77
ion-electron method of balancing redox equations, 212–215
ion-exchange resin, 314
ionic bonding
- electronegativity, 45
- ionic compounds, 41–43
- ions, 40–41
- overview, 262
- sodium + chlorine, 39–40

ions
- adding, and acid strength, 150
- defined, 39
- hydronium, 151–152, 153
- hydroxide, 152–153

isobutane, 235
isoelectric points, 266
isomers, 235
isooctane, 285
isopropyl alcohol, 239
isotonic, defined, 74
isotopes
- artificial (man-made), 291–292
- defined, 29, 290
- fissionable, 301
- neutron rich, 293

• J •

jet fuel, 282
joule (J), defined, 19, 193

• K •

K (equilibrium constant)
- acid and base equilibrium constant expressions, 334–335
- calculating with Nernst equation, 223–224
- homogeneous equilibrium constant expression, 334
- transforming into, 128–130

k (rate constant)
- defined, 103
- finding, 106–107
- orders of reaction, 104–106

K_a (acid ionization constant)
- defined, 153
- problems, solving, 160–162, 172

K_b (equilibrium constant for weak bases)
- defined, 154
- problems, solving, 163–165, 175–176

K'_c (concentration-based equilibrium constant), 129
K_d (dissociation constant), 188
Kelvin temperature
- defined, 18
- gas law problems, 84, 86

keratin, 322
kerosene, 282
ketones (R-CO-R'), 244–245
ketose sugar, 263
K_f (formation constant), 186–188
kilo-, 16
kilogram, defined, 17
kinetic energy, defined, 111
kinetic molecular theory of gases, 81–83
kinetics
- activation energy, 114–116
- catalysts, 121–123
- collision theory, 111–114

kinetics *(continued)*
integrated rate laws, 107–110
mechanisms, 117–120
overview, 11, 99
rate laws, 102–107
reaction rates, 100–101
K_{instab} (instability constant), 188
kitchen chemistry, 309
K_p (pressure-based equilibrium constant), 130
K_{sp} (solubility product constant), 180
K_{stab} (stability constant), 186–188
K_w (water dissociation constant), 159

• L •

lakes, 319
Langley, Richard H. (author)
Biochemistry For Dummies, 13
Organic Chemistry II For Dummies, 13
lanolin, 317
lanthanide series, 34–35, 36
lattice, defined, 199
laundry room chemistry
bleach, 314
overview, 309
surfactants, 310–313
water softeners, 313–314
Law of Conservation of Energy, 192, 197
Law of Conservation of Matter, 54
Law of Mass Action, 127–128
LDPE (low-density polyethylene), 251, 259
lead in gasoline, 286–287
lead storage batteries, 228
LeChatelier's Principle
catalysts, effects of, 141
common-ion effect, 184–185
concentration, altering, 138–139
Haber-Bosch Process, 142–143, 205
overview, 137–138
pressure, altering, 140–141
temperature, altering, 139–140
legal careers, 339–340
length in SI system, 16–17
"LEO goes GER phrase", 210
Lewis acid-base theory, 148, 185
Lewis structural formula, 44
ligands, 185
limiting reactant problems, 57–59
linear structure of polymers, 249
lipid interactions, 272
lipids
fats, 271–272
overview, 269
steroids, 269–271
lipstick formulation, 319
liquids
entropy, 195
properties of, 77–79
liter, defined, 18
lithium batteries, 227
London (dispersion) forces, 46, 75
lotions for skin, 317
low-density polyethylene (LDPE), 251, 259
lubricating oils, 282

• M •

M. *See* molarity
macromolecules. *See* polymers
macroscopic properties of acids and bases, 145–146
macroscopic viewpoint, 13–14
male pattern baldness, 326
malleable, defined, 37
mascara formulation, 318
mass action expressions, 127–128
mass defect, 300
mass
critical, and chain reactions, 300–302
in SI system, 17
subcritical, 301
mass number, 29, 36, 290
mass/mass percentage, 64–65
mass/volume percentage, 65–66
material scientists, 341–342
measured numbers, 23–24
mechanisms, reaction
defined, 100, 112
rate-determining step, 120
steps in creating, 117–120
melanin, 321, 323
memorization in studying chemistry, 330
Mendeleev, Dmitri (chemist), 33
meniscus, 78
mercury-201, 296
metal oxides, 154
metalloids, 37

metals
  electroplating, 226
  oxidation numbers, 211
  in periodic table, 36–37
  transition, 262
meter, defined, 16–17
methane, 238, 281
methanol (methyl alcohol), 242, 288
methyl tert-butyl ether (MTBE), 287
metric system. *See* SI system
micelle formation, 272, 310
microscopic properties of acids and bases, 146–148
microscopic viewpoint, 13–14
microstates, 195
middle notes in fragrances, 319
minoxidil, 326
molality
  calculating, 69
  defined, 64
molar solubility, 182
molarity (M)
  concentration units, 66–67
  defined, 64
  dilution process, 67–68
  reaction stoichiometry, 68–69
molarity equation, 333
mole concept
  conversions to particles, 50–52
  overview of, 47–48
  reaction stoichiometry, 52–57
  relationship to grams, 48–50
  volume of mole of gas, 90–91
mole-to-coefficient ratio, 58
momentum, defined, 30
monoatomics
  defined, 41
  oxidation numbers of, 211
monomers, 248
monoprotic acids, 151
monosaccharides, 263–264
Moore, John T. (author)
  Biochemistry For Dummies, 13
  Chemistry For Dummies, 2e, 9
  Organic Chemistry II For Dummies, 13
Motoring octane rating, 286
MTBE (methyl tert-butyl ether), 287
multiplication
  exponential or scientific notation, 20
  reporting correct number of significant figures, 25
mutations, 275

• N •

nail polish, 319
naming organic compounds
  alkanes, 234–238
  alkenes, 239
naphthalene, 241
natural gasoline, 282
Nernst, Walther (chemist), 222
Nernst equation, 222–224, 337
net-ionic equation, 147, 210–211
neutralization reactions, 147
neutron rich isotopes, 293
neutrons, 28
n-heptane, 285
n-hexane, catalytic reforming of, 283
nickel-63, 294
nickel-cadmium batteries, 227
nitrogen bases in nucleic acids, 273–274
nitrous acid, 155, 162
nomenclature, chemical, 10–11. *See also* naming organic compounds
nonbonding electrons, 44
nonmetal oxides, 154
nonmetals, 37
nonpolar covalent bonding, 45
nonpolar end of surfactants, 310
nonrenewable resources, 279
nonspontaneous processes, 191
nonstandard conditions, 202–203
normal alkanes, 232–233
notes, lecture
  recopying, 331–332
  taking in class, 331
notes in fragrances, 319
nuclear chemistry
  half-lives and radioactive dating, 296–299
  overview, 11, 289
  radioactive decay, 292–296
  radioactivity, 290–291
  transmutation, 291–292

nuclear fission
  breeder reactors, 304–305
  critical mass and chain reactions, 300–302
  power plants, 302–304
nuclear fusion
  overview, 305–306
  reactors, 306–307
nucleic acids
  components, 273–274
  mutations, 275
  roles, 274–275
nucleus, 28–29, 290
numbers, working with
  calculators, 20–21
  exponential and scientific notation, 19–20
  quadratic equation, solving, 21
  significant figures, 23–25
  unit conversion method, 22–23
  words to represent numbers, using, 47
nylon, 256–257

• O •

octane rating scale, 285–286
oils, 271
optical brighteners in detergents, 312–313
orbitals, defined, 30
orders of reaction, 103–106
organic chemistry
  functional groups, 241–246
  hydrocarbons, 232–241
  overview, 13, 231
Organic Chemistry II For Dummies (Moore and Langley), 13
osmotic pressure property, 73–74, 262
oxidation, 208–209
oxidation number method of balancing redox equations, 212
oxidation numbers, 211–212
oxidation state/number, 41
oxidation-reduction processes, 262
oxides, 154
oxidizing agents, 211
oxyacids, 149–150
oxygen
  gain of, in oxidation, 209
  loss of, in reduction, 210
  oxidation number of, 211
oxygenates, 287

• P •

P-30 (sulfur isotope), 292
PABA (para-aminobenzoic acid), 321, 322
paraffin-based waxes, 282
partial pressures, 92–93
particle size and reaction rates, 100
particles
  gas, pressure of in containers, 82
  relationship to grams, 51–52
  relationship to moles, 50–51
  subatomic, 27–28
parts per million/billion, 69–70
pascals, 18, 83
patent attorneys, 339–340
pentose sugar, 263
peptide bonds, 266
percent yield, calculating, 56–57
percentage in concentration units, 64–66
perfumes, 319–320
periodic properties, 10
Periodic Table
  overview, 33–35
  using, 36–38
periods of elements, 38
permanents for hair, 324
peroxides, 211, 245
petroleum
  alternatives, 288
  gasoline, 284–287
  refining crude oil, 279–284
petroleum jelly, 317
pH, determining, 156–158, 160–165
phaeomelanin, 323
pharmaceutical sales, careers in, 340
phenol, 241
phosphates in detergents, 312
phosphoric acid, 274
pH/pOH equations, 335
physical chemistry, 12
pine oil, 315
planetary model, 29–30
plasma, 306–307
plasticizers, 254
plastics, 250. *See also* polymers
pleated sheet structure, 268
plutonium-239, 301, 304
pOH, calculating pH with, 158

polar covalent bonding, 45–46
polar end of surfactants, 310
polarity
  dissolving process, 62
  intermolecular forces, 74–75
  liquids, 79
pollution
  air, 287
  disposal of nuclear waste, 304
  thermal, 63
polonium-218, 307
polyamides, 256–257
polyatomics
  defined, 41
  oxidation numbers of, 211
polyester, 256
polyethylene, 239, 251–252
polyetrafluoroethylene, 254
polylactic acid, 259
polymerization process, 240
polymers
  addition polymerization, 250–255
  condensation polymerization, 256–258
  disposal of, 258–260
  naturally occurring, 247–248
  polyamides, 256–257
  polyester, 256
  polyethylene, 251–252
  polyetrafluoroethylene, 254
  polypropylene, 252–253
  polystyrene, 253
  polyvinyl chloride, 253–254
  silicones, 257–258
  synthetic, 249–258
  thermoplastic, 249
  thermosetting, 250
polypeptides, 267
polypropylene, 252–253
polysaccharides, 265
polystyrene, 253
polyunsaturated fats, 271
polyvinyl chloride, 253–254
positron emission, 295
powders, body and face, 317–318
power, raising number to, 20
power plants, nuclear, 302–304
precipitation, predicting, 182–183
prefixes in SI system, 16
preignition, 284
pressure
  altering, and equilibrium, 140–141
  of gas particles in containers, 82
  Haber process, 143
  in SI system, 18
pressure-based equilibrium constant (K_p), 130
pressure-temperature relationship of gases, 87–89
pressure-volume relationship of gases, 84–85
primary structure of proteins, 267
products, 291
propane
  in crude oil, 281
  to power vehicles, 288
  structural formula, 234
propene (propylene), 239
properties
  boiling-point elevation, 71–72
  colligative, 69, 70–74
  freezing-point depression, 72–73
  of gases, 11, 83–84
  of liquids, 77–79
  macroscopic, of acids and bases, 145–146
  microscopic, of acids and bases, 146–148
  osmotic pressure, 73–74
  periodic, 10
  vapor-pressure lowering, 70–71
prostheses, silicone-based, 258
proteins
  amino acids, 266, 267
  primary structure, 267
  quaternary structure, 269
  secondary structure, 268
  tertiary structure, 268–269
protons, 28
pseudo-first order reaction conditions, 109
pure chemistry, 14

• Q •

Q (reaction quotient)
  defined, 127–128, 202
  precipitation, predicting, 182–183
quadratic equation, solving, 21
quantum mechanical model, 30
quaternary structure of proteins, 269
questions, asking in class, 332

• R •

R (gas constant), 202
radical reactive sites, 251
radioactive dating, 299
radioactive decay
  electron capture, 296
  half-life decay of isotopes, 296–299
  natural, 293–295
  overview, 292
  positron emission, 295
radioactivity, 12, 290–291
Radium-230, 292
radon, 307
rate constant *(k)*
  defined, 103
  finding, 106–107
  orders of reaction, 104–106
rate law
  defined, 103
  determining, 118
  integrated, 107–110
  rate constant, 104–107
rate law equation, 333–334
rate-determining step, 117, 120
R-COH (aldehydes), 244
R-CO-NH_2 (amides), 245–246
R-COOH (carboxylic acids), 243
R-COOR' (esters), 244, 256
R-CO-R' (ketones), 244–245
reactants
  balancing nuclear reactions, 291
  reaction rates, 100–101
reaction arrow, 291
reaction mechanisms
  defined, 100, 112
  rate-determining step, 120
  steps in creating, 117–120
reaction quotient (Q)
  defined, 127–128, 202
  precipitation, predicting, 182–183
reaction rates
  factors in, 100–101
  finding rate constant, 106–107
  formula for, 102
  integrated rate laws, 107–110
  orders of reaction, 103–106
  slowing, 102–103
reaction stoichiometry
  gases, 95–96
  overview, 11, 52, 125
  percent yield, calculating, 56–57
  reactants, limiting, 57–59
  reactants and products, calculating, 52–56
  solutions and molarity concentration units, 68–69
reactions. *See also* reaction mechanisms; reaction quotient; reaction rates; reaction stoichiometry; redox reactions
  acid-base, 155–156
  of alkene, 239–240
  chain, 300–302
  combustion, 229–230, 238
  completion of, 152
  condensation, 256
  direct electron, 210–211, 216
  electrochemical, 208
  endergonic, 114
  endothermic, 113–114, 193
  exergonic, 113
  exothermic, 112–113, 193
  fats, 272
  half-reactions, 208
  neutralization, 147
  nuclear, balancing, 291
  orders of, 104–106
  reversible, 126
  thermite, 54
  transition state of, 112
reactive site, 111
recopying lecture notes, 331–332
recycling plastics, 258–259
redox reactions
  balancing equations, 212–215
  bleaches, 314
  combustion, 229–230
  defined, 208
  electrolytic cells, 225
  galvanic or voltaic cells, 216–222
  net-ionic equation, 210–211
  oxidation, 208–209
  oxidation numbers, 211–212
  recharging cells, 227
  reduction, 209–210

reducing agents, 211
reduction
  defined, 208
  gain of electrons, 209
refining process for crude oil
  catalytic cracking, 282–283
  catalytic reforming, 283–284
  fractional distillation, 280–282
  overview, 279–280
replication process, 274–275
Research octane rating, 286
residue, 267
reverse osmosis, 74
reversible reactions, 126
R-groups of amino acids, 266
ribose, 273, 274
RNA, 273, 275
R-NH_2 (amines)
  overview, 245–246
  in sweat, 316
R-OH (alcohols), 242–243
R-O-R' (ethers), 245
rounding off numbers, 25
rubber, 250

• S •

S (entropy). *See* entropy
sales, careers in, 340
salicylic acid, 325
salt bridges, 216–217
salts, 39–40
sample size, statistically significant, 296
saponification, 311
saturated hydrocarbons, 232
saturated solutions, 63
science, defined, 14
scientific notation, 19–21
Second Law of Thermodynamics, 195, 197–199
second order rate law, 109, 110
secondary structure of proteins, 268
semimetals, 37
semipermeable membranes, 73
serendipity, 326
shampoos, 322–323
SI system
  energy, 19
  length, 16–17
  mass, 17
  overview, 15–16
  prefixes, 16
  pressure, 18
  temperature, 18
  volume, 18
significant figures
  exact, counted, and measured, 23–24
  figuring number of, in measured numbers, 24
  reporting correct number of, 24–25
silicones, 257–258
skin care chemistry
  body and face powders, 317–318
  creams and lotions, 317
  eye makeup, 318
  lipstick, 319
  nail polish, 319
  perfumes, colognes, and aftershaves, 319–320
  suntan lotion and sunscreen, 320–322
sleeping before exams, 332
slowing reaction rates, 102–103
soaps, 272, 311
sodium, 39–40
sodium fluoride, 316
sodium hypochlorite, 314
sodium metasilicate, 315
sodium perborate, 312, 314
sodium sulfate, 183
sodium tripolyphosphate, 312, 315
softening water, 313–314
solids
  amorphous, 78
  entropy, 195–196
solubility. *See also* solubility equilibriums
  defined, 63
  general rule of, 62
  molar, 182
solubility equilibriums
  common-ion effect, 184–185
  concentration of dissolved ions, 180–181
  overview, 179
  precipitation, predicting, 182–183
  sparingly soluble salts, 181–182

solubility product constant (K_{sp}), 180
solubility product equation, 336
solute component of solutions, 62–63
solutions
  colligative properties, 69, 70–74
  components of, 62
  concentration units, 64–70
  molarity, 66–68
  overview, 11, 61
  pH, determining, 156–158
  saturated, 63
solvent component of solutions, 62
sparingly soluble (insoluble), 179, 181–182
SPF (Sun Protection Factor) values, 321
spontaneity
  determining, 196–197, 198–199
  Gibb's Free Energy, 201
  predicting for enthalpy and entropy changes, 199–200
spontaneous processes, 191
spotty ignition, 284
stability, 292–293
stability constant (K_{stab}), 186–188
standard reduction potentials, 219–222
stannous fluoride, 316
starch
  as polymer, 247–248
  as polysaccharide, 265
  thermoplastic, 259
state functions, 192, 195
state of system, defined, 192
steroids, 269–271
stoichiometric ratio, 52
straight-chained alkanes, 232–233
straight-run gasoline, 282
strength, acid-base, 148–156
stress, applying to chemical systems, 137
strong acids, 151–152
strong bases, 152–153
structural formulas
  alkanes, 233–234
  covalent bonding, 44
  skin care products, 317–319
study techniques
  asking questions, 332
  checking for errors, 332
  developing and sticking to schedules, 329
  doing homework, 330
  recopying notes, 331–332
  skimming texts, 331
  sleeping before exams, 332
  striving for understanding, 330
  taking notes, 331
  using additional resources, 330–331
subatomic particles, 27–28
subcritical mass, 301
substituent groups, 234
subtraction
  exponential or scientific notation, 19–20
  reporting correct number of significant figures, 25
sulfur isotope (P-30), 292
sulfuric acid, 151
suntan lotions and sunscreens, 320–322
supersaturated solutions, 63
surface tension, 79
surfactants
  detergents, 312–313
  overview, 310
  soaps, 311
  surface tension, 79
surroundings, defined, 192
suspension agents in detergents, 312
synthesis, defined, 47
systems, defined, 192

• T •

T. *See* temperature
table salt, creating, 39–40
talc, 317
tar, 282
teaching, careers in careers, 343
technology, defined, 14
TEL (tetraethyllead), 286–287
temperature (T). *See also* Kelvin temperature
  altering, and equilibrium, 139–140
  fusion process, 306
  Haber process, 142
  reaction rates, 100–101
  in SI system, 18
  solubility of solutes, 63

as state function, 192
viscosity, 78
temperature-volume relationship of gases, 86–87
terminals, 216
ternary acids, 149–150
terpenes, 315
tertiary structure of proteins, 268–269
testosterone, 270
tetraethyllead (TEL), 286–287
textiles, 250
theoretical yield, 57
thermal pollution, 63
thermite reactions, 54
thermochemical equations, 193–194
thermochemistry, defined, 191
thermodynamics
change in energy, determining, 192–193
enthalpy, 193–195
entropy, 195–197
first law of, 192, 197
Gibb's Free Energy, 200–202
Haber-Bosch Process, 203–205
nonstandard conditions, 202–203
overview, 12
second law of, 195, 197–199
spontaneous and nonspontaneous processes, 191
third law of, 199
thermoplastic polymers, 249
thermoplastic starch, 259
thermosetting polymers, 250
Third Law of Thermodynamics, 199
thorium-230, 294
Three Mile Island power plant, 303
titration curves, acid-base, 171–177
toothpastes, 315–316
top notes in fragrances, 319
torrs, as units of measurement, 83
transcription process, 275
transition metals, 262
transition state of reactions, 112
translation process, 275
transmutation, 291–292
triglycerides, 271
tripeptides, 266
triple bonds, 43
tritium, 305

• U •

uncertainty principle, 30
unimolecular step, 117
unit conversion method, 10, 22–23
universe, defined, 192, 195
unsaturated fats, 271
unsaturated hydrocarbons, 238–240
unsaturated solutions, 63
uranium-235, 300, 301, 303
US customary system, 16
UV spectrum, 320

• V •

valence electrons, 33, 289
validity of assumptions, checking, 133, 161
vanishing cream formulation, 317
van't Hoff factor, 74
vapor pressure of water, 93
vapor-pressure lowering property, 70–71
Viagra, 326
viscosity, 78
volatility, 284
voltaic cells, 216
volume
of mole of gas, 90–91
in SI system, 18
volume/volume percentage, 66

• W •

waste products of fission reactors, 304
water
autoionization of, 159, 335
soaps, 311
softening units, 313–314
surfactants added to, 310
vapor pressure of, 93
water dissociation constant (K_w), 159
water quality chemists, 342–343
waxes, paraffin-based, 282

weak acids, 153, 160–162
weak bases, 154, 163–165
weight percentage, 64–65
Wohler, Friedrich (scientist), 231

• Z •

zeolites, 312
zero order rate law, 109, 110
zwitterions, 266

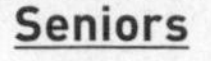

Math & Science

Algebra I For Dummies, 2nd Edition
978-0-470-55964-2

Biology For Dummies, 2nd Edition
978-0-470-59875-7

Chemistry For Dummies, 2nd Edition
978-1-1180-0730-3

Geometry For Dummies, 2nd Edition
978-0-470-08946-0

Pre-Algebra Essentials For Dummies
978-0-470-61838-7

Microsoft Office

Excel 2010 For Dummies
978-0-470-48953-6

Office 2010 All-in-One For Dummies
978-0-470-49748-7

Office 2011 for Mac For Dummies
978-0-470-87869-9

Word 2010 For Dummies
978-0-470-48772-3

Music

Guitar For Dummies, 2nd Edition
978-0-7645-9904-0

Clarinet For Dummies
978-0-470-58477-4

iPod & iTunes For Dummies, 9th Edition
978-1-118-13060-5

Pets

Cats For Dummies, 2nd Edition
978-0-7645-5275-5

Dogs All-in One For Dummies
978-0470-52978-2

Saltwater Aquariums For Dummies
978-0-470-06805-2

Religion & Inspiration

The Bible For Dummies
978-0-7645-5296-0

Catholicism For Dummies, 2nd Edition
978-1-118-07778-8

Spirituality For Dummies, 2nd Edition
978-0-470-19142-2

Self-Help & Relationships

Happiness For Dummies
978-0-470-28171-0

Overcoming Anxiety For Dummies, 2nd Edition
978-0-470-57441-6

Seniors

Crosswords For Seniors For Dummies
978-0-470-49157-7

iPad 2 For Seniors For Dummies, 3rd Edition
978-1-118-17678-8

Laptops & Tablets For Seniors For Dummies, 2nd Edition
978-1-118-09596-6

Smartphones & Tablets

BlackBerry For Dummies, 5th Edition
978-1-118-10035-6

Droid X2 For Dummies
978-1-118-14864-8

HTC ThunderBolt For Dummies
978-1-118-07601-9

MOTOROLA XOOM For Dummies
978-1-118-08835-7

Sports

Basketball For Dummies, 3rd Edition
978-1-118-07374-2

Football For Dummies, 2nd Edition
978-1-118-01261-1

Golf For Dummies, 4th Edition
978-0-470-88279-5

Test Prep

ACT For Dummies, 5th Edition
978-1-118-01259-8

ASVAB For Dummies, 3rd Edition
978-0-470-63760-9

The GRE Test For Dummies, 7th Edition
978-0-470-00919-2

Police Officer Exam For Dummies
978-0-470-88724-0

Series 7 Exam For Dummies
978-0-470-09932-2

Web Development

HTML, CSS, & XHTML For Dummies, 7th Edition
978-0-470-91659-9

Drupal For Dummies, 2nd Edition
978-1-118-08348-2

Windows 7

Windows 7 For Dummies
978-0-470-49743-2

Windows 7 For Dummies, Book + DVD Bundle
978-0-470-52398-8

Windows 7 All-in-One For Dummies
978-0-470-48763-1

Available wherever books are sold. For more information or to order direct: U.S. customers visit www.dummies.com or call 1-877-762-29
U.K. customers visit www.wileyeurope.com or call (0) 1243 843291. Canadian customers visit www.wiley.ca or call 1-800-567-4797.

Connect with us online at www.facebook.com/fordummies or @fordummies

by John T. Moore, Ed.D
Author, *Chemistry For Dummies,* Second Edition

John Wiley & Sons, Inc.

Chemistry II For Dummies®

Published by
John Wiley & Sons, Inc.
111 River St.
Hoboken, NJ 07030-5774
www.wiley.com

Published by John Wiley & Sons, Inc., Hoboken, New Jersey

Published simultaneously in Canada

For general information on our other products and services, please contact our Customer Care Department within the U.S. at 877-762-2974, outside the U.S. at 317-572-3993, or fax 317-572-4002.

For technical support, please visit www.wiley.com/techsupport.

Wiley publishes in a variety of print and electronic formats and by print-on-demand. Some material included with standard print versions of this book may not be included in e-books or in print-on-demand. If this book refers to media such as a CD or DVD that is not included in the version you purchased, you may download this material at http://booksupport.wiley.com. For more information about Wiley products, visit www.wiley.com.

Library of Congress Control Number: 2012940022

ISBN: 978-1-118-16490-7; ISBN 978-1-118-22624-7 (ePDF); ISBN 978-1-118-23946-9 (ePub); ISBN 978-1-118-26070-8 (eMobi)

Manufactured in the United States of America

10 9 8 7 6 5 4 3 2 1